Introductory Plant Pathology

H.S. Chaube
Professor
Department of Plant Pathology
G.B. Pant University of Ag. & Tech.
Pantnagar 263 145 U.S. Nagar

Ramji Singh
Scientific Officer (Plant Pathology)
U.P. Council of Agricultural Research
16, A.P. Sen Road, Lucknow 226 001

CBS

CBS Publishers & Distributors Pvt. Ltd.

New Delhi • Bengaluru • Chennai • Kochi • Mumbai • Pune
Hyderabad • Kolkata • Nagpur • Patna • Vijayawada

Introductory Plant Pathology

ISBN: 978-81-239-2670-4

First CBS Reprint: 2015

Published by:
Satish Kumar Jain for CBS Publishers & Distributors Pvt. Ltd.,
4819/XI Prahlad Street, 24 Ansari Road, Daryaganj, New Delhi - 110002
delhi@cbspd.com, cbspubs@airtelmail.in • www.cbspd.com
Ph.: 23289259, 23266861, 23266867 • Fax: 011-23243014

Corporate Office: 204 FIE, Industrial Area, Patparganj, Delhi - 110 092
Ph: 49344934 • Fax: 011-49344935
E-mail: publishing@cbspd.com • publicity@cbspd.com

Branches:
- ***Bengaluru:*** 2975, 17th Cross, K.R. Road, Bansankari 2nd Stage, Bengaluru - 70
  Ph: +91-80-26771678/79 • Fax: +91-80-26771680
  E-mail: cbsbng@gmail.com, bangalore@cbspd.com
- ***Chennai:*** No. 7, Subbaraya Street, Shenoy Nagar, Chennai - 600030
  Ph: +91-44-26681266, 26680620 • Fax: +91-44-42032115
  E-mail: chennai@cbspd.com
- ***Kochi:*** 36/14, Kalluvilakam, Lissie Hospital Road, Kochi - 682018
  Ph: +91-484-4059061-65 • Fax: +91-484-4059065
  E-mail: cochin@cbspd.com
- ***Mumbai:*** 83-C, Dr. E. Moses Road, Worli, Mumbai - 400018
  Ph: +91-9833017933, 022-24902340/41 • E-mail: mumbai@cbspd.com
- ***Pune:*** Bhuruk Prestige, Sr. No. 52/12/2+1+3/2,
  Narhe, Haveli (Near Katraj-Dehu Road Bypass), Pune - 411041
  Ph: +91-20-64704058/59, 32342277 • E-mail: pune@cbspd.com

Representatives:
- Hyderabad: 0-9885175004
- Nagpur: 0-9021734563
- Vijayawada: 0-9000660880
- Kolkata: 0-9831437309, 0-9051152362
- Patna: 0-9334159340

Printed at:
Neekunj Print Process, Delhi

Dedicated to

Our Patents

Late Sri A.N. Chaube

Late Smt. A.N. Chaube

&

Sri Ramashankar Singh

Smt. Shivaratri Devi

About the Authors

Hriday S. Chaube, Ph.D. is Professor of Plant Pathology at G.B. Pant Univ. of Agric. & Technol., Pantnagar. Dr. Chaube received his B.Sc. (Ag.) degree from the Univ. of Gorakhpur in 1964 and his M.Sc. (Ag.) Plant Pathology degree from B.H.U. Varanasi in 1966. He completed his Ph.D. (Plant Pathology) as staff candidate from G.B. Pant Univ. of Agric. & Technol, Pantnagar in 1978. Dr. Chaube has been on the staff of G.B. Pant Univ. of Agric. & Technol. Pantnagar since 1966. He was appointed Asstt. Professor in 1970, Associate Professor in 1980 and Professor in 1992.

Dr. Chaube is the member of Indian Phytopath. Soc. and Indian Soc. of Myc. & Plant Pathology. He is a regular teacher of core and elective courses offered to U.G. and P.G. students in Plant Pathology. He has published above 80 research papers, 30 research abstracts, over 50 popular articles, 10 reviews / conceptual articles, 3 bulletins and authored / edited 10 books. Dr. Chaube has been engaged in phytopathological research since 1964. His major areas of research are ecology and management of soil-borne pathogens.

Ramji Singh, Ph.D. is Scientific Officer, Plant Pathology at U.P. Council of Agricultural Research. Dr. Singh obtained his B.Sc. (Ag.) degree from University of Gorakhpur in 1984. He received his M.Sc. Ag. & Ph.D. in Plant Pathology from G.B. Pant Univ., of Ag. & Tech. Pantnagar in 1986 and 1990, respectively. He joined Central Institute of Medicinal and Aromatic Plants (CIMAP) Lucknow in 1990 as Research Associate. Later he joined UPCAR as Scientific Officer in 1992.

Dr. Singh is the member of Indian Phytopath. Soc. and Indian Soc. of Myc. & Plant Pathology. He has published above 20 research papers, several research abstracts, popular articles, reviews and chapters in books. Presently he is engaged in prioritizing, the research needs of state and developing research infrastructures in SAUs. He is also coordinating the IPM research works and IMP developmental works funded by UPCAR / UPDASP (World Bank).

Content

FOREWORD

It gives me pleasure to write the foreword of **"Introductory Plant Pathology"** by Professor H.S. Chaube, Department of Plant Pathology, G.B. Pant University of Agriculture & Technology, Pantnagar and Dr. Ramji Singh, Scientific Officer (Plant Pathology), U.P. Council of Agricultural Research, Lucknow. Progress in this branch of science has been so pronounced that such a book is most needed. It contains concise, authentic, and latest information for students who have an elementary knowledge of the subject. The book has been written keeping the interest of both pure science and agriculture graduates and would be useful from teaching point of view. While preparing the text of the book the authors have kept this in view as well as needs and course contents of the various universities in India at undergraduate level.

I hope that the text would prove to be highly beneficial to our young students and teachers.

Suresh C. Mudgal

August 26, 2002

(Suresh C. Mudgal)
Director General
U.P. Council of Agricultural Research,
16 A.P. Sen Road, Lucknow-226 001

FOREWORD

I am happy to write about the book **"Introductory Plant Pathology"** authored by "**Dr. H.S. Chaubey**, Professor, Plant Pathology" and "**Dr. Ramji Singh**, Scientist UPCAR". I, as teacher and educationist, have always felt that the students generally don't take to books particularly at UG level. They rely solely on class notes. To gain adequate knowledge and competence, books must be consulted. The students at UG level are exposed to almost every area of the subject concerned. All the chapters taught to them are never available in single volume/book. This creates a situation where students ignore reading books.

I am sure, this book containing all the chapters taught to UG students will serve the purpose and motivate them to consult books. This book would be useful to teachers as well as generate enough interest to inspire the inquisitive mind. I congratulate the authors for their efforts.

Sept., 2000

(Ranvir Singh)
Dean Agriculture
G.B. Pant University of Agriculture & Tech.,
Pantnagar-203145, Distt. U.S. Nagar

PREFACE

This would be a very long preface if we were to set down in detail the reasons why we wrote this book in the present form. Suffice it to say that it is based on a course which beginners in plant pathology study and it is an approach which we have found useful. We hope it may prove so others too. The fact that there are few introductory texts in plant pathology is also a ready excuse for writing this one. To a certain extent any course (and so any book based on it) is a reflexion of the teacher's own training and of his special interest. The senior author as a regular teacher for under graduate courses, has experienced that students don't take to books. They rely solely on class-notes. One of the reasons being the non-availability of text books covering entire course content, the way they are delivered. In the present book, we have tried to arrange the topics and the contents in such a manner that students get an easy reading with due continuity and thus understanding.

We do not claim originality in the preparation of this book and have taken help from a large number of books both foreign and Indian, dealing with various topics, we have greatly followed Alexopoulos, which in our opinion is the best at present for "biology of fungi" for beginners. We gratefully thank various authors, editors, and publishers of the books and journals we have made use of while preparing the text of this book.

The senior author (H.S.C.) is indebted to Hon'ble Vice-Chancellor (Prof. J.B. Chaudhari), Dean Agriculture, and Head Plant Pathology, G.B. Pant University of Agriculture and Technology, Pantnagar-263145, for guidance, facilities and encouragement. We also express our gratitude to Prof. S.C. Modgal, Director General, UPCAR, Lucknow for motivation and encouragement. Dr. Jyotsna Sharma, Post-Doctoral Fellow (Plant Pathology) and Ms. Deepa Khulbe, Research Scholar deserve our appreciation for help in preparation of drawings and proof reading. Miss Poonam Lohani too deserves special appreciation for providing help in computerization of the manuscript. Managerial help and proof reading by Mrs. Alka Singh (Wife of Dr. Ramji Singh) is gratefully acknowledged.

Besides our best efforts some mistakes factual or printing might have inadvertently crept in for which we beg sorry in anticipation and if pointed out to us, shall surely correct them in subsequent print/edition. We shall also welcome suggestions and healthy criticism for the improvement of this publication.

H.S. Chaube
Ramji Singh

CHAPTER I

PLANT DISEASES

I. Introduction

Plants not only sustain the man and animals, they are also the source of food for multitudes of organisms living in the ecosystem. A conflict of interest, therefore, between the contenders of suppliers from plants is inevitable. Thus, while man has been able to subjugate plants and animals for his own use, the competing microorganisms still defy their/his efforts and claim a major share of resources which man would like to use for himself. Plant diseases have been considered as stubborn barriers to the rapid progress of food production.

II. The Science of Plant Pathology

A. Concept

Plant Pathology or phytopathology (*phyton* = plant, *pathos* = ailment, *logos*= knowledge) is that branch of agricultural, botanical, or biological science which deals with cause, etiology, resulting losses and management of plant diseases.

B. Relationship with other sciences

The objectives of plant pathology are to identify the cause (s) of disease, the mechanisms of disease development, the factors affecting disease development, and finally economic and efficient management of diseases. Knowledge of basic biological and physical sciences, as well as comprehension of agricultural, environmental and social sciences are the foundation stones upon which the science of plant pathology rests. For instance, to understand and manage diseases, plant pathologists must understand the biology, physiology, reproduction, dispersal, survival and ecology of all the multiple pathogens and parasites of plants. They also must understand the concepts of stress and strains on plants and how environmental factors induce disease when the limits of tolerance are exceeded. The interrelations of different sciences with plant pathology are illustrated in the Figure-1.

III. Concept of Disease in Plants

A. An Overview

Disease is one of those terms that are very difficult to define. It is realized that disease (literally *dis-ease*) implies lack of comfort and, therefore, involves deviation from normal functioning. According to Agrios (1988), a plant is healthy when it can carry out its physiological functions to the best of its genetic potential. However, it is difficult to determine genetic potential of a plant because gene expression itself is influenced by environmental factors. Nevertheless, one can grow the plants under different sets of environmental and/or nutritional conditions, and can find out the best combination under which their growth and reproduction are optimum. This very state of plant which can not be improved

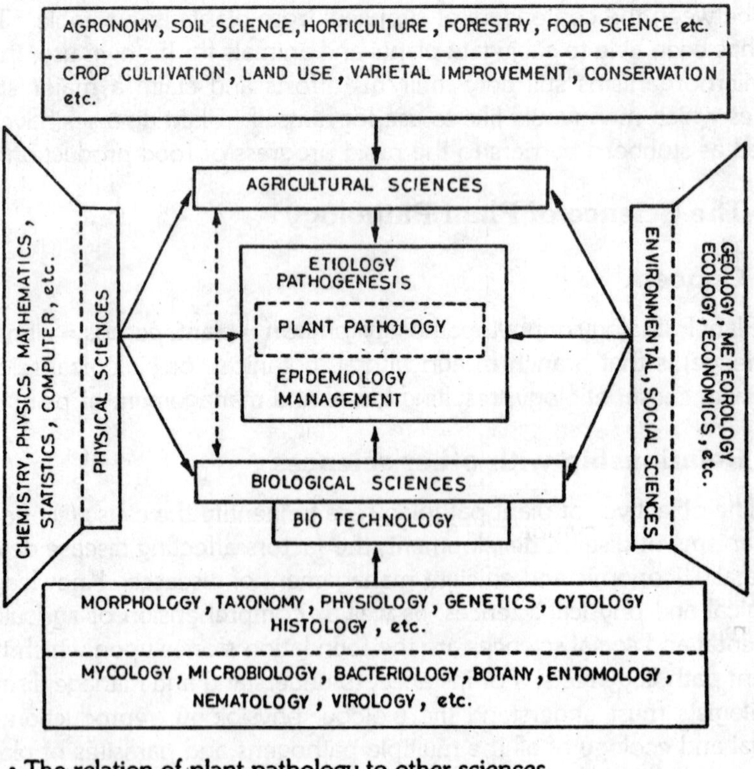
AGRONOMY, SOIL SCIENCE, HORTICULTURE, FORESTRY, FOOD SCIENCE etc.
CROP CULTIVATION, LAND USE, VARIETAL IMPROVEMENT, CONSERVATION etc.
AGRICULTURAL SCIENCES
ETIOLOGY
PATHOGENESIS
PLANT PATHOLOGY
EPIDEMIOLOGY
MANAGEMENT
BIOLOGICAL SCIENCES
BIO TECHNOLOGY
CHEMISTRY, PHYSICS, MATHEMATICS,
STATISTICS, COMPUTER, etc.
PHYSICAL SCIENCES
ENVIRONMENTAL, SOCIAL SCIENCES
GEOLOGY, METEOROLOGY
ECOLOGY, ECONOMICS, etc.
MORPHOLOGY, TAXONOMY, PHYSIOLOGY, GENETICS, CYTOLOGY
HISTOLOGY, etc.
MYCOLOGY, MICROBIOLOGY, BACTERIOLOGY, BOTANY, ENTOMOLOGY,
NEMATOLOGY, VIROLOGY, etc.

Figure-1 : The relation of plant pathology to other sciences.

further by manipulation of environmental factors, is considered as healthy and any deviation of course negative, from this state is considered as abnormal or "diseased". It is, thus, seen that any deviation from normalcy results in the diseased state.

From time to time several definitions which have been proposed, are in fact descriptive but not simultaneously exclusive. All these definitions indicate that disease; (1) is related to poor functioning of growth and reproduction in plant, (2) is malfunctioning physiology of plants, and (3) reduces Plant's ability to survive and maintain its ecological niche. Horsfall and Dimond (1959) arrived at a descriptive explanation which embodies the concept of disease. According to them disease: (1) is not a pathogen but it is caused by a pathogen, (2) it is not symptom but results in symptoms, (3) it is not a condition as the condition results from disease and is not synonymous with it, (4) is not an injury, (5) can not be catching or infectious, it is actually the pathogen which is catching or infectious, (6) results from continuous irritation, and (7) is a malfunctioning process and this must result in some suffering and hence disease is a pathological process.

Based on the concept of "Dialectical materialism" Das Gupta (1977) defined disease as a malfunctioning process involving more or less continuous interaction between host and pathogen occurring through a large number of quantitative changes between qualitative states of healthy and diseased conditions"..

Thus, most of the definitions emphasize on (1) continuous interaction/ irritation between host and pathogen and (2) disease is a malfunctioning process. These two points require in depth analysis.

B. Continuous Vs. Transient Interaction

The concept of continuous interaction, was emphasized by Horsfall and Dimond (1959) mainly for the purpose of differentiating disease from injury. Disease signifies a dynamic process, whereas injury represents a static state. However, there are situations where injury does not result in instantaneous death of the plant but it is of sufficient magnitude to lead to the malfunctioning of physiological processes of plants beyond their easy tolerance resulting in gradual development of symptoms at cellular and/or morphological level, eg. injuries caused by air pollutants, herbicides, sun scald, cold, etc. Similarly, damage to the vascular system caused by stem borer is purely an injury. If this injury is of such a magnitude that plant is unable to compensate resulting losses in its water and nutrient transports, it would suffer ultimately. The resulting effect would gradually manifest at morphological level as a wilt or dieback symptom. Therefore, there is little justification in excluding these situations from realm of disease.

Another major question is continuous interaction between whom? Of course, host is one partner but whether another partner is primary incitant or its product? What about crown gall, nutritional deficiencies, insect toximias, toxicities caused by air pollutants like OZONE, SO_2, PAN, etc., sun scald or cold injuries or symptoms induced by host- specific pathotoxins rather than the pathogen itself?. What should be considered as a threshhold point to demarcate between transient and continuous interaction?. We have already discussed the situations where injury might result in disease development. This leads to the conclusion that there is not much justification in including the concept of continuous interaction or association in definition of plant disease.

C. Physiological Vs. Biochemical Process

What are these processes where malfunctioning results in disease? What are these malfunctioning processes which constitute disease? Studies conducted in the last three decades have revealed some of the finer details of host-parasitic interaction. It is now accepted that first step in interaction, particularly in biotrophs, is mutual recognition of host and parasite as a compatible partner. This recognition is essential for establishment of genetic and subsequent physiological synchrony. Positive recognition for basic compatibility leads to a cascade of biochemical changes in the host. Several biochemical reactions are intensified, inhibited, induced, or altered in integrated manner to support growth of the pathogen. The pathogen not only draws nutrients from host to support its growth and reproduction but its interaction with the host results in synthesis and/or release of hydrolytic enzymes and toxic metabolites detrimental to the host, resulting in shift of balance more and more toward parasitism. If alterations are beyond the tolerance limit of the host, the latter suffers at its pysiological, cytological, and/or morphological level resulting in different types of symptoms and syndromes depending upon host and parasite involved. Similar to biotic causes, abiotic factors like nutritional deficiencies, toxins, pollutants, herbicides, etc. also bring about a cascade of biochemical alterations in host plant resulting in malfunctioning of its physiological processes which are gradually manifested at morphological level.

Biochemical alterations precede physiological changes which in turn may result in cytological and morphological alterations recognized as disease symptoms. Since, symptoms are not synonymous to disease but are results of disease, latter must precede the former. Hence, disease becomes synonymous to altered and induced biochemical changes in host brought about by an invading pathogen.

D. A Definition is representative of time

Lack of an adequate definition of "disease" is not at all an impediment to the progress of plant pathology. A definition does not result in an advancement of our understanding of the phenomenon. A better understanding of the phenomenon enables us to define it more accurately and precisely. Definition of any phenomenon given at a particular time must represent the status of our understanding of the same at that time. Therefore, a definition must be representative of time.

Analysis presented so far leads us to the conclusion that whenever development of symptom is gradual (as opposed to sudden death), the mechanism involved is more or less the same (i.e. alteration in biochemical reactions precedes malfunctioning of the physiological processes followed by expression at morphological level) irrespective of the nature of the cause (biotic or abiotic) or association (i.e. transient or continuous). At the same time there is growing realization that malfunctioning of the physiological processes should be considered as a symptom rather than disease itself.

Keeping in view the above discussion, a moderately precise definition of "disease" is proposed as "a sum total of the altered and induced biochemical reactions in a system of a plant or plant part brought about by any biotic, mesobiotic or abiotic factor(s) leading to malfunctioning of its physiological processes and ultimately manifesting gradually at cytological and/or morphological level. All these alterations should be of such a magnitude that they become a threat to the normal growth and reproduction of the plant".

IV. Plant Pathogens - Concepts and classification

Organisms suffer from diseases or disorders due to some abnormality in the functioning of their system. These abnormalities may be due to factors which have no biological activity of their own (abiotic factors) or those entities which show some biological activity (mesobiotic agents) and those that are established as cellular organisms. A PATHOGEN can be broadly defined as any agent or factor that induces ***pathos*** or disease in an organism, but the term is generally used to denote biotic and mesobiotic causes. PATHOGENICITY is the ability of an organism to cause disease while PATHOGENESIS is the chain of events that lead to development of disease.

Which factor, entity or agent should be called abiotic and which one biotic or living? By intution and experience, we have known that a thing that does not grow, reproduce, move, or show response to external stimuli is non-living and those that show these properties are living. However, when viruses appear in the picture, the whole concept of living vs. non-living becomes some what confused. The groups of the pathogens or causes of plant diseases are given in the chart on the next page:

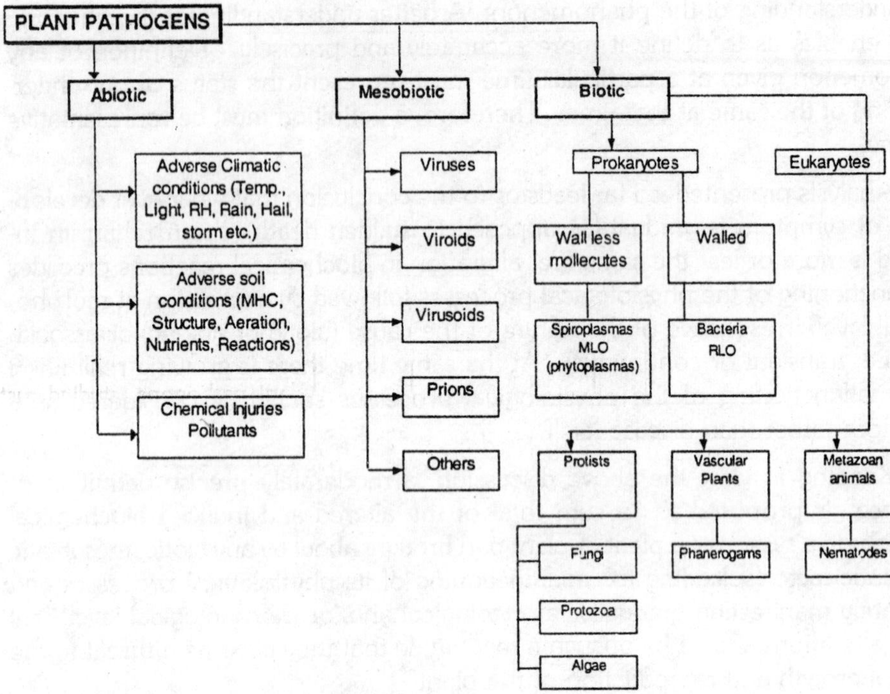
PLANT PATHOGENS
Abiotic
Mesobiotic
Biotic
Adverse Climatic conditions (Temp., Light, RH, Rain, Hail, storm etc.)
Adverse soil conditions (MHC, Structure, Aeration, Nutrients, Reactions)
Chemical Injuries Pollutants
Viruses
Viroids
Virusoids
Prions
Others
Prokaryotes
Eukaryotes
Wall less mollecutes
Walled
Spiroplasmas MLO (phytoplasmas)
Bacteria RLO
Protists
Vascular Plants
Metazoan animals
Fungi
Protozoa
Algae
Phanerogams
Nematodes

V. Classification of Plant Diseases

A. General Classification

Various schemes of classifying plant diseases have been proposed. A disease may be LOCALIZED or it may be SYSTEMIC. The diseases are SOIL BORNE, SEED BORNE or AIR BORNE. The symptoms or signs which appear on the affected plant parts, also form a basis for grouping the plant diseases. Thus, we find diseases known as RUSTS, SMUTS, WILTS, MILDEWS etc. According to host plants the diseases may be grouped as **cereal diseases, forage crop diseases, flax diseases**, etc.

Mc New (1950, 1960) proposed a system based upon seven physiological processes. Depending upon which vital functions are being adversely affected, a plant disease would be classified in one of the following groups:

(1) Soft rots and seed decays.

(2) Damping-off and seedling blights.

(3) root rots.

(4) Gall diseases and others in which meristematic activity is impaired.

(5) Vascular wilts.

(6) Diseases affecting photosynthesis (spots, blights, mildews, rusts, etc.).
(7) Diseases interfering with translocation.

B. Based on occurrence

(1) Endemic Diseases:

The word endemic means "prevalent in, and confined to a particular district or country" and is applied to disease. When a disease is more or less constantly present from year to year in a moderate to severe form, in a particular country or part of the earth, it is classed as endemic.

(2) Epidemic or Epiphytotic Diseases:

The term "epidemic" is derived from a Greek word meaning "among the people" and in true sense applies to those diseases of human being which appear very virulently among a large section of population. To carry the same sense in the case of plant diseases the term "epiphytotic" has been coined. An epiphytotic disease is one which occurs widely but periodically. It may be present constantly in the locality but assumes severe form only on occassions.

(3) Sporadic Diseases:

Those diseases which occur at very irregular intervals and locations and in relatively few instances.

(4) Pandemic Diseases:

These occur all over the world and result in mass mortality, eg. late blight of potato.

C. According to major causal agents

(1) Noninfectious or Nonparasitic or physiological diseases:

These are diseases with no biotic or mesobiotic agents associated, remain noninfectious and can not be transmitted from one diseased plant to another healthy plant.

(2) Infectious Diseases:

These are diseases which are incited by biotic and/or mesobiotic agents under a set of suitable environments. Association of a definite pathogen is essential with such diseases.

D. Based on production of Inoculum

(1) Single-cycle disease (Simple interest Disease):

When the increase of disease is mathematically analogous to simple interest of money, it is called simple interest disease. There is only one generation of disease in the course of one epidemic. Such diseases develop from a common source of inoculum, i.e. the capital is constant, and often there is one generation of infection in a season (Fig. 2.)

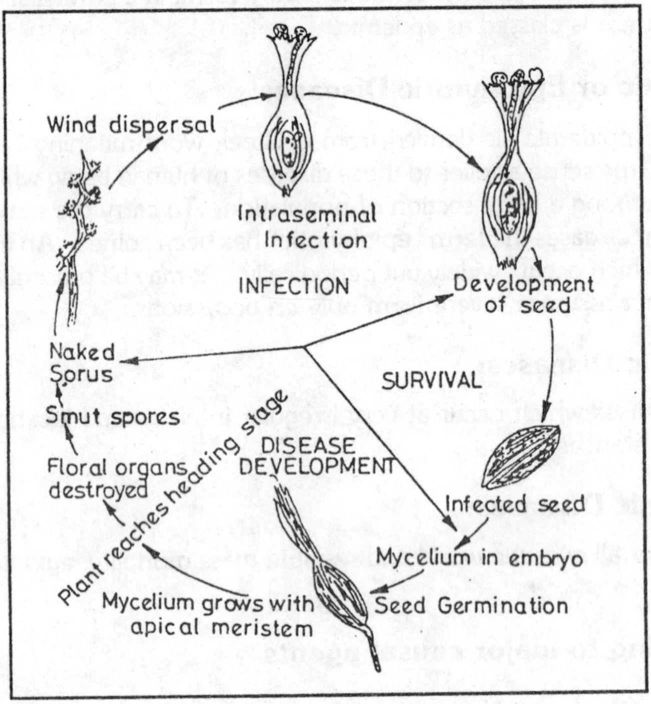
Wind dispersal
Intraseminal
Infection
INFECTION
Development
of seed
Naked
Sorus
SURVIVAL
Smut spores
Plant reaches heading stage
DISEASE
DEVELOPMENT
Floral organs
destroyed
Infected seed
Mycelium in embryo
Mycelium grows with
apical meristem
Seed Germination

Figure-2 : Simple Interest Disease (single cycle) e.g. loose smut of wheat

(2) Multiple-cycle Disease (Compound interest Disease) :

When the increase in disease is mathematically analogous to compound interest of money, the disease is called compound interest disease. There are several or many generations of the pathogen in one life-cycle of the crop, i.e. the capital is increased by the amount of interest (Fig. 3).

E. Based on Host-Pathogen Dominance System

Kommedahl and Windels (1979) have divided diseases, based on the "Host-

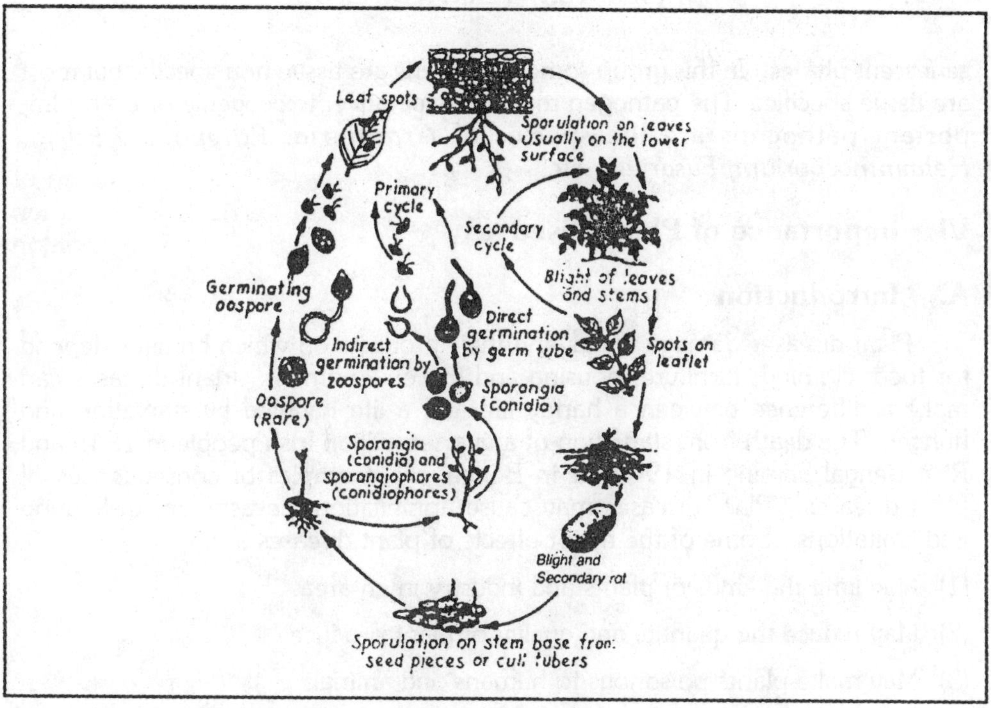
Leaf spots
Sporulation on leaves
Usually on the lower
surface
Primary
cycle
Secondary
cycle
Blight of leaves
and stems
Germinating
oospore
Direct
germination
by germ tube
Spots on
leaflet
Indirect
germination by
zoospores
Sporangia
(conidia)
Oospore
(Rare)
Sporangia
(conidia) and
sporangiophores
(conidiophores)
Blight and
Secondary rot
Sporulation on stem base from
seed pieces or cull tubers

Figure-3 : Compound interest (multiple-cycle) disease e.g. late blight of Potato.

Pathogen Dominance System" described below:

(1) Pathogen-Dominant Diseases:

The pathogen is dominant over the host, but the relationship is transitory because the resistance of the host is less initially than it becomes eventually. Such pathogens are tissue nonspecific and attack young, immature root tissues or senescent tissues of mature plant roots. They seldom damage a rapidly growing, maturing root, so the period of disease development is short. Sometimes such pathogens are macerative, sometimes they are toxicogenic, or sometimes both can occur. The pathogenesis is due primarily to the virulence of the pathogen. Physiological specialization is relatively uncommon. Important pathogens are *Macrophomina, Phytophthora, Pythium, Rhizoctonia, Sclerotium*, etc.

(2) Host-Dominant Diseases:

In host-dominant diseases the host is dominant and the pathogen is successful only when factors favour the pathogen over the host. The resistance of the host is strong enough to keep the pathogen from advancing too rapidly against the host defenses during the vegetative growth phase and the host thereby prolongs the relationship. Damage is most severe in plants in the reproductive and senescent phases. In this group some pathogens are tissue non-specific but most are tissue specific. The pathogen may be macerative, toxicogenic or both. Important pathogens are the species of *Armillaria, Polyporus, Poria. Helminthosporium, Fusarium*, etc.

VI. Importance of Plant Diseases

A. Introduction

Plant diseases damage plants and their products on which humans depend for food, clothing, furniture, housing and the environment. Plant diseases can make a difference between a happy life and a life haunted by starvation and hunger. The death from starvation of a quarter million Irish people in 1845 and Rice Bengal Famine in 1943-44 in Bengàl are examples of consequences of plant diseases. Plant diseases may cause annihilation, devastation, disfiguring and limitations. Some of the major effects of plant diseases are;

(1) May limit the kinds of plants and industry in an area.

(2) May reduce the quantity and quality of plant produce.

(3) May make plants poisonous to humans and animals

(4) May cause financial losses, and

(5) the cost of controlling is also a direct loss.

B. The Damage and Losses

Zakok (1970, 1973) has suggested a classification of losses (Table-1) that describes the complexity and inter dependence of loss at all levels of society. He also proposed three useful concepts for describing the dynamics of crop destruction. They are;

(1) Injury:

Any observable deviation from the normal crop; injury may lead to damage.

(2) Damage:

Any decrease in quantity and quality of a product; damage may lead to loss.

(3) Loss:

Any decrease in economic returns from reduced yields and cost of agricultural activities designed to reduce damage.

Table-1 : Classification of crop losses caused by injurious agents

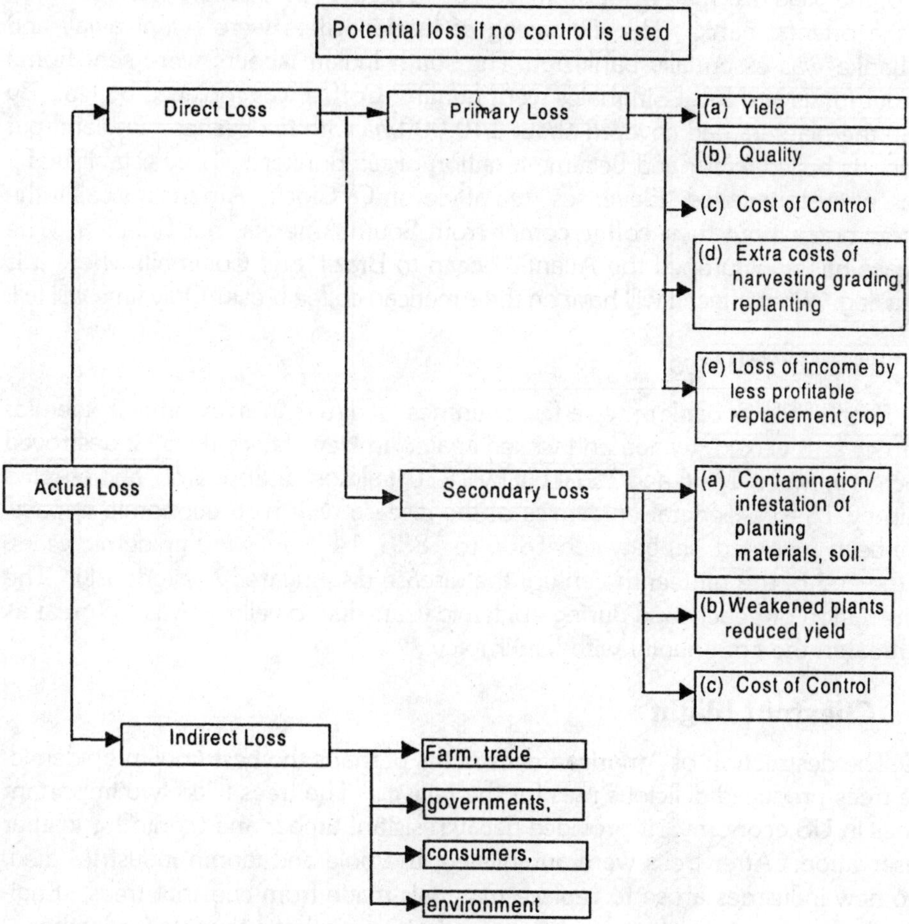
Potential loss if no control is used
Direct Loss
Primary Loss
(a) Yield
(b) Quality
(c) Cost of Control
(d) Extra costs of harvesting grading, replanting
(e) Loss of income by less profitable replacement crop
Actual Loss
Secondary Loss
(a) Contamination/ infestation of planting materials, soil.
(b) Weakened plants reduced yield
(c) Cost of Control
Indirect Loss
Farm, trade
governments,
consumers,
environment

C. Epidemics and Human Affairs

Plant disease epidemics have influenced man's food, his health, his social customs, his economics, etc.

(1) Epidemics - Human Culture

(a) Coffee Rust

From India coffee was taken to Sri Lanka in 1835. The British in Ceylon were growing only 200 ha of coffee. By 1870, they had about 2,00,000 ha and exported 50 million Kg/year. Then the devastating rust disease (coffee rust) struck.

It killed the coffee trees faster than they could be replaced. In the 19 years after 1870, the yield dropped from 50 million kg to zero. 417 planters went broke as did the oriental bank. The life saving of its depositers were swept away and Sri Lanka was essentially bankrupt. The south Indian labours were sent home without funds and local Singhales went hungry. Coffee was replaced by tea. By 1880 the planters had covered about 140,000 ha with tea bushes. England put tea in its boiled water and became a nation of tea drinkers. They established a pleasent custom called "Elevenses" tea at eleven O' Clock. Americans call it the "coffee break", but their coffee comes from South Amerika, not Srilanka. The desease has now jumped the Atlantic ocean to Brazil and Columbia where it is spreading. What affect it will have on the American coffee break? Only time will tell.

(b) Peach Yellows

Peach yellows bankrupted a few countries. It differs from all other epidemics on trees as it waxed, waned and waxed again. In New Jersey (U.S.) it destroyed peaches between 1806 and 1814 but by 1830 it almost disappeared, and erupted again by 1846. General epidemics of the disease with high economic impacts have been recorded. In between 1806 to 1886, 14 years were epidemic years. By the end of the ninteenth century the disease disappeared mysteriously. The depreciation of peach land during epidemic years due to yellows was so great as to threaten the community with bankruptcy.

(c) Chestnut blight

The destruction of American chestnut is perhaps the best known epidemic. The trees produced delicious nuts for the natives. The trees filled two important niches in US economy. It provided decay resistant timber and tannin for leather preservation. After trees were annihilated, the pole and tannin industries died. Two new industries arose to replace materials made from chestnut trees. Engineers built creosoting plants and chemist devised synthetic tannins for leather.

(2) Wars and Epidemics

(a) Late blight of potatoes - world war I

It is established that late blight of potatoes helped the British defeat the Germans in World War I. Potato growers were debarred from using the only medicine "Bordeaux Mixture". The weather in Germany during 1916 resembled that in Ireland in 1845-46 cold rain, cold rain, cold rain. The German military leaders would not release the "Cu" required to make "Bordeaux Mixture". They used "Cu" for shell casings, electric wire, etc. The result was total crop destruction. Eventually, malnutrition and starvation. The soldiers were not hungry but their family back home starved and died. This weakened soldier's moral. The military might weakened and it collasped in 1918. *P. infestans* played the role. It reminds us of the old saying that "For want of a nail the shoe was lost; for want of a shoe the horse was lost; and for want of a horse the battle was lost".

(b) Dutch Elm Disease

During World War I, the allied forces imported hundreds of chinese labourers to dig trenches in Flanders. While coming for the job, the labourers carried "wooden baskets" made from chinese elm. Some such baskets carried vector beetle harbouring the fungus. Subsequently the disease killed millions of elms in Europe. Later the disease entered the U.S. and destroyed ELM Industry. It is a significant case of a war induced epidemic.

(c) The great Bengal Famine

Dr. Padmanabhan (1973) was appointed mycologist in Bengal when the famine was at its height. While travelling to join the assignment on October 18, 1943, he writes that "he did see dead bodies, and starving and dying all along the way". The situation continued till the arrival of new crop in the summer of 1944. The war worsened the famine because the Japanese army had occupied Barma and thereby shut off the normal flow of rice from the most important source. British could not manage rice for Bengal due to their engagement in the war. Thus, you have the complex relation between epidemics and war, may be indirectly.

D. Disease Management and World Population

The human population in the world in 1993 was about 5.57 billion and became 6.00 billion on 11th of Sept., 1999, and, at the present rate of 1.70 per cent annual growth, it is expected to be 6.2 billion by the year 2000, 7.1 billion by the year 2010, and 8.5 billion by 2025 (Agrios, 1996). Currently, the world population increases by one billion every 10-11 years. Paradoxically, the developing countries, in which 56 to 80 per cent of the population is engaged in agriculture, have the lowest agricultural output, their people are living on a substandard diet and they have the highest population growth rates (2.64%). It is estimated that even today some 2 billion people suffer from hunger or malnutrition or both. To feed these people and the additional millions to come, suppression of plant diseases and reduction of yield losses due to diseases are a necessary part of increasing food supply.

E. Plant Diseases and Crop Production

Having become self sufficient in food grain production, there is no room for complacency. Notwithstanding the success, the rate of agricultural growth has to be boosted from the present 2.5% per year which is marginally above the 2.1% annual increase in population. In fact "Second Green Revolution" will have to come from making grey areas green, while simultaneously sustaining the green areas as green (Paroda, 1995). Also diversification, value addition, and commercial agriculture are to find prominent place in Indian agriculture to meet our nutritional demand and also to become globally competitive. Brownn and Kane (1994) in their book "Full House" predicted that at the current rate of population growth and environmental degradation, coupled with an improved consumption capacity of the poor, India will have to import annually over 40 million tonnes of food grains by the year 2030. This is four times more before the onset of green revolution. Swaminathan (1995) has very rightly and timely cautioned "we should examine this issue seriously and should not allow complacency overtake the farm sector just because "we have over 13 million tonnes of food grains available with government".

It is in the context, that plant protection will have to play most significant and major role in increasing food production through management of pests and diseases which have been causing a setback in agricultural production resulting in unimaginable losses. The annual loss is estimated to be over Rs. 20,000/- crores (Jayraj *et.al.*, 1994).

The crop losses caused by pests become particularly alarming when one considers their distribution among nations of varying degree of progress and development. In developed countries (Europe, U.S., Australia, New Zealand, Japan, Israel and South Africa), in which only 8.8% population is engaged in agriculture, the losses are considerably lower than those in developing countries in which over 50 per cent of the population is engaged in agriculture (Agrios, 1996). Oerke *et.al.* (1994) have indicated that the proportion of crop produce lost to diseases, insects and weeds has actually increased in all continents. It is estimated that the total annual production for all agricultural crops worldwide is about $1200 to $1300 billion (U.S. dollars, 1995). Of this, about $500 billion worth of produce is lost annually to pests. An additional loss of about $330 billion would occur annually but is averted because of plant protection practices that are employed. Approximately $ 26 billion is spent annually for pesticides alone, particularly in Europe and U.S.A. (Agrios, 1996).

F. Agricultural Technologies, Human Society and Plant Diseases

1. Crop Breeding and Plant Diseases

Dwarf and Semi dwarf wheat varieties developed at CIMMYT, Mexico, changed Mexico and some other countries in Africa and Asia from a wheat importing to a wheat exporting nation. When virulent pathogens or new virulent races/biotypes came in contact with genetically uniform crop grown extensively, rust epidemics developed and yields reduced by above 55%. Similarly *Septoria* leaf and glume blotch reduced production by over 80%. Many dwarf wheat cultivars have been found highly susceptible to powdery mildew, leaf blights, smuts, etc., as compared to local varieties. The story of male sterile corn plants that would not require detasseling is well known. This effort led to development of hybrids that were genetically uniform in carrying the cytoplasm with trait for male sterility. The same cytoplasm also carried genetic factors that made them susceptible to *Bipolaris maydis* and as a result the southern corn leaf blight destroyed more than $ 1.0 billion worth of corn in USA in 1970. Similarly, IRRI at Phillipines, several nonlodging dwarf rice varieties were developed and distributed widely. In some countries bacterial leaf blight (*X. campestris pv. oryzae*), and rice-blast (*Magnaporthe grisea*) became severe on new high-nitrogen fertilized rice varieties.

2. Cultural Practices and Plant Diseases

Irrigation, tilage, plant propagation and mechanization of agriculture have been known to affect prevalence, incidence and severity of plant diseases. For example, in Venezuela, expansion of irrigation made it possible to raise two crops of paddy annually. This change from one to two crops favoured vector of viral disease *Hoja blanca,* resulting in serious loss of the paddy crop. Irrigation by methods such as flooding, furrow, or sprinkling, have increased population and distribution of many fungal, bacterial and nematode pathogens. The use of herbicides, no doubt, have resulted in increase in areas of crops grown under reduced or no tillage system. Many pathogens survive better and cause severe losses in minimal tillage than when the crop debris is burried by ploughing and allowed to decompose.

3. Fertilizers and Plant Diseases

The increased use of fertilizers, that too in high amounts, particularly nitrogen have been recorded to increase severity of such diseases as rusts, powdery mildews, blights etc. The pathogens causing these diseases prefer young succulent tissues.

4. Pesticides and Plant Diseases

The weedicides being used increasingly in cultivated fields, may injure the plants directly and also influence soil borne pathogens and the microflora including antagonists. Similarly, other agrochemicals (fertilizers), insecticides, fungicides may also alter microbial equilibrium particularly the predators and mycoparasites. Some fungicide like benomyl if used regularly and widely will encourage *Pythium* spp. as they are insensitive and therefore may cause severe infections. The use of pesticides has been increasing steadily at an annual rate of 14% since mid 1950s. It is estimated that by the year 2000 more than 3 billion kg of pesticides will be used annually world wide. There is no doubt that pesticides have increased crop production. The cost of production, distribution, and application of pesticides is, of course, another form of financial loss caused by pests. Furthermore, these pesticides pollute the environment and ecosystems.

CHAPTER II

HISTORY OF PLANT PATHOLOGY

I. Introduction

A study of the history of any science helps to make the subject clearer. We get a better perspective of the subject; know the contributions made in that field; the problems that were encountered and the manner in which they were tackled. We can follow the guidelines set by previous scientists while studying a particular branch/ area of the science.

II. PHYTOPATHOLOGICAL TEACHING AND RESEARCH IN INDIA

The science of plant pathology in modern period in this country is rather not very old. The development of this subject started with the study of mycology. The first Indian universities that were established in 1857 at Calcutta, Bombay and Madras emphasised collection, identification, characterization and taxonomy of fungi. Upto 1930, there was more attention to the study of fungi than to diseases caused by them. Plant pathology as university subject came into being in 1930 at universities of Madras, Allahabad and Lucknow.

The study of fungi in India was initiated by Europeans in the 19th century. During 1850-1857, A.D. Cunningham and A. Berkelly started identification of fungi. Cunningham studied Indian rusts and smuts. K.R. Kirtikar was the first Indian Scientist who collected and identified fungi. Organised research on fungi and diseases caused by them was started in this country in the first decade of this century. Imperial Agricultural Research Institute was established at Pusa, Samastipur, Bihar. But after the earthquake in 1934, this institute was shifted to Delhi and now recognized as Indian Agricultural Research Institute (IARI), New Delhi. It was in this institute at Pusa that E.J. Butler (the father of Indian Plant

Pathology), before 1910, initiated scientific and exhaustive study of fungi and plant diseases. Butler stayed in India for about 20 years and besides training several Indian scientists, studied wilts of cotton and pigeonpea, diseases of rice, toddy palm, sugarcane, potato and cereal rusts. He wrote a monograph on Pythiaceous and allied fungi and a book "Fungi and Diseases in Plants" in 1918. Both publications are classics of Indian Phytopathology. Butler left India in 1920 to join as first Director of CMI in England.

J.F. Dastur (1886-1971), a collegue of Butler, studied the genus *Phytophthora* and diseases of castor and potato caused by its species. He is internationally known for establishing the species. *Phytophthora parasitica* from castor. G.S. Kulkarni published exhaustive literature on downy mildew and smuts of jowar and bajra. S.L. Ajrekar studied wilt of cotton, sugarcane smut and ergot of bajra. B.B. Mundukar, a dynamic teacher, researcher and writer will always be known for establishing "Indian Phytopathological Society:" with its journal "Indian Phytopathology" in 1948. He studied Indian smut fungi and cotton wilt. He also wrote a book "fungi and plant diseases". Prof. K.C. Mehta of Agra College discovered disease-cycle of cereal rusts, identified physiologic races and studied life-cycle of linseed rust. J.C. Luthra and his colleagues developed solar seed treatment for the control of loose smut of wheat.

Prof. S.N. Dasgupta at Lucknow and Prof. T.S. Sadasivan at Madras established strong schools of fundamental plant pathology, especially the biochemistry of host parasite relationship. Sadasivan's school contributed to the concepts of "Vivotoxins" mechanism of fungal pathogens. G. Rangaswami, B.P. Bhide and M.K. Patel did pioneering work on bacterial plant pathogens in India. Dr. M.J. Thrimulachar conducted exhaustive study of smuts and rusts. He studied their taxonomy , cytology and life-cycles. He established several new genera as well. He successfully introduced the use of antibiotics in plant disease control. Aureofungin and streptocycline are the examples of these antibiotics developed at Hindustan Antibiotic Ltd. where Dr. Thrimulachar worked.

Prof. M.S. Pavgi, a colleague of Dr. Thrimulachar, worked in the department of Mycology and Plant Pathology at B.H.U., Varanasi. He investigated several fungi particularly *Protomyces, Taphrina, Synchytrium*, etc. He studied life-cycles and parasitism of several rusts and smuts.

Teaching of plant pathology as a major subject in Indian universities started rather late. Perhaps, the university of Madras was the first to take-up plant pathology as a university science. University of Allahabad (Est. 1887) and university of Lucknow (Est. 1921) also took up the subject. Organised teaching in mycology and plant pathology (associateship) as part of agricultural sciences, was conducted by IARI, which later grew up and started degree course. Post-graduate programme was started at Agriculture College Kanpur in 1945. After the establishment of first Agricultural university in 1960 at Pantnagar, and subsequently in other states, teaching in plant pathology with its supporting courses in mycology, bacteriology, virology , nematology, biochemical plant pathology etc. became integral part of under-graduate and post-graduate programmes in agriculture. These agricultural universities have now taken up the leadership in phytopathological research, teaching and extension.

Dr. H.K. Sexena, served as Prof. and Head, Plant Pathology at Govt. Agric. College Kanpur (now C.S. Azad University of Agriculture & Technology) and trained a number of plant pathologists. His study on *R. solani* is recognised internationally. Dr. S.P. Raychaudhari, retired as Head, Division of Myc. and Plant Pathology at IARI, New Delhi. He is known for his contribution to phytovirology and mycoplasmal diseases.

Prof. R.S. Singh, served as Prof. & Head, Plant Pathology at Pantnagar. Besides being a versatile writer and teacher, he worked on biological control and ecology of soil borne pathogens. His book "Plant Diseases" is referred as Bible of Plant Pathology in India. Dr. Y.L. Nene, another former Head, Plant Pathology at Pantnagar and later Dy. Director General, ICRISAT, Hyderabad, investigated cause and cure of "Khaira Disease of Paddy". He also studied pulse viruses and chemical control of plant diseases.

II. HISTORICAL EVENTS

Significant contributions to plant pathology and allied areas in the autogenetic and pathogenetic areas are summarized in Table:

Scientist	Year	Work/Events
Micheli, P.A.	1679-1737	Studied many microscopic fungi, their spores and the growth of the same kind from such spores and of different kinds from air borne spores.
Needham, T.	1744	Obtained nematodes in wheat galls
Linnaeus, C.	1753	Described a large number of fungi in ***species plantaram***
Tillet, M.	1755	Conducted the successful inoculation of wheat by bunt, "powder" without knowing that it consisted of spores and proved the effectiveness of seed treatment.
Persoon, C.H.	1801	Described many fungi
Prevost, B.	1807	First studied the penetration of wheat by a fungus and proved that the wheat bunt was due to a fungus and could be controlled by treatment with $CuSO4$

Scientist	Year	Work/Events
Fries	1821-1832	Described many fungi
Unger, F.	1832-1847	Wrote ***treatises*** on plant diseases and emphasized the importance of ecological and physiological studies. His treatise ***Die Exantheme der pflanzen*** (1833) may be regarded as the first book on physiological plant pathology.
Anton de Bary	1831-1888	Made detailed study of the late blight fungus, proposed nomenclature and gave final proof of organisms being plant pathogens, studied rusts, smuts, downy mildews and rots, discovered heteroecious nature of rusts, life-cycle and parasitism of downy mildew and role of enzymes in pathogenesis, founder of modern experimental plant pathology.
Tulasne brothers (R.L. &C.)	1847-1865	Were with de Bary, worked on morphology and natural relationships of fungi, developed exciting methods of studying fungi.
Hartig, R.	1866-1878	Pioneer of forest pathology.
Woronin, M.	1878	Studied club root of crucifers, established slime mold as the cause of a plant disease.
Brefeld, O.	1872-1912	Introduced and developed the methods of growing microbes in pure culture.
Millardet, P.M.A.	1878-1885	Discovered Bordeaux mixture to control downy mildew of grapes which endangered grapes cultivation in Europe and threatened the wine industry in France.
Burril, T.J.	1878	Observed that fire blight of pear is caused by a bacterium
Mayer, Adolf	1886	First described a plant virus disease - the tobacco mosaic.
Biffen, H.R., Orton, W.A.	1904-1907	Studied the genetics of disease resistance and emphasized the importance of disease resistance breeding
Smith, E.F.	1888-1920	Added the fourth rule of proof to koch's postulates, did classical work on many bacterial diseases such as the wilts of cucurbits and solanaceous plants and rot of crucifers.

CHAPTER III

DEVELOPMENT OF DISEASE (PATHOGENSIS)

I. Introduction

Plant diseases result from interaction of a pathogen with its host but the intensity of this interaction is markedly affected by the environmental factors. The role of environment in pathogenesis is as important as susceptibility of the host and pathogenicity of the causal agent. Any consideration of disease in the crop, therefore, involves the DISEASE TRIANGLE.

The pathogen interacts with the host and vice-versa. Both influence each other, the host providing nutrition to pathogen and the latter causing disease in the former. The host also influences the environment through crop canopy, root exudates in soil, withdrawl of nutrients from soil, and other activities mediated through these effects. Environment affects the host through chemical, physical and biotic factors involved in plant growth and metabolism. Pathogen can also influence environment through such effects as defoliation, addition of diseased dead crop debris to soil, changing the host physiology and, thereby, host root exudation, etc. Environment may affect the pathogen in the same manner as the host but may be against the pathogen and in favour of the host or vice-versa.

Since these interactions and their effects are not spontaneous and time of interaction influences the result of interaction the "disease-triangle" can be modified as shown in the Figure-1.

II. PATHOGENESIS

Development of disease in plant is not a sudden effect. A chain of events is responsible for causation of any disease. The symptoms and manifestations of injury to the plant is the last link in the chain of events. Before symptoms appear the pathogen independently or with the host has to pass through several stages. These stages, reactions, and interactions arranged in a sequence lead to disease

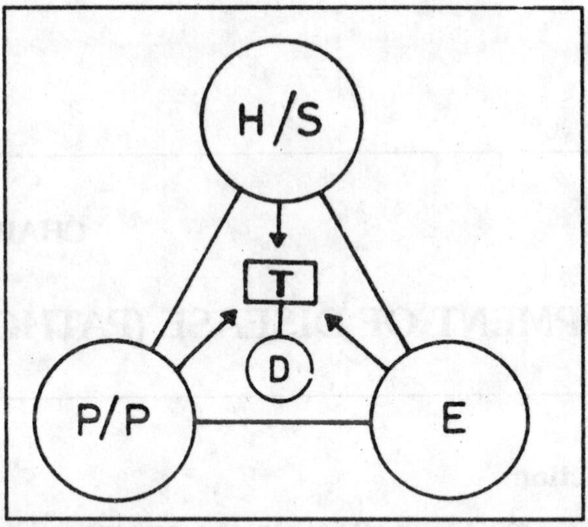
H/S
T
D
P/P
E

Figure-1 : A combination of factors required for disease appearance: H/S=Host/suscept, P/P=Pathogen/Parasite, E = Environment, D=Disease, T= Time.

development and the entire chain of events leading to disease development is known as Pathogenesis (Figure 2). The infection chain can be "continuous" or "intermittent". Further, these chains may be "homogenous" (survival on single plant species), or "heterogenous" (survival on many plant species).

III. NUTRITION LEVEL AND EVOLUTION OF PARASITISM

Fungi being devoid of chlorophyll are incapable of synthesizing their own food materials. On the basis of mode of nutrient uptake fungi are grouped into three broad biological categories. *Saprophytes, Parasites* and *Symbionts*. Those fungi which live on dead and decaying organic matter are called *saprophytes* or *saprobes* (Gr. *sapros* = rotten + *bios* = life). With further evolution, some of them have still remained incapable of growing on living organisms and survive exclusively as saprobe. They are called *obligate saprophytes* (L. *Obligare* = to bind + *sapros*). However, there are some which live as saprobe but are capable of infecting another living organism under some conditions. They are called *Facultative parasites* (L. *facultas* = ability + Gr. *parasitos*). Many fungi infect living hosts and take their food from them. They are *parasites* (Gr. *parasitos* = eating besides another). Further, some of them evolved as *obligate parasites* (L. *Obligare* = to bind + Gr. *parasitos* = table mate) and remaining as *Facultative soprophytes* (L. *Facultas* = ability+ Gr. *sapros*). Obligate parasites, in nature, obtain their food only from living protoplasm and can not be grown in culture on nonliving media. Contrary to it, facultative saprophytes are basically parasitic

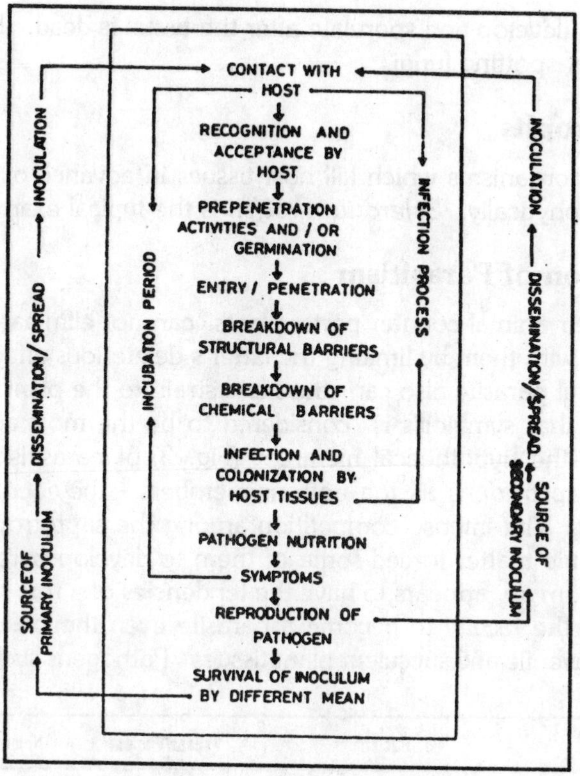
CONTACT WITH HOST
RECOGNITION AND ACCEPTANCE BY HOST
PREPENETRATION ACTIVITIES AND / OR GERMINATION
ENTRY / PENETRATION
BREAKDOWN OF STRUCTURAL BARRIERS
BREAKDOWN OF CHEMICAL BARRIERS
INFECTION AND COLONIZATION BY HOST TISSUES
PATHOGEN NUTRITION
SYMPTOMS
REPRODUCTION OF PATHOGEN
SURVIVAL OF INOCULUM BY DIFFERENT MEAN
SOURCE OF PRIMARY INOCULUM
DISSEMINATION / SPREAD
INOCULATION
INCUBATION PERIOD
INFECTION PROCESS
SOURCE OF SECONDARY INOCULUM

Figure-2 : Chain of events leading to disease development.

organisms but also capable of growing on dead organic matter under some conditions.

In fact, a number of fungi, once considered to be obligate parasites, have been cultured on artificial media. This makes the term obligate inappropriate. Luttrell (1974) recognized three broad categories of parasitism which are as follows:

A. Biotrophs

These are those organisms which, regardless of the ease with which they can be cultured, in nature obtain their food from the living tissues on which they complete their life-cycles. Some typical examples are rusts, smuts and mildews.

B. Hemibiotrophs

These are organisms which attack living tissues in the same way as biotrophs but continue to develop and sporulate after the tissue is dead. Typical examples of these are leaf-spotting fungi.

C. Necrotrophs

These are organisms which kill host tissues in advance of penetration and then live saprophytically. *Sclerotium rolfsii* is the typical example.

D. Evolution of Parasitism

Unlike their animal counter parts, plants can not eliminate their parasites; they try to live with them by limiting the latter's deleterious effects. At the same time a successful parasite also can cause less strain to the plant. It is from these considerations that symbiosis is considered to be the most advanced form of parasitism. In the hypothetical hierarchy (Fig. 3) of parasitism it is natural to consider the saprotrophs as the bottom members - they colonize only dead organic matter. The intense competition among the saprotrophs for the same source of organic matter forced some of them to develop parasitic abilities, for example *Pythium sp.* appears to have the tendencies of a true saprophyte, but it has developed the faculty to become a parasite even though restricted to only attacking the juvenile and succulent plant tissues. Pathogens such as *Rhizoctonia*

Degree of Synchronization (↑ increasing from Necrotrophs to Symbionts)

	Fungi	Bacteria	Viruses/Viroids	Nematodes	Mollecutes
Symbionts	Lichens, VAM Fungi	*Rhizobium* spp.			
Biotrophs	Powdery Mildews Downy Mildews Rusts		Viroids Viruses	Secondary Endoparasites Migratory endoparasites Ectoparasites	MLOs/ Spiro Plasma
Semi-biotrophs	*Phytophthora* spp. smuts	*Agrobacterium* spp.			
Necrotrohs	*Colletotrichum* spp. *Fusarium* spp. *Rhizoctonia* spp.	*Pseudomonas* spp. *Xanthomonas* spp. *Clavibacter*, RLB, *Xylella*, *Streptomyces*			

Figure 3 : Hypothetical hierarchical positioning of the different plant pathogens based on their parasitic advancement and theoretically measured as their ability to establish genetic and physiological synchrony with the host species.

and *Sclerotium* can be considered more hardy because in addition to attacking tender tissues, they can also exploit comperatively mature tissues. All these pathogens have a wide host range. They are necrotrophs, they cause extensive damage to the host tissues by employing chemical weapons - enzymes, secondary metabolites. On moving up the hypothetical hierarchical ladder of the parasitism (Fig. 3), necrotrophs appear showing more dependence on host and less saprophytic survival ability. They rely more on toxins than on wall degrading enzymes. Although they cause extensive tissue necrosis, little or no tissue maceration occurs. Further advancement towards semibiotrophs and biotroph has probably been guided by:

(1) Lesser dependence of the pathogens on toxins and enzymes, (2) more involvement of phytoharmones (3) decreased deleterious effects of the parasites and increased dependence of the parasite on living host cells, (4) decreased host range and, (5) more and more synchronization of the physiological processes of the host and parasite. Increased physiological synchronization guided by genetic synchronization, probably lead to the evolution of symbionts such as lichens and mycorrhizae where both host and parasite start deriving benefits from each other. Unlike biotrophic pathogens, biotrophic symbionts exhibit a comperatively wide host range. This can be explained on the basis of (1) the ability of the parasite to breach the general defense barriers of the plant, (2) the biotrophic relationship with less strain on the host plant, and (3) the benefits derived by the host due to association with these biotrophs.

A pathogen is said to have developed basic compatibility with the host if it is able to synchronize its own physiological processes with those of the host after breaching the latter's general defense barriers. However, mere establishment of basic compatibility does not guarantee that all the individuals of a pathogen would be able to infect all the individuals of a particular host species to the same extent under similar environmental conditions. This second level of interaction between host and parasite is termed as race-cultivar compatibility or specificity.

IV. STAGES IN DISEASE DEVELOPMENT

A. Inoculum and Inoculation

The infective propagules(unit of inoculum) coming in contact with the host are known as **Inoculum** and the process which ensures this contact is called **Inoculation.** An inoculum that survives dormant (overseasoning) and causes the original/ initial infections is *primary inoculum* and infections it causes as *Primary Infection*. An inoculum produced from primary infections is called a *secondary inoculum*, and it, in turn causes *secondary infection*.

B. Arrival of Inoculum (Contact)

Most of the bacterial and fungal pathogens come into contact with the hosts accidently in the form of wind-borne or water-borne propagules. Some fungi, many bacteria and most of the viruses are brought to their hosts by insect and other vectors. Motile propagules (Zoospores) of fungi are attracted to plants by root exudates (chemotaxis) and get accumulated behind the root tip (Figure 4). Plant parasitic nematodes like zoospores have also been studied for their oriented movement and accumulation at the root zone. Several plant parasitic nematodes are reported to react positively in a CO_2 gradient and move to the source of the gas. Other gradients that are produced by or associated with growing roots include water, O_2, pH, root exudates such as aminoacids, organic acids, and electrical. Among the amino acids, glutamic acid has been demonstrated to attract plant parasitic nematodes under certain conditions.

C. Recognition and Specificity

Majority of plant pathogens are known to have a well defined host-range. This type of host pathogen specificity is termed as "basic compatibility". The limited host range of most pathogens suggests that the non host defense mechanisms are not easy to overcome. It is speculated that (1) lack of specific nutrients required by the pathogen, (2) inability of the pathogen to break or by pass general defense barriers of the plant, and/or (3) non-recognition of the pathogen and host as compatible partners required for the establishment of genetic and physiological synchrony between them, may be possible factors responsible for the nonestablishment of the basic compatibility.

1. Phenomenon

In order to establish basic compatibility with the host, which is a prerequisite for the pathogenicity, a parasite must be able to breach the general defense barriers of the host and be recognized by the host as compatible partner. Host bothers about only those parasites which are able to establish basic compatibility with it. However, as discussed earlier, the establishment of basic compatibility only gives the potential, it does not guarantee that a pathogen would be able to exploit host as it is quite evident from race-cultivar type of interactions. For that a pathogen must be able to evade the recognition by the hosts as a "non-self" or if it is recognized it must be able to supress (using suppressors) or breach (using degradative enzymes) the subsequent induced reactions of the host like hypersensitive cell death, phytoalexin accumulation, cell wall barriers (HRGP, lignin, etc.) synthesis of hydrolases (chitinase, glucanase, etc.) and/or release of systemic elicitors to induce resistant reaction in uninvaded distant cells/ tissues (Figure-5).

Figure-4 : Attraction of zoospores of *P. cinnamoni* to roots (after Milholland, R.D. *Phytopathology*, 1975).

So recognition between a biotrophic parasite and host plant operates at two levels, (1) for the establishment of basic compatibility and (2) for the establishment of specificity (race-cultivar compatibility). Recognition at basic compatibility level and nonrecognition as a "non-self" at race-cultivar compatibility level would result in development of disease, while recognition as nonself at second level would trigger the defense reactions of the host (Figure-6).

Extensive studies conducted during last two decades in the field of nature of

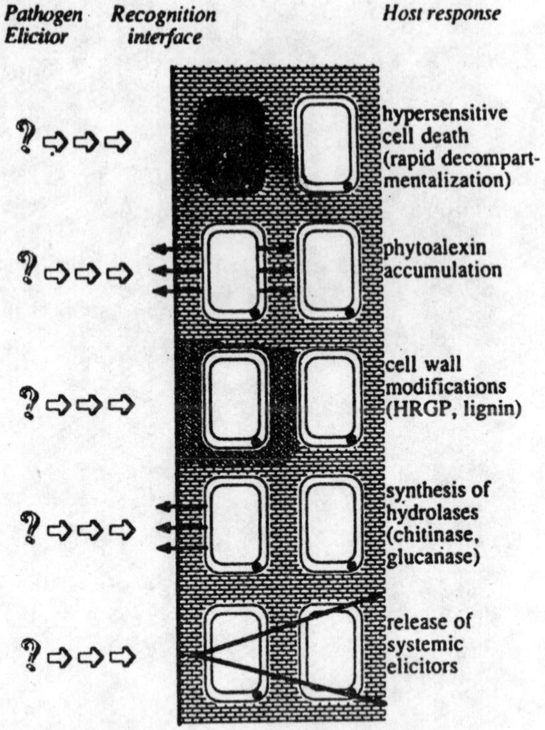
Pathogen Elicitor
Recognition interface
Host response
hypersensitive cell death (rapid decompartmentalization)
phytoalexin accumulation
cell wall modifications (HRGP, lignin)
synthesis of hydrolases (chitinase, glucanase)
release of systemic elicitors

Figure-5 : The range of host-cell responses involved in active (induced) host defense (from Callow et.al. 1987).

binding of animal hormones (insulin and glucagon, and toxins (cholera) with animal cells, pistil-pollen interaction in angiosperms in relation to inter and intraspecific incompatibility, interaction between fungal and algal components in lichens, and work on certain host-parasite interactions such as *Phytophthora infestans, Ralstonia solanacearum*-potato, *P. megasperma* var. *sojae* - soybean, and *Ralstonia solanacearum* - tomato, have brought following points to light (Singh et.al., 1988).

1. Outcome of host-parasite interaction is determined very early in the infection process - latest point being the contact of fungal haustorium to host plasmalemma (Figure 7).
2. Recognition may be either for compatibility (e.g. in *Rhizobium* legume, lichens, etc.) or incompatibility (e.g. in pistil-pollen, *R. solanacearum* tomato interactions).
3. Binding at specific site on cell surface may influence (enhance or alter) celll

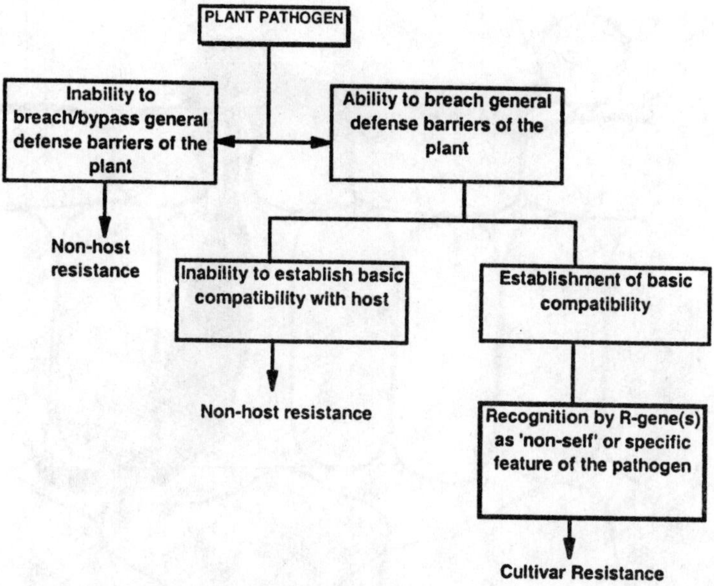
PLANT PATHOGEN
Inability to breach/bypass general defense barriers of the plant
Ability to breach general defense barriers of the plant
Non-host resistance
Inability to establish basic compatibility with host
Establishment of basic compatibility
Non-host resistance
Recognition by R-gene(s) as 'non-self' or specific feature of the pathogen
Cultivar Resistance

Figure-6 : Summary of the possible origins of non-host and cultivar resistance in terms of recognition.

metabolism (e.g. insulin and glucagon with animal cells, *Rhizobium* legume interaction) or it may lead to cell death (cholera toxin on animal cells, host-specific toxin on plant cells, *R. solanacearum* on host plant cells).

4. Recognition of host and parasite as a compatible or incompatible partners is probably determined by constitutive binding sites present on the surface of the host and the parasite. Phenomenon like hypersensitive cell death, phytoalexin accumulation, lignification, etc. may be consequence rather than cause of resistance.
5. Binding sites determining basic compatibility are probably different from those determining race-cultivar compatibility (i.e. resistance or susceptibility).
6. Binding of host with parasite (or its product) at site (s) determining basic compatibility will lead to an array of metabolic changes in the host resulting in increased synthesis of biomolecules and other nutrients required for the growth of the pathogen.
7. Binding at the race-cultivar compatibility site along with that of basic compatibility will result in resistant response by one or more of the following effects.
   (a) Shutting off of the metabolic changes induced by the binding at the basic compatibility sites.

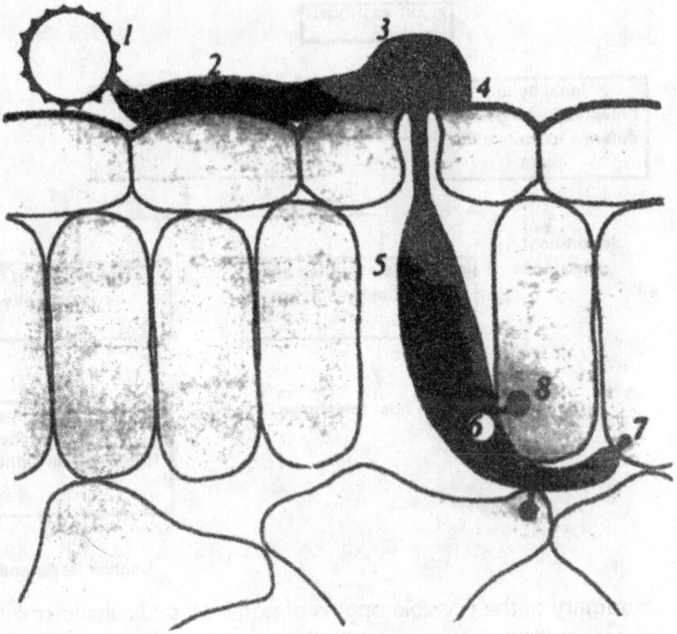
1
2
3
4
5
8
6
7

Figure-7: Diagrammatic representation of the interaction between a rust fungus and host-leaf tissue, illustrating the different levels of host-pathogen recognition: spore germination (1): directional and adhesive growth of germ tubes on host surface (2); appressorial differentiation over the vesicle in the substomatal cavity (3); and appressorial adhesion (4); expansion of the vesicle in the substomatal cavity (5); contact with mesophyll cell wall (6); initiation (7); and development (8) of haustoria in contact with host-plasma membrane (Callow et. al, 1987).

(b) Induction of death of the one or both the partners (hypersensitive response). Whether death of the pathogen precedes that of the host cell or vice-versa would depend upon the particular host-parasite system. Even both the situations may exist within the same host-parasite system depending upon the race-cultivar composition, such as in *P. infestans*-potato system.

(c) Inhibition or slowing down of furhter advance of the parasite with or without causing death of the host cell by inducing host's metabolic reactions leading to the synthesis of physical and/or chemical barriers.

Based on these observations and assumptions, a model is proposed to explain the host-parasite interaction in *P. infestans-potato* system which could also be extended to other biotrophic interactions (Figure 8) with some modifications. Instead of direct or prior to or after, surface interaction between host and para-

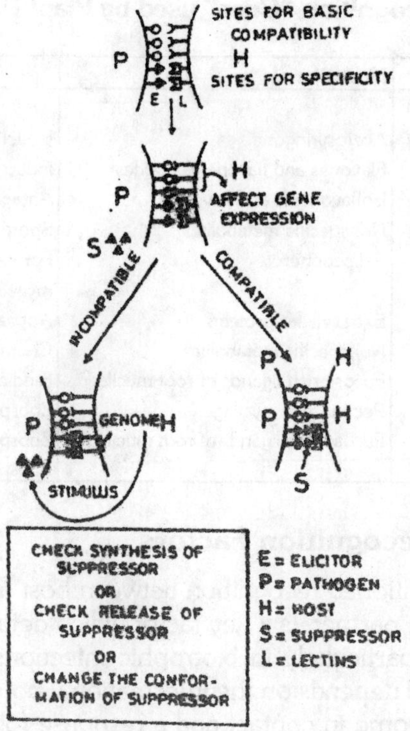
SITES FOR BASIC COMPATIBILITY
P
H
SITES FOR SPECIFICITY
E
L
P
H
AFFECT GENE EXPRESSION
S
INCOMPATIBLE
COMPATIBLE
P
H
P
GENOME H
P
H
S
STIMULUS
CHECK SYNTHESIS OF SUPPRESSOR
OR
CHECK RELEASE OF SUPPRESSOR
OR
CHANGE THE CONFORMATION OF SUPPRESSOR
E = ELICITOR
P = PATHOGEN
H = HOST
S = SUPPRESSOR
L = LECTINS

Figure-8 : A hypothetical model explaining recognition in a specialized host (H) - parasite (P) system. Recognition, determined by complimentary binding sites at the surface of the host and parasite operates at two levels-basic compatibility leads to the genetic and subsequent physiological synchronization of the host and parasite resulting in compatible response. However, simultaneous binding at the site of specificity induces host defense system leading to an incompatible reaction. Lectins (L) at the host surface and elicitors (E) on the surface of the parasite constitute the site of specificity. Supressors (S) produced by the parasite bind with host lectins constituting the site of specificity and convert the incompatible reaction into a compatible one: In order to counteract this situation the host, through its specificity binding sites tries to make it unsuitable for birding with supressors but not with elicitors present on parasite's surface or it blocks the synthesis or release of supressors in the parasite by producing some sort of inducer which ultimately leads to the incompatible response.

site, the latter may produce the chemical inducers/ suppressors which are recognized by the host or vice-versa.

Interaction of one component with the inducer/ suppressor of other component may determine the outcome of reaction. Few such chemical inducers and their effect are listed in Table-1.

Table-1 :. Host Recognition "Cues" used by Plant Pathogens and Parasites

Species	Host Signal	Process triggered
Bacteria		
Agrobacterium tumefaciens	Acetosyringone	Induction of *Vir* genes
Rhizobium meliloti	Flavones and flavone glycosides	Induction of *nod* genes
Rhizobium leguminosarum	Isoflavonoids and flavonols	Antagonism of *nod* gene
Fungi	Nonspecific metabolites	Spore germination on leaf surface
Many species	α-Tocopherol	Formation of pathogenic dikaryotic
Ustilago violacea		mycelium
Rust Fungi	Extracellular proteins	Appressorial differentiation
Zoosporic Oomycetes	Nonspecific metabolites	Chemotaxis
Phytophthora cinnamoni	Fucose-rich ligands of root mucilage	Binding to root
Phytophthora cinnamoni	Pectin, root mycilage	Zoospore encystment
Pythium aphanidermatum	Fucose-rich ligands of root mucilage	Zoospore binding
		Zoospore encystment

2. Nature of Recognition Factors

As already mentioned recognition between host and parasite as an incompatible or compatible partner, is a key factor which determines outcome of host-parasite interaction particularly in biotrophic infections. Recognition is a biochemical process and depends on the informational potential contained probably in the surfaces that come in contact and a response follows the complementary interaction of the contacting molecules. Different types of molecules which may constitute either recognition sites or may be involved in recognition are described in the following.

(a) Lectins: These are sugar binding proteins or glycoproteins of nonimmune origin which are devoid of *enzymatic* activity toward sugars to which they bind and do not require free glycoside hydroxy group on those sugars for binding. Extra or intracellular presence of lectins have been demonstrated in almost all the groups of living organisms. Lectins present on the plant surface may constitute the recognition site and bind specifically with the sugars present on the surface of the pathogen or on elicitors produced by the pathogen and this binding may determine the outcome of plant-parasite interaction. Experimental evidences are available in support of this hypothesis from several host-parasite systems including *Rhizobium*-legume (Figure 9), mycobiont - phycobiont (in lichens), *Rhizoctonia solani-Trichoderma harzianum (mycoparasite), P. infestans*-potato, *Ralstonia solanacearum*-tomato/ potato/ tobacco, and *Cladosporium fulvum*-tomato. In first three examples binding leads to compatibility and in rests in incompatibility. Interference with the bind-

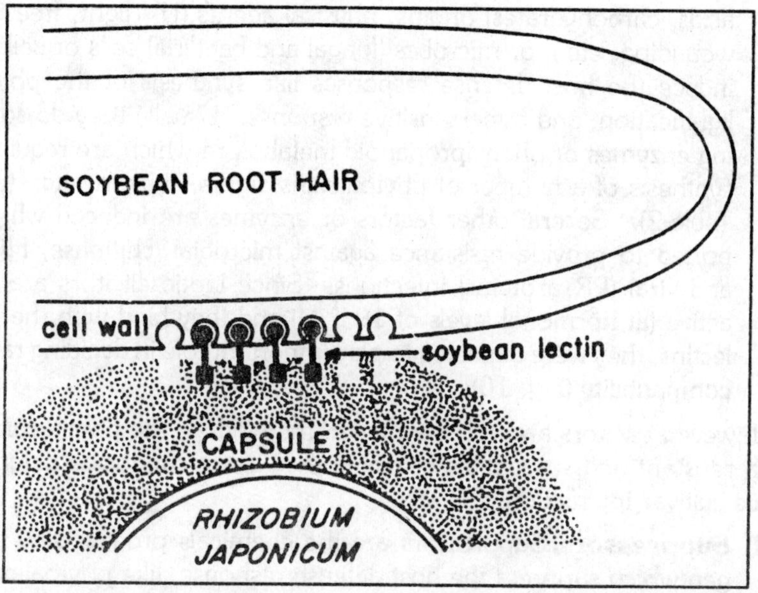
SOYBEAN ROOT HAIR
cell wall
soybean lectin
CAPSULE
RHIZOBIUM
JAPONICUM

Figure-9: A model depicting involvement of lectins in binding (recognition) of *Rhizobium japonicum* by soybean root hairs (from Bal, 1988).

ing (by saturating lectin binding sites with specific sugar moieties) may alter the course of response.

(b) Common Antigens : The presence of common antigens in partners of compatible interactions involving two taxonomically distantly related organisms, higher plant and bacteria or fungi, has been demonstrated in several host-parasite systems. Even plant roots are reported to share common antigens with their rhizosphere microflora. In several plant-parasite interactions including maize/ oat-*Ustilago maydis* degree of common antigenicity was found well correlated with the degree of susceptibility. Possible role of these common antigens in establishment of basic compatibility between host and parasite, transfer of information between interacting partners, or in the supression of resistance response has been postulated. However, most of the studies are merely correlational and little attempt has been made to determine their role in plant parasite interaction. In animal system presence of common antigens helps the parasite to avoid recognition by the immune system of the host as "non-self" but plant lacks parallel immune system.

(c) Elicitors: Several chemicals of abiotic (silver, mercury, copper salts, iodoacetate, sucrose, salicylic acid, polyacrylic acid, etc.) or biotic(i.e. microbial or plan products like proteins, complex glycoproteins, fatty acids, carbohydrates) origins, physical agents (UV light, freezing injury, wounding, etc.), or microbes (fungal and bacterial cells or cell walls) can induce the host defense responses like synthesis of the phytoalexins, lignification, and hypersensitive response. Usually they do so by inducing enzymes of phenylpropanoid metabolism which are required for the synthesis of a number of phytoalexins, lignin, suberin, etc. (Figure 10, Table-2). Several other factors or enzymes are induced which are reported to provide resistance against microbial (chitinase, HRGP, etc.) and viral (PR proteins) infections. Since biotic elicitors are extremely active (at hormonal levels of 10^{-10} M) and they bind with the plant host lectins, they were expected to play important role in deciding race-cultivar compatibility (Fig. 10).

However, elicitors are not specific i.e. they induce the phytolexin synthesis in both resistant and susceptible cultivars; they alone cannot be determinant of the race-cultivar interaction.

(d) Suppressors: Suppressors are the chemicals produced by the pathogen which suppress the host defense response, like phytoalexin synthe-

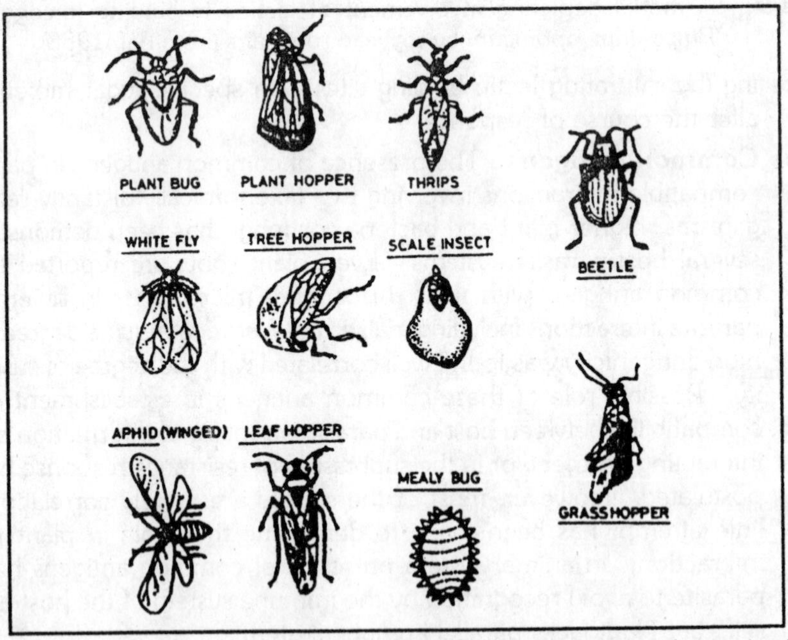
PLANT BUG
PLANT HOPPER
THRIPS
WHITE FLY
TREE HOPPER
SCALE INSECT
BEETLE
APHID (WINGED)
LEAF HOPPER
MEALY BUG
GRASSHOPPER

Figure-10: Scheme illustrating the common biosynthetic origin of various defense related phenolic materials in higher plants (C_4-H=Cinnamic acid 4-hydroxylase; 4CL= 4 coumarate; COA Ligase; PAL=phenylalanine ammonia lyase) (from Hahlbrock and Scheel, 1987).

Table-2 : Host-Gene Expression Following Interactions with Plant Pathogens and Elicitors

Genes induced	Species and tissue	Pathogen or elicitor
	Phenylpropanoid metabolism	
Phenylalanine ammonia-lyasae (PAL)	*Phaseolus* hypocotyls	*Colletotrichum*
	Phaseolus leaves	*Pseudomonas*
	Phaseolus culture	Elicitor
	Pisum endocarp	Elicitor
	Glycine culture	Elicitor
	Glycine hypocotyls	*Phytophthora*
	Petroselinum culture	Elicitor and UV
4-coumarate CoA-Ligase (4CL)	*Petroselinum* culture	Elicitor and UV
cholcone synthase (CHS)	*Phaseolus* hypocotyls	*Colletotrichum*
	Phaseolus culture	Elicitor
	Phaseolus leaves	*Pseudomonas*
	Petroselinum culture	Elicitor and UV
	Glycine culture	Elicitor
	Glycine hypocotyls	*Phytophthora*
Chacone isomerase (CHI)	*Phaseolus* hypocotyls	*Colletotrichum*
	Phaseolus culture	Elicitor
Cinnamyl alchohol dehydrogenase	*Phaseolus* culture	Elicitor
	Other proteins and enzymes	
Casbene synthetase	*Ricinus* seedlings	Elicitors
Chitinase	*Phaseolus* leaves	*Pseudomonas*
PR proteins	*Nicotiana* leaves	TMV
		Salicylic acid
	Petroselinum culture	Elicitor
	Phaseolus leaves	AMV
Thaumatin-like protein	*Nicotiana* leaves	TMV
		Salicylic acid
HRGP	*Phaseolus* cultures	Elicitors
Proline hydroxylase	*Phaseolus* cultures	Elicitors

sis, hypersensitivity etc. induced by the pathogen or elicitors produced by them. They can convert incompatible association into compatible at race-cultivar interaction level. Supressors probably act by blocking the binding of elicitor with the lectin, a complimentary binding site on the host cell surface (Figure 8). Ziegler and Pontzen (1982) found that invertases secreated by *Phytophthora megasperma f.* sp. *glycinea*, race specifically suppressed the glyceollin accumulation in soybean cultivar. Suppressive activity was imparted by the carbohydrate (rich in) moiety of glycoprotein enzyme invertase. Since supressors are race-specific, they are postulated to play more important role in deciding race-cultivar compatibility particularly in those host-parasite systems where their presence has been demonstrated. Whether common antigens also act as suppressors, is an interesting question for the scientists to answer.

V. GERMINATION AND PREPENETRATION

The pathogens in their vegetative state are capable of initiating infection. Fungal spores and seeds of phanerogamic parasites must first germinate. For that they require moisture and favourable temperature and relative humidity. According to nature of the spores, and the environmental conditions, germination of spore occurs in various ways. Sporangia and oospores of fungi produce zoospores in wet conditions. Teliospores of rust and smut fungi normally germinate by producing promycelium on which basidiospores (sporidia) are borne. Some spores germinate easily on release and sometimes even before release, while resting spores have a *dormancy*. Several environmental and nutritional factors affect spore germination.

Fungal invasion is chiefly by germ tubes or structures derived from them. There are two well known situations involving hyphae acting in a concerted way to achieve host penetration. Hyphae of *R. solani* often aggregate to form an "infection-cushion" from which multiple penetrations occur by means of appresoria and penetration pegs. Another type of penetration is exemplified by the penetration of intact root periderm by the "rhizomorphs" of *A. mellea*. In the ectotrophic infection habit among the specialized parasites such as *O. graminis*, the fungus progresses epiphytically over the roots as a sparse network of dark hyphae known as runner hyphae from which hyaline branches quickly penetrate and infect cortex tissues.

Nematodes eggs require conditions of favourable temperature and moisture to get activated and hatch. In most nematodes, the eggs contain the first juvenile stage. This immediately undergoes a molt producing the second juvenile stage. Finally the second stage larvae emerge and either they penetrate the host plant or undergo additional molts. Once nematodes are in close proximity to plant roots, they are attracted to roots by certain chemical factors (CO_2, amino acids).

VI. PENETRATION

A. Penetration Structures

In several plant parasitic fungi, the spores germinate and either germtube by itself causes penetration, directly or indirectly, or it first produces an appresorium from which infection threads develope and penetrate the host tissues. Many factors affect this process. Moisture, temperature, pH, O_2, and CO_2 concentrations must be appropriate for spore germination. The behaviour of germtube also depends on many factors. If environmental conditions are favourable, the host is susceptible and its surface is in a receptive stage, the germ tube proceeds to cause penetration.

Various explanations have been put forward as to why germ tube enters the host and how they find natural openings or wounds some germ tubes are negatively phototrophic and this may explain the entry to dark interiors of the host. Many germ tubes produce appresoria as a result of thigmotrophic response, presence of some chemicals in the germ tube, or in response to diffusable or Volatile compounds from the host surface.

In zoosporic fungi, there are different modes of penetration. Zoorpore penetration typically involves encystment of zoorpores on the outer surface of host cells and subsequently an expanding vacuole develops within the cyst. Penetration process by *Olpidium* includes growth of a penetration tube from the cyst followed by movement of the entire fungal protoplast into the host cell. Host penetration by *Phytophthora parasitica* is almost similar to that of *Olpidium* upto the point of passage of protoplast into the host cell. The formation of specialized accessory structures to facilitate passage through the cell wall has been observed in *Pythium* type of penetration. Kraft et.al. (1967), who studied the penetration of *Pythium aphanidermatum* observed that (1) the cyst germinates to produce a germ tube and (2) the germ tube differentiates into an apresorium to achieve penetration. Host penetration by members of "Plasmodiophorales" involves several unique structures including (1) the development of a tube, or *Rohr,* within the zoospore cyst (2) development of a sharply pointed rod, or stachel within the Rohr (3) Slender tubular extension of the Rohr the *Schlauch,* and (4) an adhesorium which develops during inversion of the Rohr by evagination.

At least four types of penetration are exhibited by nonmotile fungus spore. The simplest has been described for conidia of *Cladosporium* and involves growth of germ tube directly through the cuticle and into the middle lamella. Penetration by rust uredospore requires two distinct stages. The first stage involves penetration by a primary penetrating hypha from an appressorium through a stomata and into substomatal cavity. The primary hypha develops into a substomatal vesicle from which secondary hyphae arise and grow in contact with neighbouring host cells.

There are variations in the behaviour of the germ tube at the time of penetration through the stomata, as in *Peronospora destructor* infecting onion leaves, the germ tube continues to grow after the formation of the first appressorium. In *Pseudoperonospora cubensis* the *Zoospores* swim toward stomata and encyst above the line separating the guard and epidermal cells. Hyphae penetrate the stomatal aperture and swell to form a substomatal vesicle from which, in turn, other hyphae may grow to form haustoria in the adjacent cells of the leaves.

B. Mechanisms of Penetration

Pathogens may enter plants through wounds, natural openings, or by direct penetration.

In nature viruses, viroids, Phytoplasmas (MLO), RLB, etc. enter plants through wounds made by their vectors. Bacteria enter plants mostly through wounds, less frequently through natural openings, and never directly. Nematodes with the help of stylet (Figure 11) enter plant directly and, sometimes, through natural openings. Fungi enter their hosts either directly or through natural openings and wounds.

In fungal pathogens direct penetration through cutinized epidermal wall is achieved by mechanical means or enzymatic action or by both. In some fungi germ tube tip swells to form "appressoria", while in many other even hyphae as such penetrate. Appressoria formed in different species vary morphologically. They may be swollen hyphal tips to well defined melanized, thick walled structure as found in *P. oryzae* and *Colletotrichum spp.* With the latter type, the architecture and mechanical forces are certainly critical factors in penetration process, but enzymes may also have an equally important role in weakening the cuticle and in digesting pectin and cellulose. Melanization of the appressorial wall appears to be necessary for the architecture and rigidity needed to support and focus the mechanical forces involved in penetration process as the hyaline appressoria of *M. grisea* or *Colletotrichum spp.* formed in the presence of melanin biosynthesis inhibitors like tricyclazole, pyroquilone etc., fail to penetrate the host surface and cause infection.

Enzymatic penetration of host surface has now been conclusively demonstrated at least in certain host-parasite systems like papaya fruit- *Colletotrichum gloesporioides,* pea epicotyl - *Fusarium solani,* etc. In these systems cutinases are found to be essential for penetration of intact host surface and subsequently to cause infection. Cutinase inhibitors like anticutinase antibodies, DFP, (diisopropyl flurophosphate) organophosphorus insecticides, carbendazim, etc.provide protection against infection. Mutants incapable of producing cutinase are non-pathogenic. Moreover, they become pathogenic with exogenous application of cutinase. Cutinase gene is induced by cutin. Soliday et.al. (1984)used CDNA strategy; isolated in RNA for cutinase from induced cultures of *F. solani f. sp.*

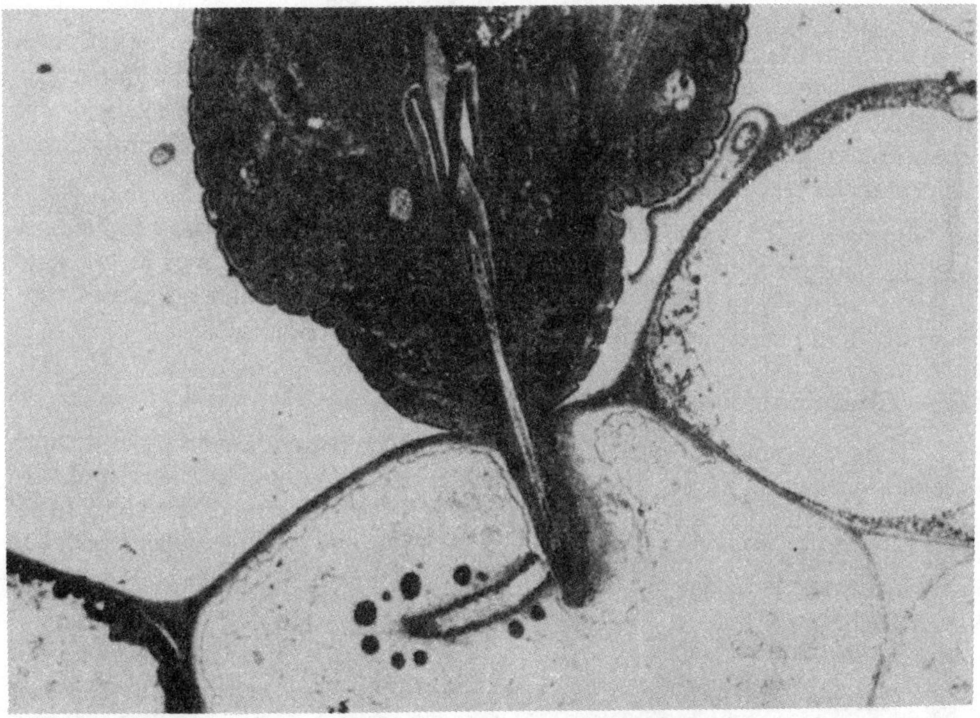

Figure-11: Electron micrograph showing penetration of host cells by plant parasitic nematode *Rotylenchulus reniformis*. (Dropkin, V.H., Introduction of Plant Nematology, John Wiley and Sons, New York, 1989, 293. With permission)

lini; prepared cDNA by reverse transcription, and cloned cutinase gene. By sequencing the gene they were able to predict the primary structure of cutinase, which, as they point out, should help in the effective development of inhibitors for use as antipenetrants. This is the first fungal gene to be cloned which is directly associated with pathogenicity.

Electron microscopic evidences suggest that in general cuticular penetration involves both mechanical as well as enzymatic activity. Cuticular membranes usually appear depressed inward during penetration suggesting involvement of mechanical force. On the other hand, clear holes produced without any sign of torn edges suggest that softening or erosion of the rather brittle cuticle probably occured prior to penetration. However, at least in those cases where direct penetration occurs and no specialized structure is formed, it would seem that enzymatic action must be almost exclusively the mode of penetration (Sisler, 1986).

C. Infection and Colonization

In viral infection the coat protein is removed during passage and nucleic acid is released in the host cell. Viruses do not absorb nutrients from cell but their nucleic acid replicates to produce more viruses nucleic acid and by using host translational machinery, form the protein coat. Assembly of nucleic acid and coat protein gives rise to virus particles. Intercellular movement of viral nucleic acid/particles takes place through plasmodesmata.

Bacteria and fungi dissolve the cell walls by enzymes after entry into the host and thus absorb nutrients. Nematodes use forces as well enzymes to break or dissolve cell walls to reach the cell protoplasm. Enzymes and hormones produced by them cause tissue disintegration and other abnormalities.

D. Chemical Weapons

The main groups of substances secreted by pathogens are enzymes, toxins, growth regulators, polysaccharides, and antibiotics. Of these enzymes and toxins play important role in necrotrophs while growth hormones are thought to play more important role in case of biotrophs (rusts, downy and powdery mildews) where pathogens obtain nutrients from living host cells.

1. Enzymes

Enzymes are the major weapons employed by necrotrophs for ingress and colonization (Table-3). Biotrophs also employ such enzymes but their deployment is highly localized mainly to facilitate their penetration. They are not involved in tissue maceration. Pectin degrading enzymes, produced either constitutively or inductively are responsible for tissue maceration. Recent studies have established that pectinases determine the pathogenicity in soft rot Erwinias. *Erwinia chrysanthemi* was found to produce five pectate lyase isozymes (PLa, PLb, PLc, PLd, and PLe). Kotoujansky (1985) observed that bacterial mutants lacking genes for PL isozymes a, d and e are avirulent on *Saintpaulia ionantha.*

This indicates involvement of these isozymes in pathogenesis. Pectate lyase encoding genes (*pel* genes) *and polygalacturonase-encoding genes* (peh *genes)* have been cloned from four different strains of *E. carotovora sp. carotovora* and *E. chrysanthemi.*

The importance of cellulases, hemicellulases, and other enzymes in pathogenesis have received scant attention. However, isolation and cloning of cellulase gene (*cel* gene) from *E. chrysanthemi* has opened up the possibilities for realistic assessment of these enzymes in pathogenesis. It is not only the production but release of wall degrading enzymes from concerned pathogens in the host tissues is essential for pathogenesis. Secretion deficient mutants of *E. chrysanthemi* and *Xanthomonas campestris* are nonpathogenic.

Plant cell wall constituents may serve as effective inducers of wall degrading enzymes in plant pathogens. Some pathogen may produce different types of wall degrading enzymes. Sequence of production and dominance of a particular enzyme may influence the nature of the symptoms. In soft rots predominant enzymes are pectinases, while cellulases and hemicellulases play important role in brown rots. Lignin degrading enzymes are reported to be employed by the white rot pathogens (Table-3).

2. Toxins

Gaumann (1954) claimed that microorganisms are pathogenic only if they are toxigenic, "Toxins" can be defined as low molecular weight, non enzymatic

Table-3 : Cell Wall and Membrane Degrading Enzymes Produced by Certain Plant Pathogens

Substrate	Enzymes	Pathogen
Cutin	Cutinases	*Colletotrichum gloeosporioides, Fusarium solani f. sp. pisi.*
Cellulose	Cellulases (C_1, C_2, C_x, and β-glucosidase)	*Ralstonia, Erwinia carotovora, E. chrysanthemi, C. lindemuthianum, F.oxysporum f. sp. lycopersici, Sclerotinia sclerotiorum, Botrytis cinerea, Ascochyta pisi, Rhizoctonia solani.*
Pectin pectic acid	Hydrolases/ Pectin methyl esterase (PME), Polygalacturonases (PG) (endo- or exo-), Pectin methyl galacturonases (PMG) (endo- or exo-) Lyases: Polygalacturonic acid *trans-eliminases* (exo- or endo-), (Pectin methyl trans-eliminases (exo- or endo-)	*E. carotovora, E. chrysanthemi, Pythium spp.: R. solani, Verticillium albo-atrum, Sclerotium rolfsii, S. sclerotiorum, B. cinerea.*
Hemicellulose	Hemicellulases	*S. rolfsii, R. solani, S. sclerotium, F. roseum, E. carotovora, E. chrysanthemi.*
Protein	Proteinases	*R. solani, Penicillium expansum, Pseudomomas lachrymans, E. carotovora, E. chrysanthemi.*
Phospholipids	Phospholipase B, Phospholipase C	*S. rolfsii, Thielaviopsis basicola, B. cinerea, E. carotovora.*
Lignin	Ligninases	*Heterobasidion annosus, B. cinerea.*
Suberin	Esterases (Cutinases) to degrade aliphatic component	*F. solani f. sp. Pisi.*

microbial products toxic to the higher plants. These are different from other microbial products like "mycotoxins" and "antibiotics", which are toxic to the animals and microbes (except producers). Toxins are classified as phytotoxin, vivotoxin, and pathotoxin. **Phytotoxins**, the toxin of microbial origin, are toxic to plants but are not regarded as of primary importance during pathogenesis. These are toxic to both host and non-host plants. **Vivotoxins** have been defined as substance produced by the pathogen in the infected host which is involved in the production of disease but not initial incitant of disease itself. Vivotoxins may also induce only a part of disease symptoms. A **pathotoxin** is defined as a host specific toxin which induces all the typical symptoms of disease at reasonable concentration, the production of which is correlated with pathogenicity.

In recent classification, toxins are divided into two categories. The first is "**host-non-specific**" which may affect many unrelated plant species in addition to main host of the pathogen producing toxin; it includes phytotoxin and vivotoxin. The second is "**host-specific**" which affects only the specific host of the pathogen; it includes pathotoxins. Toxins in general, interact with cell membrane or organelles (mitochondria or chloroplast) and alter their permeability. Host-specific and major host-nonspecific toxins are listed in Table-4 and 5, respectively.

3. Growth Regulators

Alteration in concentration of growth regulators like cytokinins, auxins,

Table-4 : Host-Specific Toxins

Toxin or organism	Pathogen	Host	Chemical nature
HV toxin	*Helminthosporium victorae*	Oats	Victoxinine (unstable and nonproteinaceous)
AK- toxin	*Alternaria kikuchiana*	Pear	Altenin(furanose ring)
PC- toxin	*Periconia circinata*	Sorghum	Proteinaceous
HC-toxin	*H. carbonum*	Corn	Cyclic peptide
AM-toxin	*A. mali*	Apple	Alternariolide
A. citri	-	Mandarin orange	Lipophilic
T-toxin	*H. maydis* race T	Corn	Linear polyketol[c]
HS-toxin[a]	*H. sacchari*	Sugarcane	α-D-galactoside of 1, 2 dihydroxy cydo propane[c]
A. alternata	*A. alternata*	Tomato	Cationic compound
P. teres[b]	*Pyrenophora teres*	Barley	Peptide
A. citri	-	Rough lemon	Lipophillic
AB-toxin	*A. brassicae*	Brassica spp.	Destruxin B (cyclopeptide)

a Report indicates host-specific toxin but data incomplete.

b Host selective not host specific

c Not confirmed

Table-5: Some Important Host-nonspecific Toxins

Toxin	Produced by	Chemical nature
Tentoxin	*Alternaria alternata*	Cyclic tetrapeptide
Alternaric acid	*A. solani*	Hemiquinone derivative
Cercosporin	*Cercospora beticola* *C. kikuchi* *C. personata*	Benzoperylene derivative
Cerato-ulmin	*Ceratocystis ulmi*	Large M, carbohydrate
Diaporthin	*Endothia parasitica*	Isocoumarin
Fusaric acid	*Fusarium spp.*	5-n-butylpicolinic acid
Lycomarasmin and Lycomarasmic acid	*F. oxysporum f. sp. lycopersici* and *vasinfectum*	Amino acid derivative
Tabtoxin	*Pseudomonas tabaci/ coronafaciens/ garcae*	Amino acid derivative
Pyricularin	*Pyricularia oryzae*	Nitrogen containing compound
Picolinic acid	*P. oryzae*	picolinic acid
Malformin	*Aspergillus niger* *F. moniliforme f.sp. subglutinans*	Cyclic peptide

gibberellins, ethylene and abscisic acid, have been found to be associated with several plant diseases particularly in those resulting in abnormal plant/ organ/ cell growth like, tumors, galls, knot, stunting, epinasty, curling, hypertrophy, hyperplasia, etc. (Table-6). Increase in level of a growth regulator may be either due to its induced production (by the pathogen or host) or inhibited degradation. Decrease in concentration is due to the enhanced degradation by the host's or pathogen's enzymes. Biotrophs like powdery mildews employ these hormones to draw their nutrients from the host cell.

Alterations in the level of growth regulators seems to be a consequence of pathogenesis.However, at least in three diseases hormones have been demonstrated to be determinant of pathogenicity. These are crown gall caused by *Agrobacterium tumefaciens,* (cytokinin and auxin), oleander and olive knot caused by *Pseudomonas syringae* sp. *savastanoi* (auxin), fasciation disease caused by *Corynebacterium fascians* (cytokinin). In all these three bacteria genes have been identified and cloned which are responsible for the hormonal production. Deletion of these genes result in conversion of virulent strains into avirulent.

VII. GENETICS OF HOST-PARASITE INTERACTION

No information is available on the genetics of basic compatibility. However, a good amount of information is available regarding genetics of specificity (race-cultivar compatibility). The studies on the genetics of resistance in the host and pathogenecity in the pathogen are important to understand the basic aspects of

Table-6 : Certain Plant Diseases Involving Altered Concentration of Growth Regulators

Growth Regulators	Diseases (pathogens)
Cytokinins	**Increased concentration:** crown gall (*Agrobacterium tumefaciens*). fasciation disease (*Corynebacterium fascians*), white rust of crucifers (*Albugo candida*). bean rust (*Uromyces phaseoli*). club roots of crucifers (*Plasmodiophora brassicae*), western pine blister rust (*Cronartium fusiforme*), white pine blister rust (*C. ribicola*), root knot of tobacco (*Meloidogyne incognita*)
	Reduced concentration: *Verticillium* wilt of tomato and cotton, root knot of tomato (*M. incognita)*
Auxins	**Increased concentration:** crown gall (*A. tumefaciens*), peach leaf curl (*Taphrina deformans*), wheat stem rust (*Puccinia graminis f. sp. tritici*), wheat powdery mildew (*Erysiphe graminis*), safflower rust(*Puccinia carthami*) white rust of crucifers (*A. candida*), downy mildew of crucifers (*Peronospora parasitica*), tomato wilt (*Verticillium albo-atrum*), banana wilt(*F. oxysporum f. sp. cubense*) bacterial wilt of solanaceous crops (*Pseudomonas solanacearum*), Oleander knot (*P. syringae Pv. savastanoi*)
	Reduced concentration : Mangomalformation: tobacco mosaic (TMV), Potato leaf roll (PLRV), Curly top of sugarbeet (SBCTV)
Ethylene	**Increased concentration:** fruit rot of citrus (*Pencillium digitatum*), tomato wilt (*F. oxysporum* f. sp. *lycopersici)*, wilt of tulip (*F. oxysporum* f. sp. *tulipae*), Verticillium wilt of cotton (*V. albo-atrum*).
Gibberellins	**Increased concentration:** "Bakanae" disease of rice (*Gibberella fujikuroi*), creeping thistle rust (*Puccinia punctiformis)*
	Reduced concentration: Anther smut of sea campion (*Ustilago violacea)*

host and parasite. Depending on the type of interaction, the breeding/ management strategies of resistance could be framed.

A. Genetics of Resistance

The rediscovery of Mendel;s law of inheritance in 1900 provided the foundation necessary for the analysis of the differential reaction of varieties to diseases. The first reported genetic study of resistance to a disease was published in 1905 by Biffen in England. He obtained a 3 (susceptible):1 (resistant) ratio in the F_2 populations of crosses between wheat variety (Ribet) resistant to *Puccinia striiformis* (yellow rust) and susceptible variety(Michigan, Bronza, or Red King). The F_3 families appeared in the ratio of 1:4 true breeding susceptible lines, 1:2 segregating lines and 1:4 true breeding resistant lines. Between 1912 to 1970 numerous research papers on the inheritance of resistance were published and most of them confirmed Biffen's findings.

Pearson and Sidhu (1971) reviewed the work carried out on the inheritance of resistance since Biffen's time. They concluded that (1) regardless of the species that was involved in host-parasite interaction, resistance generally segregated in Mendelian ratios. Resistance was usually found to be determined by "major" rather than by "minor" genes. The alleles for resistance were dominant over those for susceptibility, (2) in a relatively small number of studies the two factor genetic interaction was found, (3) in a relatively small number of studies evidence has been found for linkage of resistant genes, and (4) alternate resistance alleles were distinguished according to specific types.

The information published after 1971 on the mode of inheritance of resistance of diseases against some of the important crop plants against important pathogens (host-parasite system) have been compiled by Singh (1986). The mode of inheritance could be monogenic, oligogenic, or polygenic.

B. Genetics of Pathogenicity

Pearson and Sidhu (1971) reviewed the literature on the genetics of pathogenicity and made the following generalizations (1) the virulence/ avirulence was usually under Mendelian control, (2) the genes which induced susceptible reaction were usually inherited as recessive, and (3) linkage of genes for virulence was reported only occasionally and there was no report of allelism. Van der Plank (1975) observed that virulence and avirulence in the pathogen are the counterparts of vertical susceptibility and resistance in the host. Aggressiveness and nonaggressiveness in the pathogen are the counterparts of horizontal susceptibility and resistance in the host, respectively.

Genetics of pathogenicity can be best explained by citing Flor's findings. The two varieties Ottawa 770B and Bombay, each having one dominant gene for resistance, and their hybrid segregated in the F_2 generation to give a digenic ratio when tested with two rust races 22 and 24. Similarly, the F_2 segregates of the hybrid between race 22 and 24 showed digenic ratio when tested on the varieties Ottawa 770B and Bombay.

Nelson and Kline (1969) studied the pathogenicity of 291 ascospore isolates obtained from different crosses between isolates of *Cochliobolus heterostrophus* to 9 different gramineous species. A minimum of 13 different genes for pathogenicity was identified. Segregations and comparisons of responses of paired differential species indicated that pathogenicity to 4 host species is controlled by 5 different genes, and the pathogenecity to 4 species is conditioned by 5 different sets of 2 genes each. All pathogenic capacities are inherited independently.

C. The Gene-for-Gene Hypothesis

If either the virulence of the pathogen or the resistance of the host increased unopposed, it would have led to the elimination of either the host or the pathogen, respectively, which obviously has not happened, This shows that evolution of virulence and resistance are stepwise which can be explained by "gene-for-gene" concept according to which for each gene that confers resistance to the host there is a corresponding gene that confers virulence to the pathogen or vice-versa.

The gene-for-gene hypothesis was proposed by Flor as the simplest explanation of results of studies on the inheritance of pathogenicity in the flax rust fungus, *Melampsora lini*. On the varieties of flax (*Linum usitatissimum*) that had one gene for resistance to the parent race, F_2 cultures of the pathogen segregated into monofactorial ratios. On varieties which had 2, 3, or 4 genes for resistance to the parent race, the F_2 cultures segregated into bi-, tri-. or tetrafactorial ratio. This suggested that for each gene that conditions resistance in the host there is a corresponding gene in the parasite that conditions pathogenicity.

The gene-for-gene hypothesis, in its simplest form, is illustrated in Table-7, in which 'A', represents the dominant gene for avirulence and 'a', the recessive gene for virulence in the pathogen and where 'R', represents the dominant gene for resistance and r, the recessive gene for susceptibility in the attacked plant.

According to gene-for-gene hypothesis, only one possible gene combination, A-R, would result in resistance (Table-7). In all other gene combinations, the susceptible reaction would result because the host plant is susceptible (r), the parasite is virulent (a), or both conditions (a-r) are fulfilled in the same host-parasite interaction.

Table-7 : Gene-for-Gene Interaction

	Genes in the plant	
	R	r
Genes in the pathogen		
A	A-R (Resistant)	A-r (Susceptible)
a	a-R (Susceptible)	a-r (Susceptible)

The significant aspect of the gene-for-gene concept can be further illustrated by considering interactions in which the corresponding pairs of genes occur at two or more different loci on the chromosomes. With two loci, four different gene combinations are possible. Notations for genes at two loci in the parasite would be $A_1 A_2$, $A_1 a_2$, $a_1 A_2$, and $a_1 a_2$. For corresponding genes at two loci in the host, the notations would be $R_1 R_2$, $R_1 r_2$, $r_1 R_2$, $r_1 r_2$. All possible interactions between corresponding pairs of genes, along with the disease reaction that would occur from each interaction, are given in Table-8.

Because of the specificity of interaction, resistance is expressed only when combinations $A_1 R_1$ or $A_2 R_2$ occur in the same host-parasite system. That is, A_1, recognizes only R_1 and A_2, recognizes only R_2.

Table-8 : Disease Reactions Following Interactions Between Corresponding Genes at Two Different Loci

Genes of the parasite	Genes of the host	Disease reaction
$A_1 A_2$	$R_1 R_2$	Resistant
$A_1 A_2$	$R_1 r_2$	Resistant
$A_1 A_2$	$r_1 R_2$	Resistant
$A_1 A_2$	$r_1 r_2$	Susceptible
$A_1 a_2$	$R_1 R_2$	Resistant
$A_1 a_2$	$R_1 r_2$	Resistant
$A_1 a_2$	$r_1 R_2$	Susceptible
$A_1 a_2$	$r_1 r_2$	Susceptible
$a_1 A_2$	$R_1 R_2$	Resistant
$a_1 A_2$	$R_1 r_2$	Susceptible
$a_1 A_2$	$r_1 R_2$	Resistant
$a_1 A_2$	$r_1 r_2$	Susceptible
$a_1 a_2$	$R_1 R_2$	Susceptible
$a_1 a_2$	$R_1 r_2$	Susceptible
$a_1 a_2$	$r_1 R_2$	Susceptible
$a_1 a_2$	$r_1 r_2$	Susceptible

"Gene-for-gene" type of associations have been demonstrated in several host-parasite systems (Table-9).

VIII. SURVIVAL OF PLANT PATHOGENS

A successful infectious pathogen must be able to bridge the discontinuities in

Table-9 : Host Parasite Systems for Which Gene-for-Gene Relationship Has Been Demonstrated or Suggested

Disease	System
Apple scab	*Malus-Venturia inaequalis*
Bacterial disease	*Gossypium – Xanthomonas campestris pv.malvacearum* *Leguminosae- Rhizobium (symbiosis)*
Blight	*Zea-Drechslera turcica*
Bunts	*Triticum- Tilletia caries, T. controversa*
Late blight	*Solanum - Phytophthora infestans*
Leaf mold	*Lycopersicon- Cladosporium fulvum*
Mildews	*Hordeum - Erysiphe graminis* *Triticum- E. graminis*
Potato wart	*Solanum - Synchytrium endobioticum*
Rust	*Zea- Puccinia sorghi* *Linum - Melampsora lini* *Triticum-Puccinia graminis* *Triticum - P. striiformis* *Triticum P. recondita* *Avena - P. graminis* *Coffee - Hemileia vestatrix* *Helianthus-P. helianthi*
Smuts	*Avena - Ustilago avenae* *Triticum U. tritici* *Hordeum - U. hordei*
Viruses	*Lycopersicon - Tobacco mosaic virus* *Lycopersicon -Tomato spotted wilt virus* *Solanum - Potato viruses* *Phaseolus - Bean common mosaic virus*

infection-chain or disease-cycle due to gaps between successive host crops and cyclic unfavourable season. It must be able to survive during unfavourable environments. Human activities aimed at interfering with the ability of the pathogen to bridge the discontinuities in the infection chain constitute the most important approach to disease management. Discontinuity in growth causes reduction in the amount of inoculum.

A. Sources of Survival

The sources of survival of plant pathogens are outlined in the following chart:

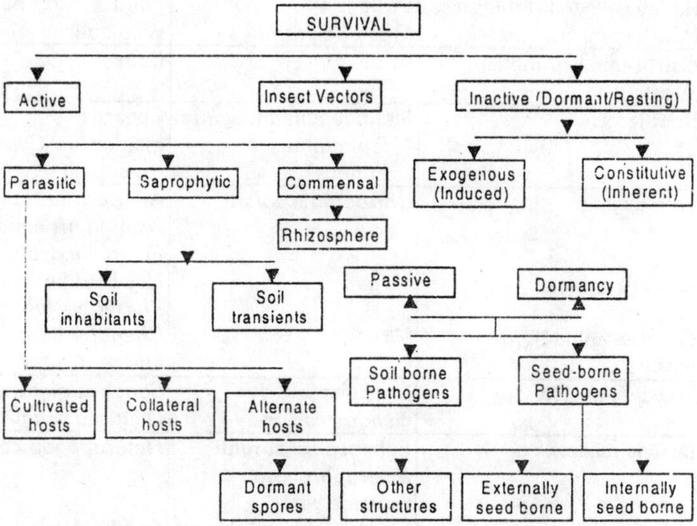
SURVIVAL
Active
Insect Vectors
Inactive (Dormant/Resting)
Parasitic
Saprophytic
Commensal
Exogenous (Induced)
Constitutive (Inherent)
Rhizosphere
Soil inhabitants
Soil transients
Passive
Dormancy
Soil borne Pathogens
Seed-borne Pathogens
Cultivated hosts
Collateral hosts
Alternate hosts
Dormant spores
Other structures
Externally seed borne
Internally seed borne

1. Infected Plants as Reservoir of Inoculum

a. The Cultivated Host

Perennial hosts are easy source of survival of plant pathogens which attack them. In temperate regions, bacterial pathogens remain in the living but dormant tissues of the perennial hosts during the off season. Certain pathogens, such as *X. campestris pv. citri, X.campestris pv. jugalandis, X. campestris pv. pruni, P. syringae pv. morsprunorum* and *Erwinia amylovora* are known to be carried over perennially in holdover cankers or blighted twigs which may produce bacterial ooze in favourable weather to furnish fresh inoculum.

Similarly, several fungal pathogens like *Colletotrichum gloeosporioides (Glomerella cingulata)* (anthracnose of mango), *Oidium mangiferae* (powdery mildew of mango), *Podosphaera leucotricha* (powdery mildew of apple), *Botryosphaeria ribis* (canker of apple), *Phyllosticta* spp. (apple blotch), etc. survive on infected organs of the plants.

b. Collateral and Other Weed Hosts

Along with cultivated crops, several undesirable plants (weeds, etc.), both annual and perennial, grow independently in the nature. Such plants which belong to the same botanical family to which cultivated hosts belong, are known as collateral hosts. Some important examples of pathogens surviving on collateral and weed hosts are given in Table-10.

Table-10 : Collateral and Other Weed Hosts

Cultivated host	Disease	Pathogen	Wild or weed hosts
Barley	Barley yellow dwarf disease Powdery midew	*Virus* *Erysiphe graminis*	*Lolium perenne* *Wild species of Hordeum*
Carrot	Carrot thin leaf motley dwarf	*Virus*	*Daucus spp.*
Cotton	Root knot	*Meloidogyne incognita Hoplolaimus columbus*	*Cyperus esculentus Cyperus roundus*
Cowpea	Web blight Cowpea mosaic Cowpea aphid borne mosaic	*Rhizoctonia solani* *Virus* *Virus*	*Amaranthus spinosus* *Aspillia africana* *Fleurya extruans* *Newbauldia laevis* *Phaseolus tahyroides* *Chenopodium amaranticolor*
Cucurbits	Powdery mildew	*Erysiphe cichoracearum*	*Sonchus oleraceus* *Digitaria sanguinalis*
Maize	Downy mildew Maize dwarf mosaic Maize chlorotic dwarf Maize rough dwarf	*Sclerospora sorghii Sclerophtohora rayssiae var. zea Virus Virus Virus*	*Heteropogon contortus* *Sorghum spp.* *Halepense spp.* *Digitaria sanguinalis*
Oat	Cyst nematode	*Heterodera avenae*	*Avena spp.(wild oats)*
Pea	Pea seed borne mosaic	*Virus*	*Vicia villosa*
Pear	Fire blight	*Erwinia amylovora*	*Cratagus spp.*
Rice	Bacterial leaf blight Hoja blanka Tungro	*X. campestris pv. oryzae* *Virus* *Virus*	*Leersia oryzoides* *Echinocloa spp.* *Panicum spp.* *Oryza spp.* *Echinocloa spp* *Leersia hexandra*
Soybean	Rhizoctonia blight	*Rhizoctonia solani*	*Cynodon dactylon*
Sugarbeet	Curly top	*Virus*	*Sophia spp.*
Sugarcane	Red rot Downy mildew	*Colletotrichumn falcatum* *Peronosclerospora sacchari*	*Sorghum vulgare* *S. halepense* *Saccharum spontaneum* *Zea mays*
Water-melon	Watermelon mosaic	*Virus*	*Momordia dioica* *M. charantia* *Coccinia grandis*
Wheat	Stem rust Stripe rust	*Puccinia graminis* *P. striiformis*	*Wild species of Triticum: Aegilops; Hordeum* *Agropyron repens*

c. Alternate Hosts

One of the two kinds of plants on which a parasitic fungus (e.g., rust) must develop to complete its life-cycle is known as "alternate host". The role of such hosts in perpetuation is not as important as of collateral hosts. In temperate regions the alternate host of *Puccinia graminis* (black or stem rust of wheat), the barbery bush, grows side by side with the cultivated host. In such areas this wild host belonging to a different family may be of some importance for survival of the fungus. It helps in completion of the heterogenous infection chain of the rust fungus. A list of some weeds and wild plants that serve as alternate hosts for rusts affecting field or fruit crops is given in Table-11.

2. Saprophytic Survival

Garrett (1970) has characterized many of the root pathogens ecologically as root inhabitants or soil inhabitants. Root inhabitants are considered to be ecologically obligate parasites, whereas soil inhabitants grow well saprophytically and can survive longer in soil in the absence of a susceptible host plant.

Fungal pathogens like *Pythium, Rhizoctonia, Sclerotium,* etc. survive as soil inhabitants for considerable period by the virtue of their ability to attack and

Table-11 : Alternate Hosts on Which Rust Fungi Complete Their Life Cycle

Cultivated hosts	Rust fungus	Alternate host
Barley	*Puccinia hordei*	*Ornithogalum spp.*
Cherry	*Puccinia cerasi*	*Eranthis spp.*
Cotton	*Puccinia stakmanii*	*Bouteloua spp.*
Gooseberry	*P. caricis var. grossulariata*	*Carex spp.*
Oat	*Puccinia coranata*	*Rhamnus spp.*
Pear	*Gymnosporangium spp.*	*Juniperus spp.*
Plum	*Tranzschelia discolor*	*Anemone coronaria*
Red currant	*Cronartium ribicola*	*Pinus spp.*
Sorghum	*Puccinia sorghi*	*Oxalis stricta*
Wheat	*Puccinia graminis*	*Berberis spp.*
	Puccinia recondita	*Thalictrum spp.*
		Isopyrum spp.
		Anchusa spp.

colonize dead plant materials and thus, remain active as saprophytes. However, the chances of very prolonged active survival of even these pathogens has been doubted by many workers. Antagonism by other soil microflora reduces their ability to continue saprophytic activity unless they have strong competitive saprophytic ability.

Another category of saprophytic survival is of those fungi that are slightly more tolerant to antagonism and usually predominate in the rhizosphere. In this region these pathogens are restricted to pioneer colonization of dead substrates and are tolerant to competition by other soil microflora.

The third category of saprophytic survival in soil is of those pathogen that have low competitive saprophitic survival ability and survive saprophitically for only a short time. These are described as "root inhabiting" fungi. Many vascular wilt causing species of *Fusarium (F. oxysporum f. sp. udum*, *F. oxysporum f. sp. vasinfectum*), *Verticillium,* the root rot pathogen *Phymatotrichum omnivorum,* etc. fall in this category.

Besides fungi, the bacteria form another important group of plant pathogens to survive in soil. Plant pathogenic bacteria have been grouped in relation to survival in soil; (1) transient visitor, (2) resident visitor, and (3) saprophytes. Crosse(1968) subsequently proposed a slightly modified scheme consisting of four groups; (1) with permanent soil phase (2) with protected soil phase, (3) with transitory soil phase, and (4) with no soil phase. Phytopathogenic bacteria characterized as transient visitors consist of species whose populations are developed almost exclusively in the host plant where maximum number of generations are produced. When such bacteria reach soil, their populations decline rapidly and do not remain important source of primary inoculum for the next season. Being poor competitors, most phytopathogenic bacteria fall under this category. Examples are most species and pathovars of *Xanthomonas,* nonsoft rot species of *Erwinia,* and pathovars of *Corynebacterium michiganense.* However, in temperate regions and under conditions not favourable for intense soil microbial activities, these bacteria may perpetuate in soil in association with crop debris. Investigations have shown that *Xanthomonas* spp. do not multiply or survive in soil in the free state. *X. campestris* pv. *phaseoli, pv. translucens, pv. malvacearum, pv. citri, pv. campestris, pv. vesicatoria*, and pv. oryzae *are known to decline rapidly* and reach extinction within few days or weeks after their introduction into soil. Certain species of *Pseudomonas* are incapable of persisting in free state in natural soil for extended periods. The corynebacteria though not soil borne, can survive in soil in safely placed crop debris.

The resident visitors also have their maximum generations in the host but their numbers gradually decline in soil. If populations enter the soil at a sufficiently high rate, as happens in case of bacterial diseases of underground parts, the slow decline would permit net increase of such bacteria from season to season. The long term persistence of such bacteria in soil is host dependent; increase, decrease or total extinction depending on the cropping practices. Bacterial pathogens with an extended soil phase include crown gall bacterium (*Agrobacterium tumefaciens)* and *Ralstonia solanacearum* race 1 (bacterial wilt of solanaceous crops). These can be considered true soil-borne pathogens or soil inhabitants. *R. solanacearum* race 1 can survive in soil in free state. This pathogen survives under bare fallow for 4 to 6 years and up to 10 years in soil cropped with non host or nonsusceptible crops. Susceptible weed hosts prolong its soil survival. *Agrobacterium tumefaciens* can be reisolated from artificially infested soil without crop cover but gradually declines in natural soil. *Streptomyces scabies* and *S. ipomoea* are other examples of soil inhabiting bacteria. Their primary inoculum comes from soil.

There are very few phytopathogenic bacteria which are truely saprophytes with permanent soil phase. The group of Buddenhagen's classification is typified by bacteria whose populations are largely produced in the soil including the rhizoplane and whose relation to plant disease is only-ephimeral. This group includes the true soil saprophytes, the rhizoplane bacteria, the green fluorescent *Pseudomonas* causing soft rots, the species of *Bacillus* and the pectolytic soft rotting *Clostridium* species which are opportunistic pathogens.

Survival of soft rot causing bacterium *Erwinia species* has always been a matter of controversy. *E. carotovora* considered a soil inhabitant, was found in subsequent studies that, after being introduced into soil, its populations reach nondetectable level rapidly. Populations of *E. carotovora* sub sp. *carotovora* in potato root zone soil increase as seed tubers decay but subsequently decline to very low levels after the plants are removed.

3. Survival Through Dormant/ Resting Structures (Organs)

Among infectious plant pathogens fungi, nematodes and phanerogams survive through their resting or dormant structures. Phanerogams produce seeds just like any other flowering plant and through these seeds they can live in dormant stage, sometimes for years. Majority of phytophagous nematodes, survive through their dormant structures (eggs, cysts, galls formed from host tissues). The eggs are mainly present in soil. Cysts or similar structure such as galls are also present in soil or sometimes in seed lots. Plant parasitic fungi are the only organisms that produce spores, analogous to eggs of nematodes, and other structures for their inactive survival. In most fungal pathogens these dormant stages are the major sources of survival.

Dormancy: It is "reversible interruption of phenotypic development of an organism". Viewed in this way, dormancy can be of two types: "exogenous or induced" dormancy and "constitutive or inherent" dormancy. Exogenous dormancy or induced dormancy is a condition wherein development is delayed because of unfavourable physical or chemical conditions of the environment. On the other hand, constitutive or inherent dormancy is a condition wherein development is delayed due to an innate property of the dormant stage, such as a barrier to the penetration of nutrients, a metabolic block or self inhibiotory substances produced by the spores.

In soil the presence of widespread "fungistasis" has been studied in detail since 1953. This phenomenon is described as failure of propagule to grow even if necessary physical conditions for germination are favourable. It is responsible for induction of dormancy in fungal spores especially those that depend on some external source of nutrients for germination. Fungistasis is of various types but the most important mechanism is one which has a biotic origin.

Thus, the asexual spores (conidia, chlamydospores), sexually produced spores (oospores, ascospores), fruiting bodies (acervuli, pycnidia, sporodochia, cleistothecia, perithecia, etc.) and other dormant structures like thickened hyphae, sclerotia, rhizomorphs, etc. are the main structures for dormant survival. When the hemibiotrophs/ facultative saprophytes fail to continue as saprophytes in host plant debris or in soil, they produce resting structures. Wilt causing *Fusarium* spp. (vascular pathogens) perpetuate in the form of chlamydospores formed due to conversion of conidia and vegetative hyphae. In addition to nutritional factors and competition, in *Fusarium oxysporum* chlamydospores are reported to be formed under the influence of metabolites from certain bacteria. The formation of such structures ensures survival of these fungi under the conditions of intense antagonism in soil because dormant structures are not easily affected by antagonistic activities of other soil microflora.

4. Survival with Seeds

Seed can harbor a wide range of microflora, viruses and other causal agents of plant diseases. When a pathogen is carried with or within seed without necessarily producing disease in the offspring, it is "**transport**" of the pathogen by the seed. If the seed-borne inoculum necessarily produces disease in the offspring it is "**transmission**". A pathogen is "**externally seed borne**" when it is external to the functional seed or fruit parts essential to production of a new plant. When embedded in the tissues of the seed, the pathogen is "**internally seed-borne**".

Table-12: Longevity of Some Important Seedborne Fungi

Fungus	Crop	Viability in years
Alternaria sp.	*Hordeum vulgare*	10.6
A. brassicicola	*Brassica oleracea*	7
Ascochyta phaseolorum	*Phaseolus vulgaris*	2.5
A. pisi	*Vicia faba*	9
	Pisum sativum	7
A. rabiei	*Cicer arietinum*	2-3
Botrytis allii	*Allium cepa*	3.5
B. cinerea	*Linum usitatissimum*	3.33
B. fabae	*Phaseolus vulgaris*	0.75
Cercospora beticola	*Beta vulgaris*	2.5
C. kikuchii	*Glycine max*	2
C. gossypii	*Gossypium sp.*	13.5
Dreschslera avenae	*Avena sativa*	7
D. graminea	*Hordeum vulgare*	5
D. oryzae	*Oryza sativa*	10
F. moniliforme	*Gossypium sp.*	13.5
F. udum	*Cajanus cajan*	2
Macrophomina phaseolina	*Phaseolus spp.*	2.42
Phoma betae	*Beta vulgaris*	5
Sclerospora graminicola	*Pennisetum typhoides*	2
S. nodorum	*Triticum aestivum*	7
Tilletia caries	*T. aestivum*	9
Ustilago nuda	*Hordeum vulgare*	11
U. tritici	*Triticum aestivum*	7

i. Fungal Pathogens: Neergaard (1979) reviewed the data available on maximum longevity of seed borne fungi. The longevity of some important seed-borne fungal pathogens are given in Table-12.

ii. Bacterial Pathogens: A large number of bacterial pathogens have been demonstrated as seed-borne. Some important phytopathogenic bacteria and their longevity are listed in Table-13.

iii. Viruses: About 100 viruses have been reported to be transmitted by seed. However, only a small portion (1 to 30%) of the seeds derived from virus infected plants transmit the virus and the frequency varies with the host-virus

Table-13:Longevity of Some Important Seed Borne Bacteria

Bacterium	Crop	Viability in years
Corynebacterium flaccumfaciens var. aurantiacum	*Phaseolus vulgaris*	15
C. flaccumfaciens pv. flaccumfaciens	*P. vulgaris*	24
C. flaccumfaciens var. violaceum	*P. vulgaris*	8
C. michiganense pv. insidiosum	*Medicago sativa*	3
C. michiganense pv. tritici	*T. aestivum*	5
E. carotovora pv. carotovora	*Nicotiana spp.*	0.66
P. syringae pv. apii	*Apium grave-olens*	1
P. syringae pv. glycinea	*Glycine max*	1.33
P. syringae pv. lachrymans	*Cucumis sativus*	2
P. syrigae pv. pisi	*Pisum sativum*	0.84
P. syringae pv. sesami	*Sesamum indicum*	0.9
P. syringae pv. tabaci	*Glycine max*	1.5
	Nicotiana spp.	2
Xanthomonas campestris pv. campestris	*Brassica oleracea*	3
Xanthomonas campestris pv. malvacearum	*Gossypium sp.*	4.75
Xanthomonas campestris pv. manihotis	*Manihot esculenta*	1.5
Xanthomonas campestris pv. oryzae	*Oryzae sativa*	0.16
X. campestris pv. phaseoli	*Phaseolus vulgaris*	15
X. campestris pv. sesami	*Sesamum indicum*	1.33
X. campestris pv. glycinea	*Glycine max.*	2.5
X. campestris pv. tomato	*Lycopersicon esculentum*	20
X. campestris pv. vesicatoria	*Capsicum frutescens*	10
X. campestris pv. zinniae	*Zinnia elegans*	4
X. campestris pv. phaseoli var. fuscans	*Phaseolus vulgaris*	3

combination. In most seed transmitted viruses, the virus seems to come primarily from the ovule of infected plants, but several cases are known in which the virus in the seed seems to be just as often derived from the pollen that fertilized the flower. A list of seed-borne viruses and their longevity is presented in table-14.

Table-14: Longevity of Some Important Seed Borne Viruses

Virus	Crop	Viability in years
Alfalfa mosaic	*Medicago sativa*	3.5
Barley stripe mosaic	*Triticum aestivum*	3 19
Bean common mosaic	*Phaseolus vulgaris*	3
Bean southern mosaic	*Phaseolus vulgaris*	0.6
Bean western mosaic	*P. vulgaris*	3
Cowpea aphid-borne mosaic	*Vigna unguiculata*	2
Cucumber mosaic	*Phaseolus vulgaris Stellaria media*	2.25 1.75
Muskmelon mosaic	*Cucumins melo*	3
Prune dwarf	*Prunus spp.*	3.5
Raspberry ringspot	*Capsella bursa-pastoris Stellaria media*	6 6
Sowbane mosaic	*Chenopodium murale*	6.5
Squash mosaic	*Cucurbita pepo*	3
Tobacco mosaic	*Lycopersicon esculentum*	3.9
Tobacco ring spot	*Petunia violacea Glycine max.*	5

iv. Plant Parasitic Nematodes: Survival and seed transmission of plant parasitic nematodes have been found in the species of genera: *Anguina, Aphelenchoides, Ditylenchus, Heterodera,* etc. Seed-borne nematodes known to cause some important plant diseases along with their viability are given in Table-15.

5. Survival in Association with Insects. Survival of plant pathogens in association with insects is not uncommon in the nature. The bacterium associated with corn wilt (*E. stewartii)* is present in the intestinal tract of its vector, *Diabrotica*

Table-15: Longevity of Some Important Seed Borne Nematodes

Nematode	Crop	Viability in years
Anguina tritici	*Triticum aestivum*	14-28
Aphelenchoides besseyi	*Oryza sativa*	3
Ditylenchus dipsaci	*Avena sativa*	8
	Medicago sativa	5
Heterodera glycines	*Glycine max*	1.8

undecimpunctata (spotted cucumber beetle) and *Chaetocnema pulicaria* (the corn flea beetle). It not only survives in these beetles but is distributed over long distances also. The cucurbit wilt bacterium (*E. tracheiphila*) is totally dependent on cucumber beetles for its survival between seasons. The hibernating adult striped cucumber beetle (*Acalymma vittatum*) and the spotted cucumber beetle (*D. undecimpunctata*) harbour the pathogen during off-season in t-heir intestinal tract and transmit it during feeding in the crop season.

Similarly, the potato black leg organism, *E. carotovora* sub sp. *atroseptica* can live in all stages of the seed corn maggot (*Hylemya platura*) and may persist in the intestinal tract of both adult flies and larvae. Since the pathogen survives pupation, the emerged adult may contaminate eggs as they are laid.

6. Survival in Vegetatively Propagated Plant Parts Several horticultural and cash crops, and most of those crops where modified roots or stems happen to be the actual edible parts, are raised from vegetatively propagated organs such as tubers, suckers, cutting, runners, etc. In majority of such cases the plant organs used for raising new crops, are the major sources of survival for plant pathogens. Out of many examples, the survival of pathogens in two such crops - potato and sugarcane where tubers and sugarcane pieces are used for raising new crops, are given in Table-16.

IX. DISSEMINATION OF PLANT PATHOGENS

Knowledge of modes by which pathogens are dispersed is essential for devising suitable and effective disease management practices. Several terms such as **distribution, dissemination, dispersal, spread, and transmission** have been used invariably and rather loosely. **Distribution** implies the spread of a pathogen into new geographical areas and its establishment there. **Dispersal** is what happens between take-off of a spore and its deposition - it does not include its germination or infection of the plant whereas **spread** implies that the pathogen reaches and infects plants. Spread can be used to describe progressive colonization of the infected organs or plants, passage of the pathogen from infected plants to others in the same field or crop area, or long distance spread of pathogens between plants which are widely separated, as in intercontinental spread.

A. Modes of Dispersal

1. Direct (Active)

a. Through Seed:

The survival of plant pathogens with seeds has been explained in the preceding chapter. Since most of the cultivated crops are raised from seed, the

Table-16: Survival of Plant Pathogens in Potato Tubers and Sugar cane Pieces

Pathogens (Sugarcane)	Pathogens (Potato)
Pernosclerospora sacchari	*Colletotrichum atramentarium*
Colletotrichum falcatum	*Rhizoctonia solani*
(Physalospora tucumanensis)	*Phytophthora infestans*
Ustilago scitaminea	*Spongospora subterranea*
Fusarium moniliforme	*Helminthosporium solani*
Acremonium furcation	*Synchytrium endobioticum*
Xanthomonas vasculorum	Alfalfa mosaic virus
Xanthomonas albilineans	Corky ring spot virus
Grassy shoot (MLO)	Potato virus A
White leaf disease (MLO)	Potato virus M
Mosaic of sugarcane (virus)	Potato virus Y
Spike of sugarcane (virus)	Potato aucuba mosaic
	Potato leaf roll
	Potato mop-top
	Potato spindle tuber
	Potato witches broom
	Tobacco black ring
	Yellow dwarf virus

transmission of diseases and transport of pathogens by seeds, has much importance for plant pathologists and farmers. Dispersal of inoculum through seed is accomplished either as mixture and contaminant dormant structures of the pathogen (e.g. sclerotia of ergot fungus, galls containing nematode larvae, cysts, smut sori, etc.) or through presence of propagules on the seed coat (externally seed-borne-covered smut of barley), or as dormant mycelium in the seed (internally seed borne-loose smut of wheat). Agrawal and Sinclair (1987) have made a thorough review of seed borne plant pathogens that cause major diseases of major crops.

b. Transmission by Vegetative Propagation Materials:

Over 40% of the bacterial plant pathogens are transmitted on vegetatively propagated material. Among the pertinent examples are *X. campestris pv. citri* (citrus canker), *Agrobacterium tumefaciens* (crown gall) and pathogens of potato such as *Erwinia carotovora sp. carotovora* and *spp. atroseptica, Corynebacterium sepidonicum*, and *R. solanacearum*. Dispersai of bacteria to healthy plants often occurs during cultural practices (cutting, pruning, grafting, etc.) Vegetative propagation is most important, often the only method of transmission of Phytoplasmas and RLB.

c. Through Soil:

As discussed in the preceding paragraphs (survival of plant pathogens), it has been conclusively demonstrated that a large number of plant pathogens survive in soil. Thus, soil as such becomes an important means for their short or long distance dispersal. Zoosporic fungi, flagellate bacteria and nematodes are capable of active mobility. Several others are passively transported.

i. Active Motility: Hickman and Ho (1966) observed that zoospores of *Pythium aphanidermatum* moved at speeds upto 14.4 cm/h over short periods of chemotactically directed movement. However, Lacey (1967) noted that zoospores of *Phytophthora infestans* moved about 1.3 cm in 2 weeks in wet soils.

Autonomous dispersal takes place by active growth of hyphae or hyphal strands. *Phytophthora cinnamoni* moved at least 4.5 m uphill in 22 months. According to Shipton (1972) runner hyphae of *Gaeumannomyces gramminis* grew for a radius of 1.5 m in a growing season. Wehrle and Ogilvie (1956) also observed that mycelial growth of *G. graminis* was about 1.5 m in a season. The rate of spread of *Rhizoctonia solani* has been estimated to be 25 cm/month; 21.2 cm in 23 d, and 2.5 cm/d. The rhizomorphs of *Armillaria mellea* grew through field soil at about 1 m/year. The rate of such spread in *Phymatotrichum omnivorum* (root rot of cotton) is estimated as 3 to 30 ft. per season and 2 to 8 ft/month in alfalfa crop. *Foma, Ganoderma, Polyporus, Sclerotium*, etc. also move independently.

Active movement of plant parasitic nematodes has also been studied. According to Wallace(1978) spread rate in soil of 0.1 to 0.5 cm/d appear to be reasonable approximation for nematodes. *Pratylenchus zeae* travelled about 0.1 cm/d in sandy loam soil and *Globodera rostochiensis* at about 0.3 cm/d. Tarjan (1971) found that *Radopholus similis, Pratylenchus coffeae* and *Tylenchulus semipenetrans* moved at about 0.4, 0.2 and 0.1 cm/d, respectively. The average speed of *Pratylenchus penetrans* in soil under optimum conditions was about 0.3 cm/d. *Ditylenchus dipsaci* moved at 0.5 to 1.0 cm/d and *Tylenchulus semipenetrans*, 0.1 cm/d.

ii. Passive Dispersal in or alongwith Soil: Pathogens perpetuating in soil either saprophytically on plant debris or in dormant stage are transported from one place to another within a plot or from one plot to another or even from one area to another area. During the cultural operations in á field, soil is moved from one point to another within the field through agricultural implements, workers' feet, erosion, etc. Movement of farm equipments and animals from one farm to another may likewise transport the pathogen. Irrigation or rain water, wind with high velocity, hail storms, etc. also transport soil particles and plant debris from one location to another and thus short or long distance dispersal of pathogen propagules takes place in nature.

2. Indirect (Passive)

a. Animal Dispersal and Other Agencies

i. Insect: Since the discovery that bees and wasp can transmit *Erwinia amylovora* (fire blight of apple and pear), much information has been accumulated on the role of insects and other small animals in the dissemination of plant pathogenic fungi, bacteria and viruses.

ii. Fungi: Spread of fungal pathogens by insects, although accidental, is fairly widespread especially for those disease in which pathogens produce sugary and sticky substances. About 66 species of fungi causing plant diseases are transmitted by more than 100 species of insects belonging to at least 6 orders. Spores of *Ustilago violaceae* (smut of Caryophyllaceae) and *Botrytis anthophilla* (mold of clover) are carried from the infected anthers to healthy flowers by pollinating insects. The conidia of *Claviceps* are disseminated from infected ovaries to healthy ones by insects which feed on the sugary honeydew.

(iii) Bacteria: A wide variety of insects act as vectors of bacterial pathogens either through incidental association or through intimate biological association. The best example is *Erwinia amylovora* (fire blight of apple and pear). More than 100 insects that visit apple blossoms easily spread this bacterium to healthy blossoms. Close biological association includes transmission of bacteria with a limited host range by insects restricted to a small group of plants. Examples include transmission of *E. stewartii* (corn wilt) by corn flea beetle (*C. pulicaria)* and *E. tracheiphila* (cucurbit wilt) by cucumber beetle (*A. vittatum)*

3. Viruses

Japanese workers were the first to have established the transmission of rice dwarf disease by the leaf hopper. The virus vector relationship is specialized phenomenon. This relationship is usually classified according to the length of time the virus can persist in and transmitted by the vector. Insects with sucking mouth parts carry plant viruses on their stylets - *Styletborne* or *nonpersistent viruses* or they accumulate the viruses internally, and after passage of the virus through the insect system, they introduce the virus again into plants through their mouthparts - *circulative* or *persistent viruses*. Some circulative viruses may multiply in their vectors and are then called propagative viruses.

Members of relatively few groups of insects as sown in Figure 12 are known to transmit viruses.

i. Mites: Agents of at least 14 diseases are transmitted by *Eryophid mites*. These include *Agropyron* mosaic, cherry leaf mottle, currant reversion, fig mosaic, rye grass mosaic, peach mosaic, sterility mosaic of pigeonpea, wheat spot mosaic and wheat streak mosaic viruses.

ii. Nematodes: Hewitt et.al. (1958) were the first to demonstrate that nematodes can transmit a virus when they proved that fanleaf disease of grapevine was transmitted by the nematode *Xiphinema index*. Nepoviruses are transmitted by species of related nematode genera *Longidorus, Paralongidorus,* and *Xiphinema,* while tobra-viruses are transmitted by species of *Trichodorus* and *Paratrichodorus.* A list of virus diseases transmitted by these nematodes is given in Table-17.

iii. Fungi: A number of soil-borne fungi of the groups of Chytridiomycetes

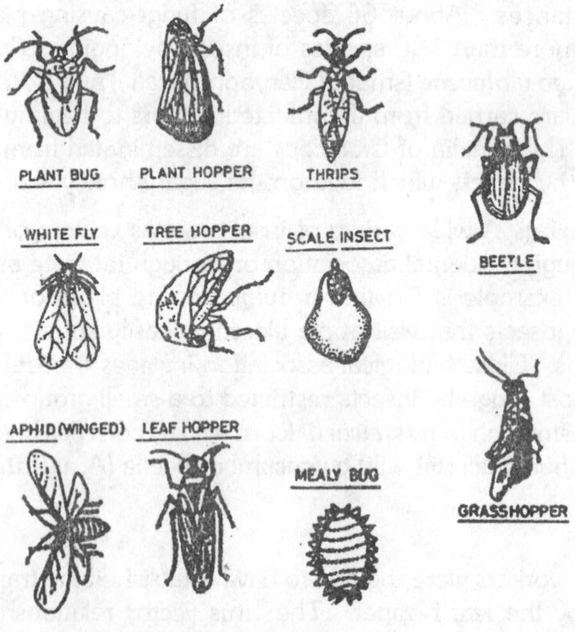
PLANT BUG
PLANT HOPPER
THRIPS
WHITE FLY
TREE HOPPER
SCALE INSECT
BEETLE
APHID (WINGED)
LEAF HOPPER
MEALY BUG
GRASSHOPPER

Figure-12: Insect vectors of plant viruses.

Table-17: Nematode Vectored Plant Viruses

Type of virus	Vector	Hosts
	Neporiruses	
Arabis mosaic	*Xiphenema diversicaudatum, X. coxi*	Cherry, cucumber, grapevine
Cherry leaf roll	*X. diversicaudatum* *X. coxi, X. vuittenzi*	Black berry, cherry, dog-wood, elm, rhubarb, walnut
Grapevine fan leaf	*X. index* *X. italiae*	Grapevine
Mulberry ring spot	*Longidorus martini*	Mulberry
Raspberry ring spot	*L. elongatus* *L. macrosoma* *X. diversicaudatum*	Blackberry, raspberry, redcurrants, Strawberry
Strawberry latent ring spot	*X. coxi* *X. diversicaudatum*	Black current, cherry, celery rose, strawberry
Tobacco ring spot	*X. americanum* *X. coxi*	Bean, blueberry, gladiolus, grapevine, tobacco
Tobacco black ring	*L. attenuatus* *L. elongatus*	Celery, potato, strawberry, tomato
Tomato ring spot	*X. americanum*	Blackberry, cherry, grapevine, peach, tobacco
	Tobraviruses	
Pea early browning	*Paratrichodorus spp.* *Trichodorus spp.*	Pea, alfalfa
Tobacco rattle	*Paratrichodorus spp.* *Trichodorus spp.*	Potato, lettuce, tobacco
	Other viruses	
Brome mosaic	*X. diversicaudatum* *L. macrosoma*	Grasses
Carnation ring spot Prunus necrotic ring spot	*X. diversicaudatum* *L. macrosoma*	Carnation Plum, sour cherry

and Plasmodiophoromycetes transmit viruses. Viruses transmitted by soil-borne fungi are listed in Table-18.

iv. Transmission by Dodder: Many plant viruses are transmitted from one plant to another through establishment of parasitic relationship between two plants by the twining stems of the parasitic plant dodder (*Cuscuta* spp.). The tobacco rattle virus is transmitted by at least 6 species of *Cuscuta*. The cucumber mosaic virus (Cucumis virus 1) is transmitted by at least 10 species. Some impor-

Table-18 : Virus Disease Tansmitted by Fungi

Fungal vector	Virus disease
Olpidium brassicae	Lettuce big vein and tobacco necrosis on tobacco, bean, potatoes and tulips
	Tobacco stunt
Olpidium cucurbitacearum	Cucumber necrosis
Polymyxa betae	Beet necrotic yellow vein
Polymyxa graminis	Barley yellow dwarf mosaic
	Oat mosaic
	Rice necrosis
	Wheat spindle
	Streak mosaic
	Wheat soil-borne mosaic
Pythium ultimum	Pea false leaf roll
Spongospora subterranea	Potato mop-top
Synchytrium endobioticum	Potato virus X

tant viruses transmitted by dodder are summarized in Table-19.

4. Wind Dispersal (Anemochory)

Some plant pathogenic bacteria, seeds of some angiospermic plant parasites, eggs and cyst of nematodes and majority of fungal pathogens such as downy mildew fungi, powdery mildew fungi, rusts, smuts, sooty molds and leaf spots, and/or blight causing fungi which produce abundant spores of extremely varied morphology and origin (conidia, uredospores, basidiospores, ascospores, etc.) on the host surface, are disseminated by winds. Air dispersal of inoculum is accom-

Table-19 :Some Virus Diseases Trnasmitted by Dodder

Virus	*Cuscuta sp.*
Arabis mosaic	*C. subinclusa and C. californica* frequently and *C. campestris* occasionally
Barley yellow dwarf	*C. campestris*
Citrus tristeza	*C. americana*
Citrus exocortis viroid	*C. subinclusa*
Cucumber green mottle mosaic virus (Cucumis virus 2)	*C. subinclusa*
	C. lupuliformis
	C. campestris
Potato leaf roll	*C. subinclusa*
Tobacco etch virus	*C. californica*
	C. lupuliformis

plished through rain drop splashes, sprinkler irrigation, dew or fog drip, air-borne debris, air-borne dust, aerial strands, aerosols, etc.

The number of spores produced can be astronomical. Powdery mildews may produce several thousands conidia per square centimeter of infected leaf surface, a fairly modest out put compared with 10^5 or more spores in some downy mildews. A single smut sorus may contain millions of spores, a heavily infected barberry bush is said to produce up to 70×10^3 million aeciospores of *Puccinia graminis* and a relatively small apothecium of *sclerotinia* can produce about 30 million ascorpores.

In general, the fungus spores are more abundant in the air close to the earth/ plant surface than at higher altitudes. However, several investigators have encountered clouds of spores, numerous bacteria and other minute objects several thousand meters above the earth. Uredospores of *Puccinia graminis* f. sp. *tritici* have been caught as high as 14,000 ft. above infected fields, living spores of various fungi have been caught from aeroplanes above the carribean sea 600 mi. from the nearest source and living spores of several molds were caught at 72,500 ft. and set to close at 36,000 ft.

5. Water Dispersal (Hydrochory)

Rain, flood and/or irrigation water may carry the propagules especially that in or near the soil. The water separates and distributes spores in the microenvironment. The splashing and spattering of water during heavy rains may result in distribution of inoculum to plant parts near the soil and may distribute propagules to different parts of the same plant or to neighbouring plants.

6. Human Dispersal (Anthropochory)

Man is to a large extent responsible for dissemination of pathogens which he does in two ways-through his person and through the objects he transport from one place to another.

CHAPTER **IV**

SYMPTOMATOLOGY

I. INTRODUCTION

Plant pathogens induce different reactions in the body of their hosts. This results in creation of abnormalities which appear on the plants. Any condition resulting from disease that indicates its occurrence is called SYMPTOMS. Generally, a distinction is made beteween **symptoms** and **signs**. A symptom of disease is expressed as a reaction of the host to a causal agent, where as, a sign is evidence of disease other than that expressed by the host. Signs are usually the structures of the pathogens. A disease is first noticed by the presence of symptoms and/or signs, and recognition of specific type of symptoms or sign aid in the eventual diagnosis of the disease.

II. SYMPTOMS

Symptoms are divided into three general categories—

(A) Necrotic symptoms: Those symptoms that result from cessation of function leading to death.

(B) Hyperplasia/Hypertrophy symptoms: Hyperplasia (Gr. hyper = over + plasis = molding, formation), excessive multiplication of cells; abnormal rate of cell division. Hypertrophy (Gr. hyper = over + trophe = food), excessive enlargement of cells.

(C) Hypoplastic - under development or retardation of function.

(A) NECROTIC SYMPTOMS (PLATE-I)

1. **Blight:** Rapid killing of foliage, blossoms, and twigs.
2. **Blotch:** Large, irregular lesions on leaves, shoots and stems.

3. **Canker:** Necrotic, often sunken lesions in the cortical tissues of stems and roots.
4. **Decay:** Disintegration of dead tissues
5. **Dieback:** Progressive death of twigs and branches from their tips toward the trunk.
6. **Hydrosis:** A water soaked, translucent condition of the tissue due to cell sap passing into intercellular spaces.
7. **Scald:** Blanching of the epidermis and adjacent tissues.
8. **Scorch:** Browning of leaf margins resulting from death of tissues
9. **Shot hole:** Circular hole, in leaves resulting from the dropping out of the central necrotic areas of spot.
10. **Spot:** Lesions, usually defined, circular or oval in shape, with a central necrotic area surrourded by variously coloured zones.
11. **Wilt:** Leaves or shoot loose their turgidity and droop.
12. **Yellowing :** Leaves turn yellow due to degeneration of chlorophyll.

(B) HYPERPLASTIC / HYPERPLASIA SYMPTOMS (PLATE-I)

1. **Anthocyanescence:** Purplish or reddish colouration of leaves or other organs due to over development of anthocyanin pigment.
2. **Callus:** Over growth of tissues at the margins of wounds or diseased tissues.
3. **Curl:** Rolling or folding of leaves due to localized overgrowing tissues.
4. **Fasciation:** Flattened condition of a plant part that is normally round.
5. **Fasciculation or witches broom:** Broom like growth of densely clustered branches.
6. **Sarcody:** Abnormal swelling of tissues above girdled branches or stems.
7. **Scab:** Roughened, crust like lesions.
8. **Tumefaction:** Tumor like or gall like or knot like over growth of tissues.
9. **Virescence:** Development of chlorophyll in tissues where it is normally absent.

(C) HYPOPLASTIC SYMPTOMS (PLATE-I)

1. **Chlorosis :** Failure of chlorophyll development in normally green tissues.
2. **Dwarfing :** Sub normal size of an entire plant or some of its parts.
3. **Etiolation :** Yellowing due to lack of light
4. **Rusetting :** Crowded condition of foliage due to lack of internode elongation.

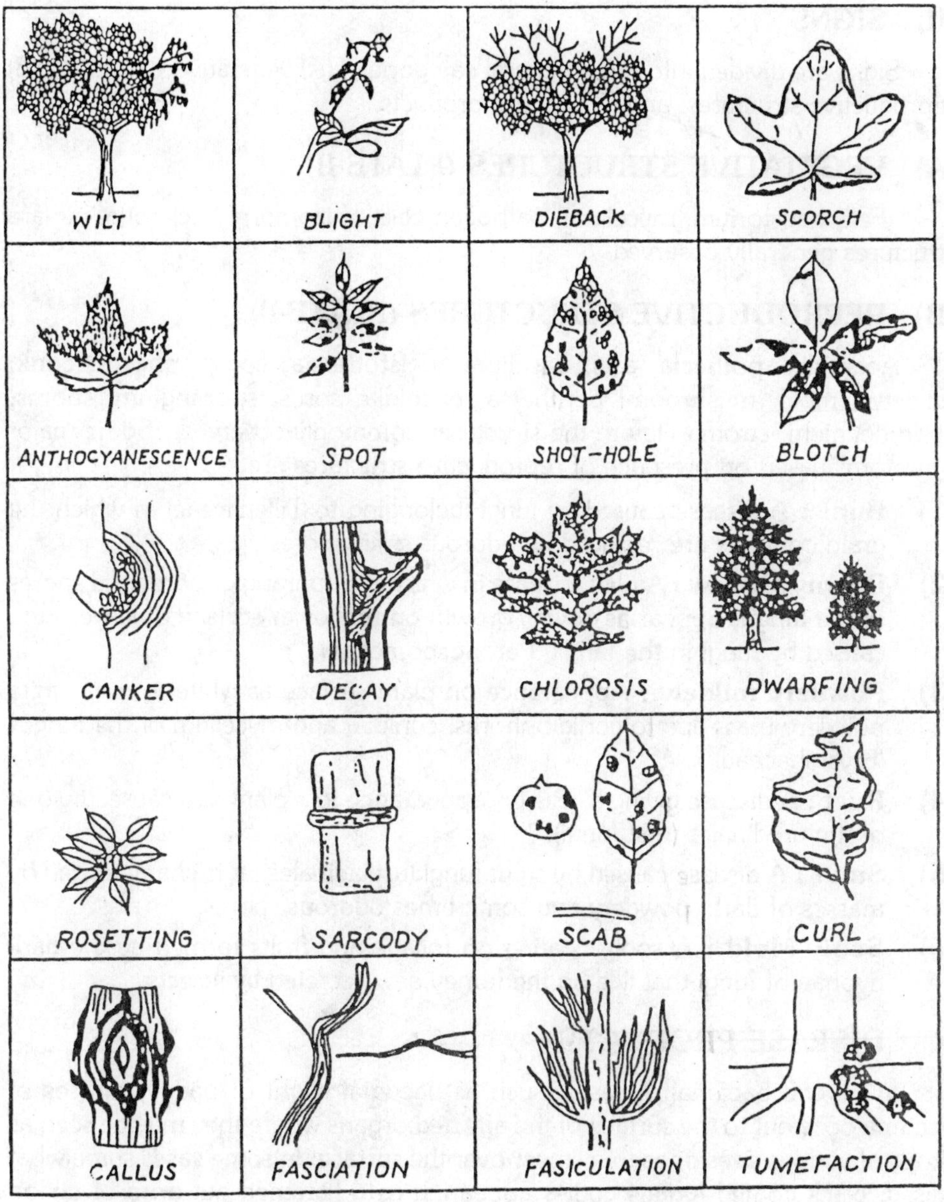
WILT
BLIGHT
DIEBACK
SCORCH
ANTHOCYANESCENCE
SPOT
SHOT-HOLE
BLOTCH
CANKER
DECAY
CHLOROSIS
DWARFING
ROSETTING
SARCODY
SCAB
CURL
CALLUS
FASCIATION
FASICULATION
TUMEFACTION

Plate-I : Various Symptoms of Disease

5. **Suppression :** Prevention of the development of certain organs.

III. SIGNS

Signs are divided into three general categories : (A) vegetative structures, (B) reproductive structures, and (C) disease products.

(A) VEGETATIVE STRUCTURES (PLATE-II)

Felt, haustorium, mycelium, pathogen cells, rhizomorph, sclerotia, etc are structures generally observed.

(B) REPRODUCTIVE STRUCTURES (PLATE-II)

Acervuli, apothecia, asci, basidium, cleistothecia, conidiophores, conk, mildews, mold, mushroom, perithecia, pycnidia, sorus, sporangium, spores, sporodochium, stroma etc are the structures commonly observed. Some major symptoms based on presence of reproductive structures are:

(1) **Bunt :** A disease caused by fungi (belonging to Tilletiaceae) in which the grain contents are replaced by odorous smut spores.

(2) **Downy Mildew :** A plant disease in which the sporangiophores and spores of the fungus appear as downy growth on the lower surface of leaves etc., caused by fungi in the family Peronosporaceae .

(3) **Powdery mildew :** Appearance on plant surface as white to dirty white powdery mass due to conidiophores, conidia, and mycelium of the fungus (Erysiphaceae)

(4) **Rust :** A disease giving a "Rusty" appearance to a plant and caused by one of the uredinales (rust fungi).

(5) **Smut :** A disease caused by smut fungi (ustilaginales); it is characterized by masses of dark, powdery and some times odorous spores.

(6) **Sooty Mold :** A sooty coating on foliage and fruits formed by the dark hyphae of fungi that live on the honey dew secreted by insects.

(C) DISEASE PRODUCTS

In several bacterial diseases, such as bacterial blight of paddy, masses of bacteria ooze out to the surface of the affected organs where they may be seen as drops of various sizes or as thin smeer over the surface. In some cases some what raised, black coated fungus bodies appearing as a flattened out drop of tar on leaves appear.

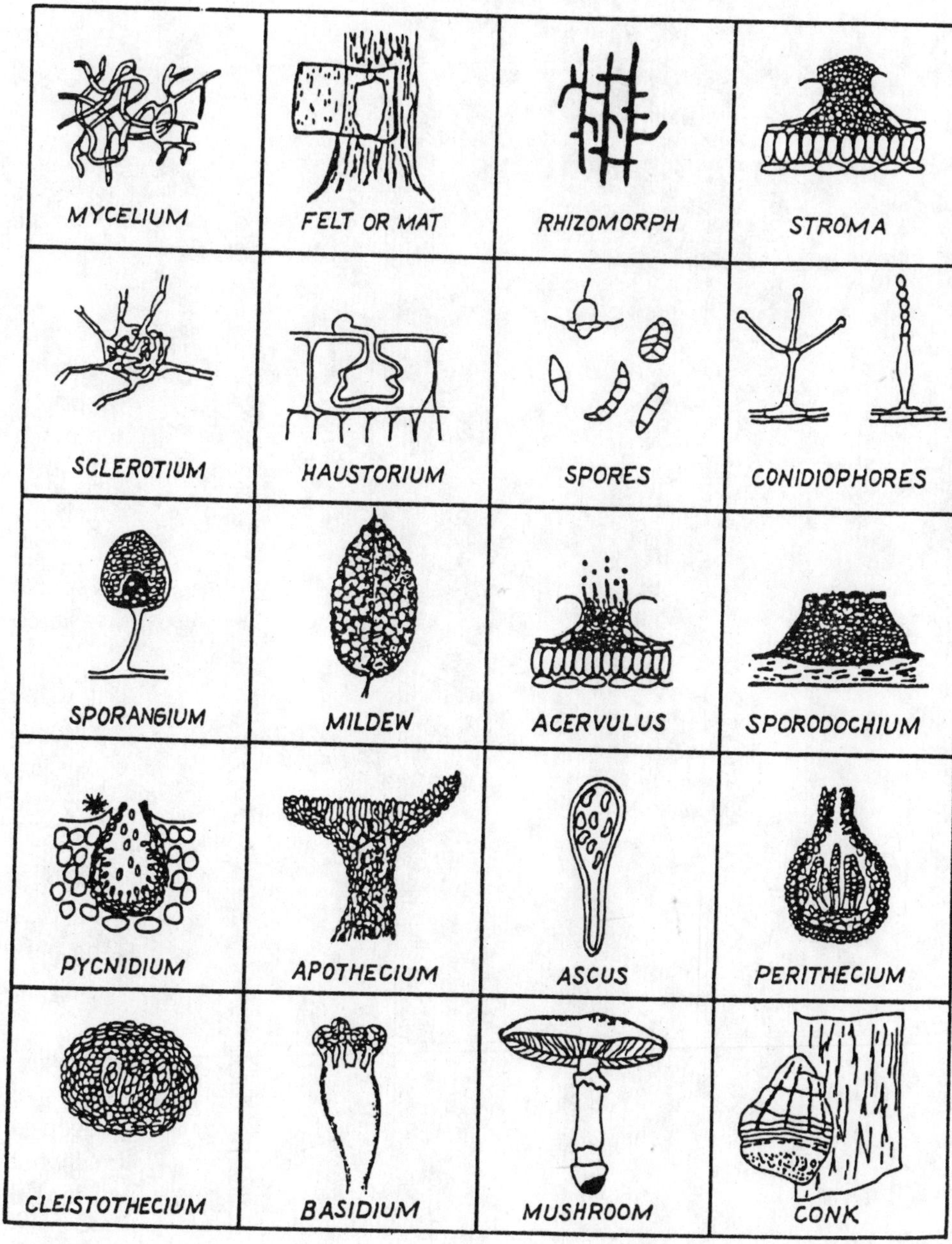
MYCELIUM
FELT OR MAT
RHIZOMORPH
STROMA
SCLEROTIUM
HAUSTORIUM
SPORES
CONIDIOPHORES
SPORANGIUM
MILDEW
ACERVULUS
SPORODOCHIUM
PYCNIDIUM
APOTHECIUM
ASCUS
PERITHECIUM
CLEISTOTHECIUM
BASIDIUM
MUSHROOM
CONK

Plate-II : Various Signs (Pathogen Structures)

CHAPTER V

PATHOGENESIS AND HOST-PHYSIOLOGY

I. INTRODUCTION

After establishment of physiological and genetic synchrony between the host and parasite, host tissues show various responses. Normal physiological activities are disturbed resulting in anatomical and morphological changes. Notable physiological changes include altered cell permeability, tissue disintegration, changed growth, reproduction, absorption and translocation of water and nutrients, respiration and photosynthesis.

II. TISSUE DISINTEGRATION

Plant pathogens are known to employ chemical weapons during their growth and development inside the host tissues. Among them, the enzymes (cuticular, pectic, cellulytic, lignolytic, etc) bring about destruction of cell walls leading to plasmolysis of protoplasts, death of cells, and ultimately utilization of cell contents by the parasite. Besides enzymes, toxins also affect cellular integrity and disrupt cell functioning and ultimately cellular structures. There are very few diseases that do not show tissue disintegration and cell death. As a result of tissue disintegration or rot, the symptoms known as blight, canker, anthracnose, etc appear on plants.

III. EFFECT ON PLANT GROWTH

Plant pathogens affect host's meristematic activities. This results in over or under growths, leaf curls, club root, knots, galls, etc. The changes in growth pattern of the host are caused by imbalance in production, accumulation, and translocation of growth regulators in plants. The normal plant synthesises growth promoting substances in quantities just enough for its normal growth. The plant also produces growth inhibitors to regulate the activities of growth promoters and other chemical substances. During pathogenesis, this regulatory mechamism is altered, disturbed or destroyed and as a consequence there is unregulated synthesis of growth hormones and other substances, and therefore, changes in plant's habit are seen. The increase in amount of IAA has been observed in many diseased conditions of plants. This could be due to excessive production of IAA by the diseased plant or due to production by the pathogen in clolonized tissues or it could be due to its reduced destruction in diseased tissues. Plants contain IAA oxidase enzymes which degrade IAA and keep the amount of IAA under check to permit Plant's normal growth. Pathogens through their metabolites inactivate IAA oxidising enzymes which results in rise in the level of IAA. This condition has been proved in maize smut (*U. maydis)* and stem rust of wheat (*P. graminis)*. Excess of IAA (hyperauxinity) has been detected in white blister of *B. napus (A. candida)*, and wheat powdery mildew (*E. graminis)*. Similarly, gibberellins, normal constituents of green plants' perform numerous functions. They function as chemical signals which activate cell extension, dormancy breaking, flowering, etc. Gibberellins have strong growth promoting qualities. These include elongation of roots and shoots, excessive flowering, fruiting, etc. These compounds are suspected to be operating in many host-parasite system such as downy mildew of sugarcane (*S. sacchari)* and pea rust *(U. pisi)*. The *bakanae* or *foolish seedling* disease of rice *(G. fujikuroi)* is characterized by abnormal elongation of stem due to excessive elongation of internodes. Gibberellins were isolated from these seedlings.

The significance of kinins in pathogenesis is rather uncertain. It is, suspected that these substances affect host-parasite interaction in bean rust, root knot, and some bacterial diseases. Many growth inhibitors are also synthesised in plants. Excess of these substances may react with growth promoters and render them ineffective. Thus, normal plant growth may be arrested. Dormin and ethylene have been studied. Dormin induces dormancy and functions as antagonist of gibberellins. This compound also masks the effects of IAA.

Ethylene (C_2H_4) is highly active growth regulator and plays important role in pathogenesis. Some prominent effects of ethylene are epinasty, tissue proliferation, increased respiration, premature senescence, shedding of leaves, and stimulation of root formation.

IV. EFFECT ON REPRODUCTION AND FOOD STORAGE

When pathogenesis reaches a particular stage, the reproductive process of the plant is affected either directly or indirectly. Direct effect includes partial or complete destruction of flowers, fruits, seeds, etc. Indirect effects are the results of weakening of the plants. The powdery mildews do not show serious effect on reproductive organs or seeds but the seeds are weak, undersized, shrivelled, etc. or pod formation and development affected due to distrubed host physiology and photosynthesis. The increased transpiration in plants affected by rusts also results in shrivelled grains of low viability.

V. EFFECT ON PHOTOSYNTHESIS

Carbohydrates are synthesized by chloroplasts in green parts of the plants through the process of photosynthesis. This is a basic function of all green plants that enables them to convert light energy (from sun) into chemical bonds of energy for utilization by the cells. In photosynthesis $C0_2$, H_20 and light energy combine in chloroplasts to form carbohydrates $6C0_2 + 6H_20 + \text{energy} = C_6H_{12}0_6+60_2$.

Since photosynthesis provides the basic material for synthesis of all the organic compounds in plant tissues, it is apparent that any condition that obstructs this process is pathogenic. In a large number of plant diseases the symptoms on leaves clearly suggest that photosynthetic processes are impaired due to infection. Such symptoms include chlorosis, necrotic lesions, leafspots, blights, etc. The effects of pathogenesis on photosynthesis can be attributed to causes grouped under two categories : (a) the destruction of chlorophyll and chloroplasts, and (b) decreased efficiency of the photosynthetic process per mole of chlorophyll.

VI. EFFECT ON HOST RESPIRATION

Respiration is that process by which the cells through enzymatic oxidation of organic materials, produce energy for various activities and carbon skeletons. The energy finallly produced by various oxidative reactions is used for maintenance, growth and synthetic processes. In plants the process of respiration takes place in two major steps; (a) in the first step known as glycolysis, hexose sugars are degraded to form pyruvates, and (b) in the second step, the terminal phase, pyruvates are oxidised to produce CO_2, H_2O and energy. Thus, the two steps can occur simultaneously only in presence of oxygen and can be expressed as $C_6H_{12}O_6+6O_2=6CO_2+6H_2O+\text{Energy}$.

The diseased plants generally show increased rate of respiration. This increase significantly alters metabolic processes in the host. Level of many enzymes associated with respiratory process is increased. Accumulation and oxidation of phenols also increases. Two mechanisms have been proposed —

1. Uncoupling of oxidative phosphorylation that causes enhanced O_2 uptake. Certain substances such as 2, 4 - dinitrophenol (DNP) prevent formation of ATP from ADP. As a result of this uncoupling a high level of ADP accumulates in the cells and causes increased respiration.
2. Stimulation of metabolism in diseased plant is a more convincing mechanism. In many plant diseases, growth is first stimulated, protoplasmic streaming

becomes faster, synthetic processes such as protein, nucleic acid, and carbohydrate synthesis, stimulated and translocation of synthesized products from healthy to diseased areas occurs. All these processes require energy, and therefore, accelerate breakdown of ATP with increased utilization of ATP, the level of ADP and PO_4 rises in the cells. This causes increased respiration.

VII. EFFECT ON UPTAKE AND TRANSLOCATION OF WATER AND NUTRIENTS

The water uptake capacity of roots can be affected in three ways; 1. roots are injured, 2. permeability of root cell walls is altered, and 3. development of roots is checked. The pathogens causing different root diseases/injuries/wounds reduce the density of active roots and thus water uptake is reduced in the same proportion. Roots absorb water and nutrients through the process of osmosis. If the pathogens or metabolites produced by them alter cell wall permeability, the osmosis is affected. The decreased osmotic activity causes decrease in water uptake.

The plant pathogens causing several diseases (damping off, stalk rot, cankers, etc.) can enter the xylem vessels. In the affected vessels pathogen propagules or metabolites produced by them may be present and cause obstruction. Thus, disintegration as well as obstructions both reduce the water conducting capacity of the root system, plugging of xylem vessels has atleast partly been attributed to presence of mycelium, spores or slimy bacterial mass. Wilt causing vascular pathogens produce pectolytic and cellulolytic enzymes. These enzymes degrade middle lamilla and release substances that form gels and gums which form plugs and obstruct the passage. Browning of xylem vessels is a common feature of vascular wilts. This is due to formation of melanin. This pigmant is formed due to oxidation of phenols.

Bacteria and fungi produce polysaccharides that induce wilt symptoms. The entrapping of such materials in the cell wall openings causes obstruction in the flow of water from one vessel to another and laterally to other cells. Abnormal development of xylem vessels, often occurs in vascular wilts. These changes reduce water transport in plants. Many pathogens produce tyloses which are outgrowth of parenchyma adjacent to the xylem and appear as peg-like structures. They obstruct passage of water in the same manner as the gels and gums.

VIII. EFFECT ON TRANSPIRATION

Increased transpiration has been observed in several diseases such as rusts, powdery mildews etc. The main cause of increased transpiration is the disintegration of cuticle. Increased cell wall permeability and malfunctioning of stomata are also contributory factors.

CHAPTER VI

EPIDEMIOLOGY OF PLANT DISEASES

I. INTRODUCTION

Epidemiology is primarily concerned with epidemics (epiphytotics) but the term has a wide meaning and has come to include most field aspects of plant diseases. It is the interaction of crop pathogen and environment, populations of plants and pathogens rather than individuals being involved. Epidemiology covers the effects of environmental factors on disease prevalence, incidence and severity, in addition to survival and spread of plant pathogens. A proper understanding of epidemiolgy is necessary for prediction of plant diseases and formulation of effective control measures. In the recent years, especially after the application of statistics and mathematical analysis by Van der Plank (1963) the subject has received serious considerations.

II. CONCEPT

The word "epidemic" stems from the Greek, *epi* = (on) and *demons* = (people). Literally epidemic is defined as "affecting or tending to affect many individuals within a population, a community, or a region at the same time an outbreak, a sudden and rapid growth, spread and development".

In recent years the term "epidemiology" has come to have a broad meaning within plant pathology. The term has been variously defined as the study of diseases in populations; the study of environmental factors that influence the amount and distribution of diseases in population, and the study of rates of change (either increase or decrease) in the amount of disease in time, in space, or both. (Horsfall and Cowling, 1978)

III. THE ELEMENTS OF AN EPIDEMIC

Disease results from the interaction of a pathogen with its host but the inten-

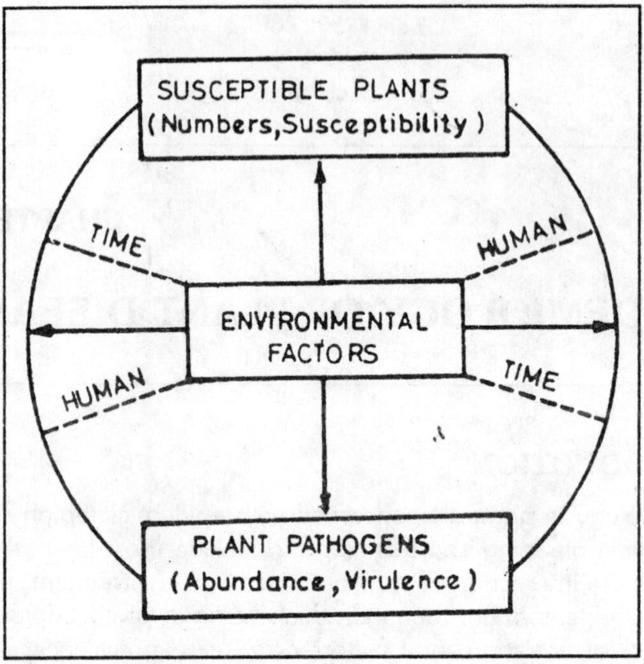
SUSCEPTIBLE PLANTS
(Numbers, Susceptibility)
TIME
HUMAN
ENVIRONMENTAL FACTORS
HUMAN
TIME
PLANT PATHOGENS
(Abundance, Virulence)

Figure 1 : A schematic diagram of the interrelationships of the factors involved in plant disease epidemics.

sity and extent of this interaction is markedly affected by the environmental factors. Although these factors are not the causal agents of infectious diseases, they are the final determinants of almost all events that constitue the infection chain leading to pathogenesis and also the events that follow, viz., spread of the disease in the population. The interaction of these three components of disease have been visualized as triangle, generally referred to as the "disease triangle". However, to understand an epidemic, a fourth dimension —time—must be added to the "disease triangle" to give a "disease pyramid" or "disease cone". Agrios (1988) has added another dimension, i.e. , the activities of humans (Figure 1). Fry (1987) has explained the complex nature of disease involving interactions between host x pathogen x environment through an equation:

$$D_t = S \sum_{i=0}^{t} f(p_i, h_i, e_i)$$ where D_t is disease at time t

and is the sum of interaction of pathogen (p_i includes inherent ability to induce disease and population size), host (h_i includes susceptibility, distribution, and population size), and environment (e_i includes physical, biological, and chemical factors) over time (from i =0 to t).

A. Host

Levels of genetic susceptibility or resistance, degree of genetic uniformity of host plants, distance of susceptible host from the source of primary inoculum, abundance, and distribution of susceptible host, disease proneness in the host due to environment, and availability of alternate or collateral hosts are the factors which play important role in the development of epidemics involving host.

The host plant having high levels of resistance (vertical resistance) do not allow some races and/or biotypes of a pathogen to become established in them and thus, no epidemic can develop unless and until there is evolution of new races and/or biotypes that can attack that resistance and thus, making the host susceptible. Plants with horizontal or general resistance will become infected, but the rate of development will be slow. On the other hand, susceptible host plants lacking genes for resistance against existing races of the pathogen provide suitable and adequate substrate for establishment and development of new infections.

The disease in an area is initiated by primary inoculum surviving at some source. Longer the distance from the source of primary inoculum, longer will be time required for build up of epidemic in a susceptible crop. During dispersal in different directions the density of primary inoculum is diluted and as the distance increases fewer propagules are likely to reach the susceptible surface. Continuous cultivation of a susceptible variety in a given area, large areas under a similar susceptible variety and distribution of the variety over large continuous areas help in build up of inoculum and improve the chances of epidemic. Under these conditions, the pathogen is able to use maximum number of its propagules effectively, increase the rate of multiplication many times and repeat the disease-cycle quickly. This phenomenon has been observed in case of *Helminthosporium* blight of victoria oats and Southern maize leaf blight with Texas male sterile cytoplasm.

Plants change in their susceptibility to disease with age. For example, in root rots, downy mildews, smuts, rusts, etc., the hosts or their target organs are susceptible only during the growth period and become resistant during the adult period. In some other diseases such as flower or fruit blights caused by fungal pathogens like *Botrytis, Alternaria, Glomerella, Penicillium* and *Monilinia,* and in all post harvest infections, plant parts (fruits) are resistant during growth and early adult period but become susceptible near ripening.

For pathogens spreading through heterogenous infection chain, presence of an alternate host is necessary for providing primary inoculum. The amount of inoculum thus available will determine the intensity of primary infection and subsequent spread. Presence of collateral hosts plays the same role for pathogens of homogenous infection chain. For example, grass hosts of *Sclerospora philippinensis* (downy mildew of maize and sorghum), *Ustilago scitaminea* (sugarcane smut) and *M. grisea* (rice blast), may produce abundance of inoculum aiding in build-up of epidemics. In annual or seasonal crops, such as maize, wheat, barley, vegetables, chickpea, cotton, tobacco, etc., the epidemic generally develops much faster than do in perennial woody crops such as fruit and forest trees. For instance, epidemics of fruit and forest trees like pear decline, Dutch elm, chestnut blight, citrus tristeza, etc., take years to develop.

B. The Pathogen

For any epidemic rapid cycles of infection are essential and successful infection can be caused only by virulent or aggressive isolates of the pathogen. High birth rate or fast reproductive cycles is another important contributory factor for epidemics. The pathogens that assume epidemic form invariably have the capacity to produce enormous quanitiy of spores that are adapted to quick and long distance dispersal in a short time, so that they can take advantage of favorable weather conditions during that short period. In most cases, these spores are asexually produced usually on the exposed surfaces of the host for quick dispersal by wind, water, and insects. These are polycyclic pathogens that usually cause leaf spots, blights, rusts, mildews, and are responsible for most of the catastrophic plant disease epidemics in the world. Soil-borne pathogens like *Fusarium, Phymatotrichum, Verticillium,* etc., and most nematodes usually have 2 to 4 reproductive cycles per growing season. Since, the dispersal of such pathogens is limited both in space and time, only localized and slower developing epidemics are caused. On the other hand, monocyclic pathogens like smut fungi require an entire year to complete a life cycle. In such diseases, inoculum builds up from one year to the next, and the epidemic develops over several years. Similarly, epidemics caused by pathogens that require more than one year to complete a reproductive cycle are slow to develop. Examples are cedar apple rust (2years), white pine blister rust (3 to 6 years), and dwarf mistletoe (5 to 6 years). Such pathogens produce inoculum and cause series of infections each year only.as a result of overlapping of the polyetic generations.

Among fungi the vicissitudes of dispersal by wind, minute size of unprotected spores, possible chances of falling on wrong plants, etc., are many factors that cause high death rate among the propagules. However, this weakness is offset by extremely high birth rate. Epidemics attributed to low death rate of pathogens are those in which the causal organism is systemic and protected by the plant tissues. Thus, the chances of high mortality are considerably reduced. The chief source for accumulation of inoculum for epidemics of such diseases is the diseased plant organs used for vegetative propagation and, therefore, the build-up of epidemic is comparatively slow. When a particular area becomes saturated with diseased planting materials chances of occurrence of epidemics are very high.

Adaptability of pathogen is another factor vital to development of epidemics. For pathogens having capacity to adapt to adverse conditions, the occurrence of epidemics is almost certain. The units of propagules produced by the pathogen are dispersed by external agencies which must be available if epidemics are to develop.

C. Environmental Factors

The effect of weather on disease development has been discussed elsewhere. Optimum moisture, temperature, light, etc., are necessary for activities of biotic pathogens. Assuming that a particular fungal pathogen meets all the conditions discussed above for causing epidemic; high reproductive cycles, high aggressiveness, ensured and effective dispersal, susceptible host; even then development of epidemic may not ocour if weather is not favorable for germination of spores or in the absence of light stomata have not opened to permit entry of the infection thread or when the stomata open the moisture is so deficient that the germtube has dried. The weather also affects the activity of the pathogen on the host surface. It may not permit sporulation on the host surface thus reducing amount of inoculum for secondary spread.

D. Activities of Humans

Many activities of farmers, such as selection of sites, propagative materials and various disease management practices including cultural practices, have a direct or indirect effect on plant disease epidemics. The use of infected seeds, nursery stocks, and other propagative materials, increase the amount of primary inoculum within the crop and thus, greatly favor the development of epidemics. Contrary to it, the use of healthy, pathogen free or suitably treated planting materials can greatly reduce the chance of epidemic. Continuous monoculture, large acreages planted to the same or related variety of crop, high levels of nitrogen fertilization, overhead (sprinkler) irrigation, herbicide injury, and poor sanitation all increase the possibility and severity of epidemics. The use of certain chemical or planting of a certain variety may lead to selection of virulent strains that are either resistant to the chemical or can attack the resistance of the variety and thus lead to epidemics.

IV. Decline of the Epidemics

In epidemics of crop plants, no epidemic remains for ever in a population. In an area where a disease assumes epidemic proportion, majority of the plants get infection, and therefore, there is saturation of the pathogen in the host population as nonavailability of more uninfected plants limits the pathogen. All these result in production of less inoculum, fewer secondary infections, and finally no new infections. The plants that escape infection are those that possessed resistance or in which resistance developed during the epidemic. It is possible that in future only these plants will be grown in that area. therefore, one of the positive effects of an epidemic is that it eliminates susceptible individuals and permits only the resistant individuals of the population to survive and breed.

With passage of time, decline of proneness in the host also accounts for decline of the epidemics. Most pathogens attack the plant at a particular stage of its growth. Once the plant has crossed this stage, its proneness for contacting infection is reduced or completely lost. Under these conditions the epidemic will automatically decline. When the plant is receptive for infection throughout its life and its population has been affected by an epidemic, the weather conditions may not remain always congenial for disease development. This will result in reduced spread of the inoculum and the epidemic will decline. The aggressiveness of the pathogen may also be reduced. After the population of susceptible host has been destroyed, the pathogen may try to parasitize the remaining resistant individuals of the same species. In these adverse conditions it may lose its power of successful infection, its reproduction may slow down, and thus it may not remain as aggrressive as when the conditions were favorable (Kranz, 1978)

V. Patterns of Epidemics

The pattern of an epidemic in terms of disease incidence or severity is expressed in curves that show the progress of a disease over distance or over time. This curve is called "disease-progress curve". The shape of disease-progress curve may reveal informations about the onset of the disease, amount of infective propagules, changes in host susceptibility and proneness over time, recurrent weather events and the effectiveness of disease management practices. Disease-progress curves are generally characteristic of some group of disease, though they vary somewhat with location and time due to influence of weather, crop variety, etc. For example, a saturation-type curve is characteristic for monocyclic diseases (Figure 2), a sigmoid curve is characteristic for polycyclic diseases (Figure 2), and a bimodal curve (Figure 2) is characteristic for diseases affecting different organs of the plant. There is another type of disease curve called "disease-gradient curve" (Figure 3). Since the amount of disease is generally higher near or at the source of inoculum and decreases with increasing distance from the source, most disease curves are hyperbolic and quite similar, at least in the early stages of the epidemic. The incidence and severity of disease decrease steeply within short

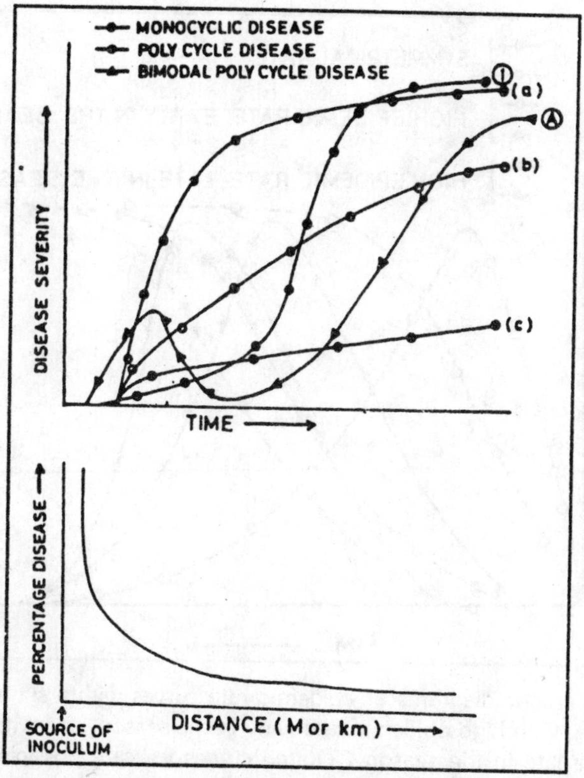
MONOCYCLIC DISEASE
POLY CYCLE DISEASE
BIMODAL POLY CYCLE DISEASE
(a)
(b)
(c)
DISEASE SEVERITY
TIME
PERCENTAGE DISEASE
SOURCE OF INOCULUM
DISTANCE (M or km)

Figure-2 : Schematic diagrams of disease — progress curves of some basic epidemic patterns (a-c). Three monocyclic diseases of different epidemic rates (I) a polycyclic disease, (A) a biomodal polycyclic disease.

Figure-3 : Schematic diagram of disease-gradient curve.

distance of the source and less steeply at greater distances until they reach zero or a low background level of occasional diseased plants. Disease gradients, however, are sometimes flattened near the source as a result of multiple infections and may become more flat with time as secondary spread occurs.

The rate of growth of the epidemic can be obtained by plotting the disease-progress curve from the informations obtained at various time intervals. The rate of epidemic is the amount of increase of disease per unit of time in the plant population under study. The patterns of epidemic rates are given by curves called "rate curves" and these curves are different for various groups of diseases (Figure 4). These rate curves identify at least three major classes of epidemic patterns : symmetrical (bell -shaped, Figure 4), for example, in late blight of potato; or asymmetrical curves (Figure 4) with epidemic rate greater early in the season due to greater susceptibility of young leaves—for example, in apple scab or most

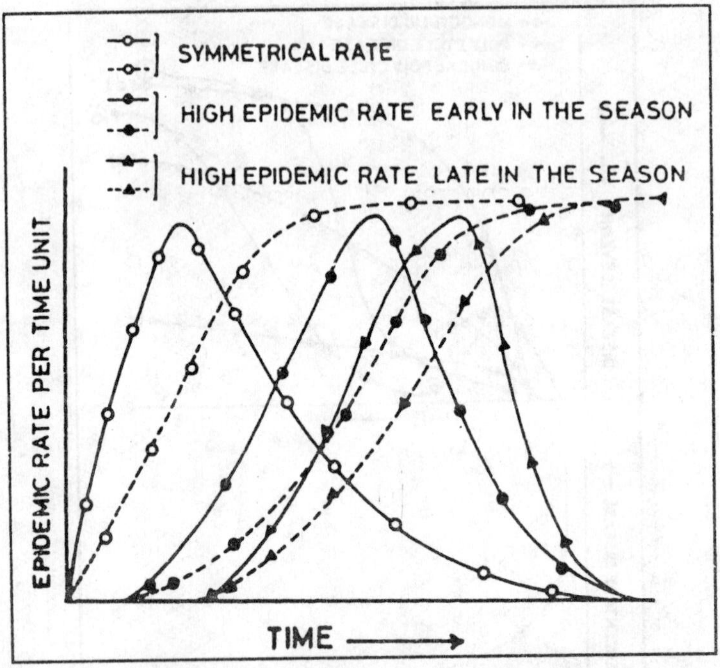
SYMMETRICAL RATE
HIGH EPIDEMIC RATE EARLY IN THE SEASON
HIGH EPIDEMIC RATE LATE IN THE SEASON
EPIDEMIC RATE PER TIME UNIT
TIME

Figure-4 : Schematic diagrams of epidemic rate curves with a symmetrical epidemic rate, with high epidemic rate early in the season, and with a high epidemic rate late in the season. Dotted curves indicate possible disease-progress curves that may be produced in each case from the accumulated epidemic rate curves.

downy and powdery mildews; or asymmetrical with the epidemic rate greater late in the season (Figure 4) as observed in many diseases which start slowly but accelerate markedly as host susceptibility increases late in the season.

VI. Models of Epidemics Development

An epidemic is a dynamic process. Researchers have used mathematical models as a tool in analysis of disease dynamics. Van der Plank was instrumental in establishing mathematical models for analysing epidemics when he published his book *Plant Disease Epidemic and Control* in 1963.

A. Monocyclic Pathogens

The amount of disease caused by a monocyclic pathogen in a season is the function of several interacting factors. The pathogen factors include size and distribution of the pathogen population and the genetic potential of the pathogen to induce disease. Host factors are genetic potential of the host as well as its size and distribution. Environmental factors include both biotic and abiotic factors. The equation $x_t = QRt$ explains how disease develops from the interaction of host x pathogen x environment through time. X_t is the amount of disease at time t, Q is the size of initial inoculum, R is the efficacy of inoculum and t is the length of time that host and pathogen interact in their enviroment (Fry, 1987).

This explains several important characteristics of the interaction between the host population and monocyclic pathogen. The size of initial inoculum (Q) does not increase during the season as pathogen produces no additional inoculum which could induce disease in the same season. R (pathogen efficacy) may range from zero to some positive value. If Q or R is zero, no disease will occur, t (duration of host x pathogen interaction) may influence the amount of disease.

Fry (1987) with the help of Van der Plank's equation [1] $d_x/d_t = QR$ where increase in disease (dx) during a short time period (dt) is a function of initial inoculum (Q) and its efficacy (R, a rate), explains that this equation neglects an important factor (the amount of healthy tissues) that can have a greater influence on disease increase. He explains that for a given pathogen population in a given environment, the rate of disease increase is likely to be greater when there is availability of large amount of healthy tissues than when there is a small amount. Thus, he corrects the equation and describes it as $dx/dt = QR\,(1 - x)$ where $(1 - x)$ describes the amount of available tissues or the proportion of the host population which is not yet infected. To make it more useful he rearranges this equation as $dx'\,(1 - x) = QR\,dt$ and then after integration another equation $\ln\{1/(1 - x)\} = Q\,Rt + k$ is developed. The symbol ln indicates natural logarithms (to the base e). Thus, the left-hand side of this equation is natural logarithm of $1/(1 - x)$. The letter k is the constant which results from integration ($k = \ln\{1/(1 - x_0)\}$ where x_0 = the amount of disease at $t = 0$). This equation can be used to predict the required effect of disease management to achieve a desired degree of disease suppression (Fry, 1987).

B. POLYCYCLIC PATHOGENS

Diseases caused by polycyclic pathogens are influenced by the size and distribution of initial pathogen population, the inherent ability of the pathogen to induce disease, host reaction, environmental factors including cultural manipulations, time during which host and pathogen interact, and rate of reproduction of the pathogen.

To explain the increase of disease induced by polycyclic pathogen Fry (1987) described an equation $dx/dt = xr\,(1 - x)$, where dx/dt is the instantaneous rate of disease increase at a specific time, x is the proportion of tissue diseased, r is the rate at which new infections occur (apparent infection rate), and $(1-x)$ is the proportion of tissue available for infection. He proposed a simplified equation if the amount of diseased tissue is very small (i.e., <0.01). Then (1 - x) is approximately 1.0 and the equation becomes $dx/dt = xr$. Upon rearrangement it becomes $dx/x = r\,dt$ and then integrated to obtain $\ln x = rt + k$, where t = 0, the constant of integration = x_0, the value of x at the beginning of the time period of concern. After taking antilogs it was found that $x = x_0e^{rt}$, where x is the amount of disease at time t, x_0 is the amount of initial disease (at t = 0), e is the base of natural logarithm, r is the exponential rate of disease increase, and t is the interval during which host and pathogen interact. This equation is the description of exponential growth. At low levels, disease induced by a polycyclic pathogen increases exponentially. At higher levels of disease, the rate of the increase is limited by the diminishing supply of uninfected tissues.

VI. COMPUTER SIMULATION OF EPIDEMICS

In the last three decades, the use of computers has helped pathologists to write programs that allow simulation of several plant disease epidemics. The computer is given data pertaining to various subcomponents of the epidemic and control practices at specific points in time. The computer then provides continuous information concerning not only spread and severity of the disease over time but also the economic losses likely to be caused by the disease under the conditions of the epidemic as given to the computer.

The first computer simulation program called EPIDEM, was written in 1969 to simulate *Alternatia* early blight of potato and tomato. Subsequently, computer simulations were written for *Mycosphaerella* blight of chrysanthemums (MYCOS), for southern corn leaf blight caused by *Helminthosporium maydis* (EPICORN), and for apple scab caused by *Venturia inaequalis* (EPIVEN).

CHAPTER VII

FUNGI

I. INTRODUCTION

Man's interest in fungi started with the observation of the umbrella shaped mushrooms and toadstools growing naturally on soils. Since these grew attached to the soil like plants, they were regarded as plants. It is assumed that MYCOLOGY after being derived from the Greek word (*Gr. mykes* = mushroom+logos= discourse) etymologically is the study of the mushrooms. Although much before the invention of the microscope, mushrooms largest of the fungi attracted the attention of naturalists but it was only in the 17th century after the invention of the microscope, that the actual systematic study of the fungi began. It is virtually extremely difficult to define the exact limits of this group. Infact, fungi constitute a most fascinating group comprising of more than 1,00,000 species (Ainsworth and Bisby, 1983).

II. DYNAMICS AND SPECTRUM OF ACTIVITIES

Fungi occur in almost all type of habitats. Dynamics and spectrum of their activities are summamrised in the chart given on the next page:

III. FUNGI AND PLANTS

Fungi comprise a separate major group of organisms which differ from plants in origin, evolution and organization due to adaptation to different mode of primary nutrition (absorptive). In adaptation to absorptive mode of nutrition, the higher fungi have evolved : (1) a non-motile life embedded in food supply and (2) a mycelial organization that combines maximum surface of contact with the free movement of food and protoplasm through the mycelial system. This difference of fungi from plants and animals has been supported by studies of cytochrome C, a component of the terminal respiratory chain of enzymes in aerobic organisms.

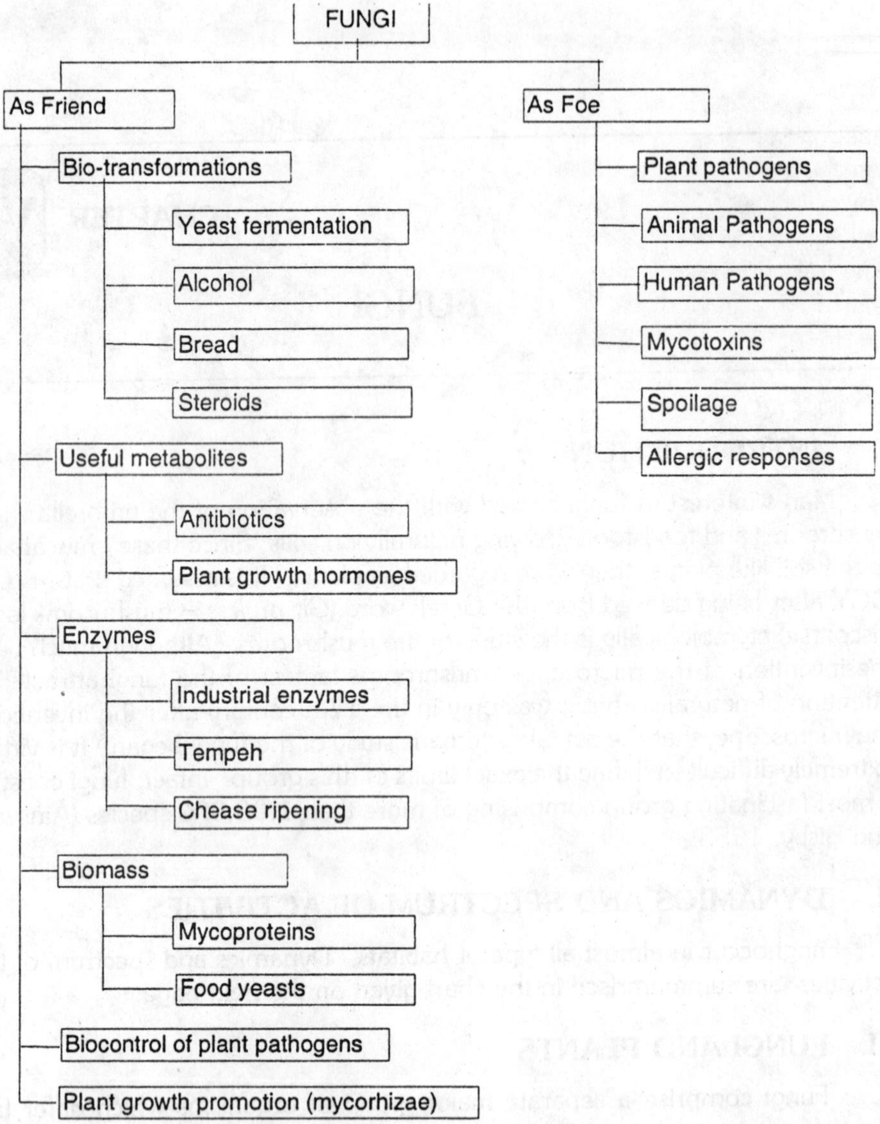
FUNGI
As Friend
Bio-transformations
Yeast fermentation
Alcohol
Bread
Steroids
Useful metabolites
Antibiotics
Plant growth hormones
Enzymes
Industrial enzymes
Tempeh
Chease ripening
Biomass
Mycoproteins
Food yeasts
Biocontrol of plant pathogens
Plant growth promotion (mycorhizae)
As Foe
Plant pathogens
Animal Pathogens
Human Pathogens
Mycotoxins
Spoilage
Allergic responses

The fungi differ from plants as they lack chlorophyll and they can not synthesise their own food from Co_2 and water and are forced to live as parasites or saprophytes. Although fungi resemble green plants in having their protoplasts encased in cell walls, they differ basically in their mode of nutrition. Fungi do not possess stems, roots, or leaves as do plants, and have not developed a complex vascular system. Another feature that separates fungi from the plants is the fact that their primary carbohydrate storage product is glycogen rather than starch.

IV. DIVISION OF ORGANISMS

When the living organisms were divided into plant and animal kingdoms, fungi were placed among the plants (cryptogams-Thallophyta). The complex and heterogenous behaviour of many lower organisms created limitations in this system. To solve this problem, E. Haeckel, a German biologist proposed a third kingdom Protista (*Protisto* = Very first) in 1866 to accommodate organisms that lacked tissues. Protists themselves were later divided into lower and higher protista. The lower protista which included bacteria and blue-green algae was named the fourth kingdom, Monera. Thus, the living world came to have four kingdoms, i.e. Plantae, Animalia, Protista and Monera. Whittaker (1969) proposed a five kingdom system and placed fungi in a separate kingdom, coordinate with higher plants and animals.

The development of electron microscope and associated preparative techniques for biological materials made it possible to study the ultrastructures in the cell such as intranuclear intranucleolar structures, cell organelles other than nuclei, membrane of the cell and intracellular organelles, etc. These observations led to the recognition of two kinds of cells in the living systems. In the more complex *eukaryotic cells,* which is the unit of structure in plants, metazoan animals, protozoa, fungi and most algae, the nucleus is bound by a nuclear membrane and divides by mitosis.

The less complex *prokaryotic cells* are the unit of structure in two groups: the bacteria including mycoplasmas and rickettsiae, and the blue-green algae or blue-green bacteria. The prokaryotic nucleus (nucleoid) is not membrane bound and its division is nonmitotic. This unambiguous bipartite division of organisms, exclusively based on cellular properties still retains algae, fungi and protoza in the protists as eukaryotes distinct from plant and animals on the basis of little or no differentiation of cells and tissues as given below:

The Primary Subdivison of Cellular Organisms

Eukaryotes	1. Multicellular extensive differentiation of cells and tissues	(A) Plants (B) Animals
	2. Unicellular, Coenocytic, or multicellular with little or no differentiation of cells and tissues	(C) Protists Algae Protozoa Fungi
Prokaryotes	Bacteria including photobacteria or blue-green algae and mollecutes	

V. GENERAL FEATURES OF FUNGI

Fungi are too diverse to be easily defined. The main characteristics are summarized in Table-1.

Table-1 :

Characters	Remarks
Distribution	Cosmopolitan
Habitat	Ubiquitous as saprobes, symbionts, parasites, hyperparasites
Sporocarps	Macroscopic or microscopic, showing limited tissue differentiations
Nutrition	Heterotrophic, absorptive
Thallus	Plasmodial, pseudoplasmodial, pseudomycelial or mycelial
Cell Wall	Well defined, typically cutinized but cellulose in oomycetous fungi
Nuclear status	Eukaryotic, multinucleate, homo or heterokaryotic, haploid, diploid
Life-cycle	Simple to complex
Reproduction	Asexual, sexual (homo or heterothallic), parasexual

VI. SOMATIC STRUCTURES

Somatic (*Gr. Soma* = body) refers to the assimilative phase in fungi; a structure or function as distinguished from the reproductive. It is also referred to as vegetative.

A. The Thallus

Thallus (pl. thalli; *Gr. thallos* = shoot) is a relatively simple plant body devoid of stems, roots and leaves. It may be PLASMODIAL, PSEUDOPLASMODIAL, PSEUDOMYCELIAL or MYCELIAL. Plasmodium (pl. plasmodia, Gr. *plasma* = a molded object) is a naked, multinucleate mass of protoplasm that moves and feeds in an amoeboid fashion such as the somatic phase of *Myxomycota* (slime molds). Pseudoplasmodium (*Gr. pseudo* = false+ plasmodium) is a sausage shaped structure consisting of many amoebae aggregation in the cellular slime molds (*Dictyosteliomycota* or *Acrasiomycota*). Pseudomycelium (*pl. pseudomycelia*; *Gr. pseudo=false+mycelium*) is a series of cells adhering end to end due to incomplete wall separation after budding in some yeasts. It deffers from true mycelium in which compartments or cells are formed by septum formation just behind the apex of an extending hyphal tip. Mycelium (*pl. mycelia; Gr. mykes*=mushroom, fungus) is the mass of hyphae constituting the body (thallus) of a fungus.

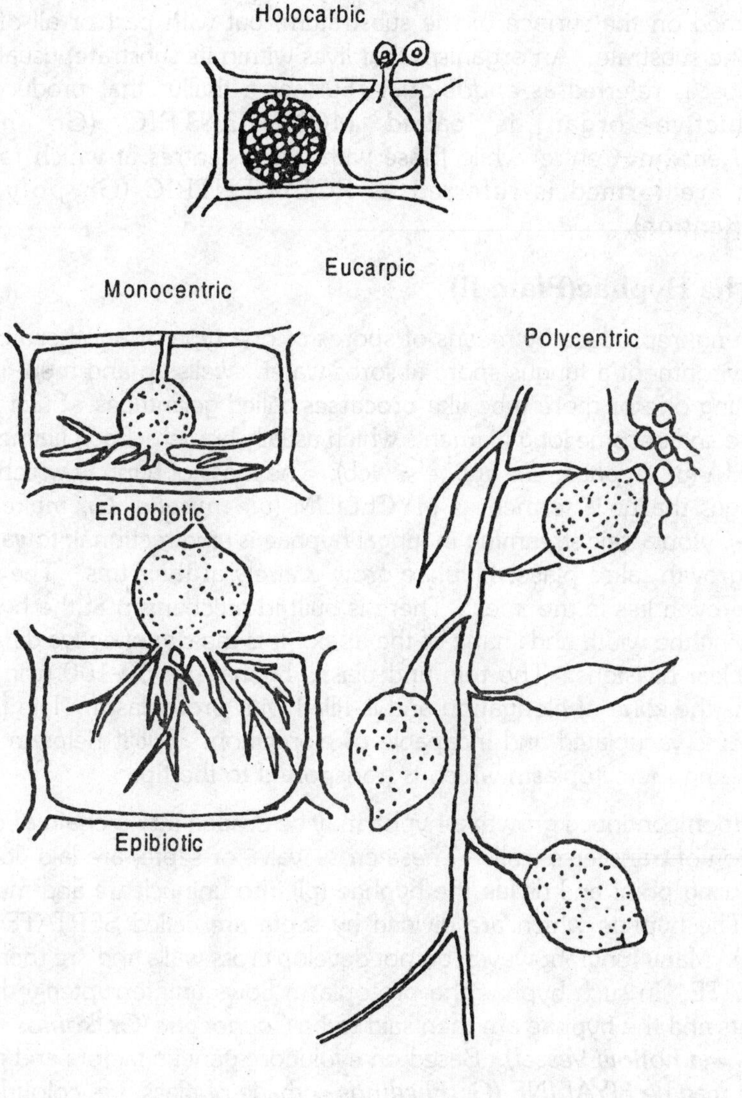

Holocarbic
Eucarpic
Monocentric
Polycentric
Endobiotic
Epibiotic

Plate-I : Types of thallus development in Chytridiales (after Webster, 1970).

The most simple types of thalli (**plate -I**) are wholly converted into reproductive cells and are termed HOLOCARPIC (*Gr. holos*=entire + *Karpos*=fruit) (**plate-1**). In majority of fungi, however, only parts of the thallus become reproductive forming differentiated or undifferentiated reproductive cells. Such thalli are EUCARPIC (*Gr. eu= good +Karpos=fruit). In relation to host, the thalli may be EPIBIOTIC (Gr. epi=On+bios=life) or ENDOBIOTIC (Gr. endos=inside+bios=life).* In fungi with epibiotic thalli, the reproductive organs are formed on the surface of the substratum, but with part or all of the soma within the substrate. An organism that lives within its substrate, usually the cells of its host is referred as *endobiotic*. Further a thallus that produces a single reproductive organ is called MONOCENTRIC (*Gr. monos* = single+*Kentron*=Centre) while those with many centres at which reproductive organs are formed is referred as POLYCENTRIC (*Gr. poly* = much, many+*Kentron*).

B. The Hyphae(Plate-II)

Fungi reproduce by means of spores of one type or another. In a favourable environment a fungus spore absorbs water, swells-up and then germinates by sending one or more tube like processes called germtubes. Such germtubes elongate and become long filaments which usually branch. Each filament is called a HYPHA (*pl. hyphae, Gr. hyphe* = web). The mass of hyphae which constitute the fungus thallus is termed as MYCELIUM (*pl. mycelia, Gr. mykes* = mushroom). Cytoplasmic streaming in fungal hyphae is unidirectional, towards the tip, where growth takes place, Hyphae grow entirely at their tips. The key to the fungal growth lies in the apex. There is built in mechanism at the hyphae apex for varying the width and shape of the apex. It is the site of active differentiation and nuclear divisions. The thin and plastic hyphal tip, 50-100 μ in the apical region is the zone of elongation and is filled with protoplasm. The portion behind this is vacuolated and incapable of elongation. But it helps in growth by synthesizing the cytoplasm which is transported to the tip.

Upon continued growth a hypha may be divided into a chain of cells by the formation of transverse walls. These cross walls, or septa, are laid down behind the growing point and divide the hyphae (pl) into uninucleate and multinucleate cells. The hyphae which are divided by septa are called SEPTATE HYPHAE (plate II). Many fungi, however, do not develop cross walls and are then said to be ASEPTATE. In such hyphae the protoplasm flows uninterruptedly through the filaments and the hyphae are then said to be *Coenocytic* (Gr. *koinos* = common + kytos = *a hollow Vessel).* Based on evolution, genetic factors and nutrition, a hyphae may be *HYALINE (Gr. Hyalinos* = made of glass, i.e. colourless/ transparent) or coloured/ pigmented. Many fungi while parasitizing host plants grow superficially. Such growth is ECTOPHYTIC while others grow inside the host tissues and are referred as ENDOPHYTIC. In plant tissues they are inter-cellular (lying in between the cells) or intra-cellular (lying in the lumen of the cell) (Plate II).

Branching of the hyphae especially those bearing spores, is of special significance since it aids in identification and classification of fungi. Branching is *dichotomous* when the apex of a hyphae ceases elongation and forks into two lateral branches. It may be *lateral* when main hyphae is left to grow and subapical lateral branch is given off. Two lateral branches opposite to each other, are also produced. In many fungi, a whorl of three or more branches is produced from the same point. This is known as *verticillate* branching. In *sympodial branching*, each successive leading apex becomes restricted in growth and is overtaken by a lateral branch from below. In monopodial branching the apex of leading hyphae is not suppressed but keeps pace in growth with the most active of the lateral branches from below. The sympodial branching may form a *cymose* while the monopodial branching forms a *racemose*.

C. The specialised somatic structure

1. **Rhizoids** - (Gr *Rhiza* = root+*oeides* = like)

A short, root like filamentous out growth of the thallus at the base of thalli/ sporophores, serving as an anchoring/attachment organ and also absorption of nutrients from the substrate.

2. **Rhizomycelium** (Gr. *Rhiza*+Mycelium)

A rhizoidal system extensive enough to resemble myceliumn superficially.

3. **Appresorium [pl. appresoria) L= appremere = to press against]**

A flattened, hyphal, pressing organ, from which a minute infection peg usually grows and enters the epidermal cell of the host (Plate II).

4. **Haustorium [(pl. Haustoria) L. *Haustor* = Drinker]**

An absorbing organ originating on a hypha and penetrating into cells of the host (Plate II)

D. Snares or Traps of Predacious Fungi

Many fungi trap and kill small animals, such as nematodes, to feed on their body contents. Such fungi develop special trapping or catching branches from the hyphae (Plate III).

1. Sticky Branches : Short lateral branches often only a few cells long are formed. These may anastomose to form loops.

2. Sticky network: The branches curl around and anastomose with

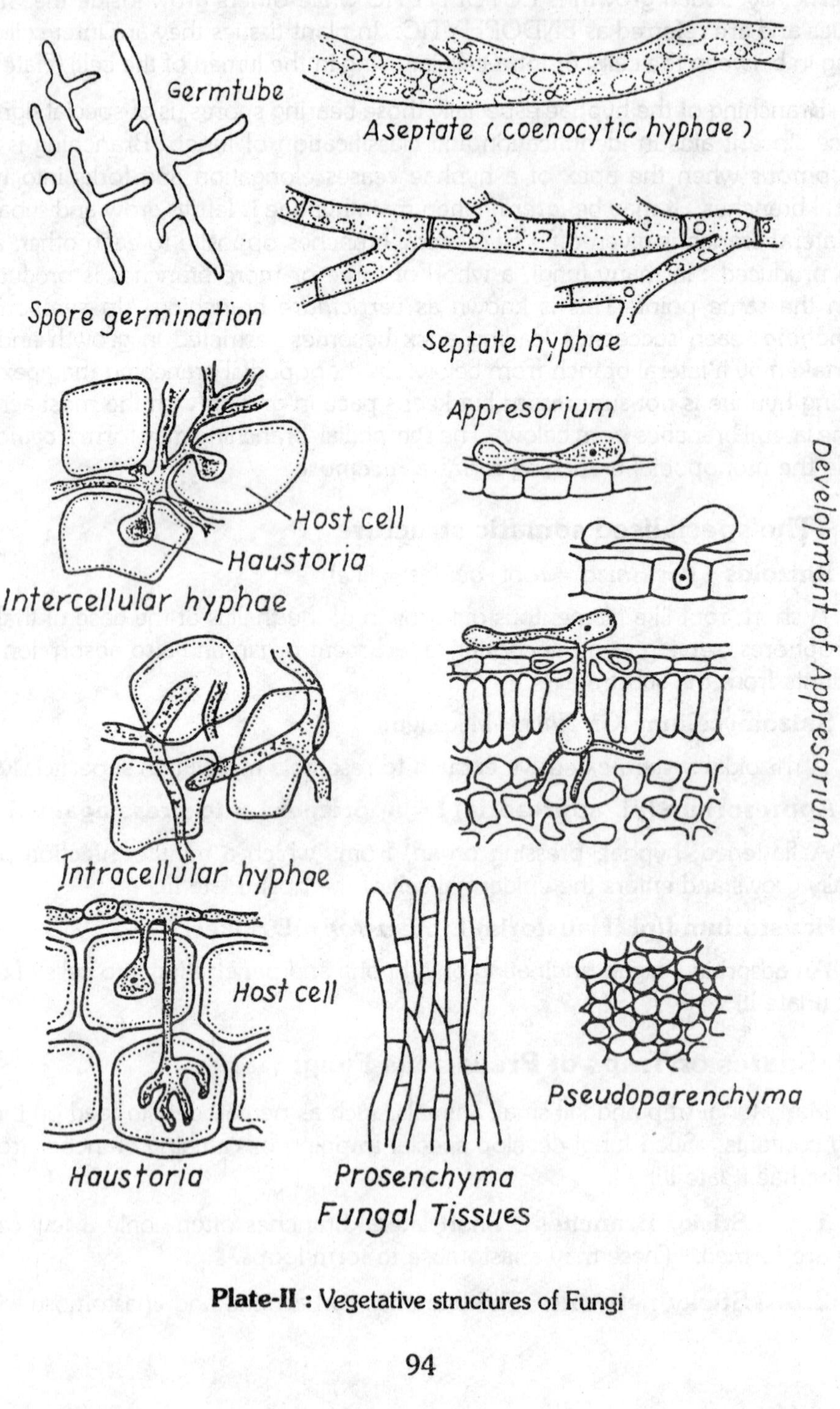
Germtube
Aseptate (coenocytic hyphae)
Spore germination
Septate hyphae
Appresorium
Development of Appresorium
Host cell
Haustoria
Intercellular hyphae
Intracellular hyphae
Host cell
Haustoria
Prosenchyma
Pseudoparenchyma
Fungal Tissues

Plate-II : Vegetative structures of Fungi

similar branches or with parent hyphae. These loops produce complex network. The adhesive surface of the network helps to hold the prey.

3. Sticky knobs : Small subspherical or spherical lobes on one or two celled lateral hyphae are formed. Only the terminal cell is sticky.

4. Constricting Rings: The trap is formed when a short hyphal branch curls back on itself and anastomoses, forming a 3 or 4 celled ring. The ring is borne on a 1 or 2 celled lateral hypha. When a nematode enters and contacts the inner walls of the ring cells, the cells bulge inward filling the lumen of the ring and holding the prey fast.

5. Non-constricting rings: These structures are as above but are neither constricting nor sticky. The nematodes get entrapped by the rings.

E. Hyphal Aggregations and Tissues (Plate-II)

1. Rhizomorph (Gr. *Rhiza*=Root, *morph* = shape)

The hyphae become root like or string like and attain great length. The mycelium aggregates to form thick strands in which hyphae lose their individuality and the entire strand acts as a unit.

2. False Tissues: The mycelium of fungi is made up of loosely interwoven hyphae which may be organised in various ways. They are detailed as under:

3. Plectenchyma (Gr. *pleko*= I weave, *enchyma* = infusion) Mycelium of many fungi organises into a loosely or compactly woven tissue. It may be of two types:

(a) Prosenchyma (Gr. *pros* = towards + *enchyma* = infusion) A type of plectenchyma in which the component hyphae lie parallel to one another and are easily recognized as such:

(b) Pseudoparenchyma: [(pl. pseudoparenchymata) Gr. pseudo- false + *parenchyma* = a type of plant tissue] A type of plectenchyma consisting of oval or isodiametric cells, the component hyphae having lost their individuality.

3. Stroma [(pl. stromata) Gr. *stroma* = mattress] A compact stomatic structure, much like a mattress, on which or in which fructifications are usually formed.

4. Sclerotium (pl. sclerotia; Gr. *skleron* = hard) It is another modified form of plectenchyma and form a resting body. These are filled with food materials and the wall either remains thin or becomes thick forming a protective rind which may be brown or black in colour.

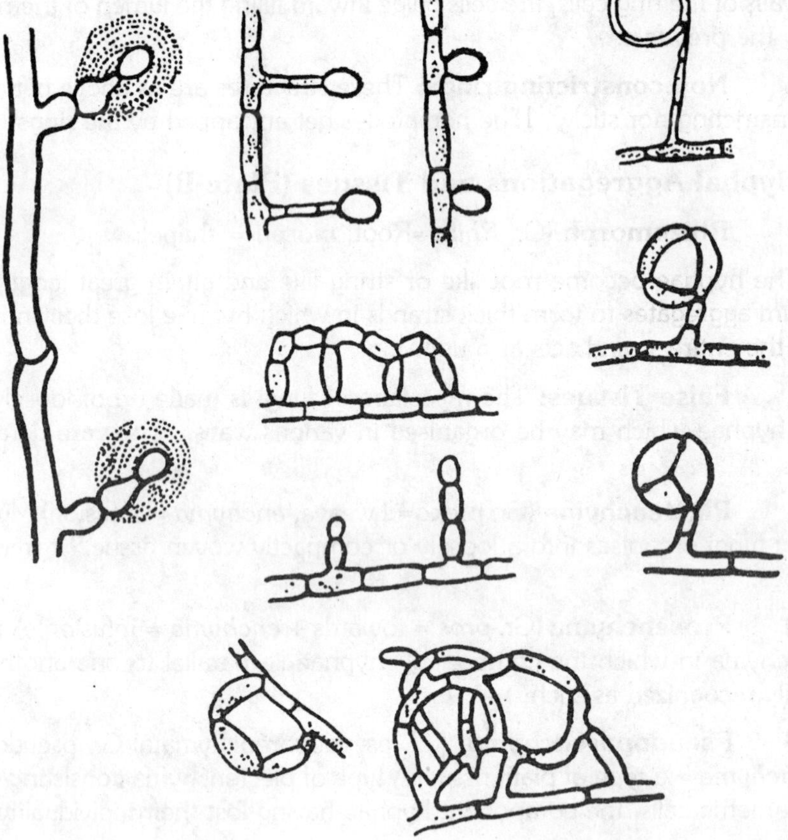

Plate-III : Snares or traps of predaceous fungi

F. The Fungal Cell

Except for a few primitive fungi, the vegetative cells of all other fungi are enclosed in a cell wall. The cell wall is a dynamic structure that is subject to change and modification at different stages in the life of a fungus (Peberdy, 1991). It is composed basically of a skeleton or microfibrillar component located to the inner side of the wall and usually embedded in an amorphous matrix material that extends to the out surface of the wall. The skeleton component consists of highly crystalline, water insoluble materials that include β-linked glucans and chitin, while the matrix consists mainly of polysaccharides that are mostly water soluble. These latter polysaccharides include α- glucans and glycoproteins. Miscellaneous components that may be present in fungal cell walls include lipids, melanins, D-galactosamine polymers, and polyuronids (Peberdy, 1991). Although there are scattered reports of the presence of cellulose in the walls of a few fungi, this compound generally is thought to be absent from most of the true fungi. On the other hand, cellulose is a characteristic component of the walls of the stramenopiles, including oomycetous fungi.

It has been known since begening that hyphae grow at their tips. How this growth occurs is still not understood fully? One theory of hyphal tip growth is the "Steady-state" hypothesis (Wessels, 1986, 1988). This hypothesis suggests that the hyphae apex is inherently viscoelastic and expandable and that the newly synthesized wall at the apex consists of a mixture of noncrystalline chitin and β-glucan. As a result of subsequent cross linking of the polymers of the wall the viscoelastic mixture then gradually develops rigidity. The second hypothesis (Barthicki-Garcia, 1973) suggests that the wall is inherently rigid and that for growth to occur there must be a permanent delicate balance between the lysis of the wall followed by synthesis of wall polymers and the pushing out and mending of the wall. In either of these two hypotheses, it is clear that the subapical region of a growing hyphae provides the energy; enzymes, wall precursors, and membranes necessary for hyphal tip growth. At this point there is strong evidence indicating that many of the raw materials needed by the growing hyphal tip are delivered to the tip by membrane bound vesicles.

G. Fungal Organelles

The fungal hyphae almost invariably contain large numbers of nuclei. The nuclei of most fungi are quite small, generally spherical to ovoid in shape and extremely plastic and capable of squeezing through tiny septal pores. The typical fungal nucleus usually contains a prominent nucleolus that is often centrally positioned. Aside from nuclei, perhaps the most conspicuous fungal organelle is the mitochondrion. Mitochondria are numerous in hyphae and at ultra structureal level they usually appear as electron-dense structures oriented parallel to the long axis of a hyphae.

Other cytoplasmic components of fungi include ribosomes, strands of endoplasmic reticulum, vacuoles, lipid bodies, glycogen storage particles, microbodies, golgi bodies, filasomes, multivesicular bodies and the microtubules and microfilaments that comprise the fungal cytoskeleton. Spherical structures known as woronin bodies also are present in certain types of fungi and are associated with septal pores.

H. The Septa

Septa are formed either to cut-off reproductive cells from rest of the hypha or to separate-off the damaged parts or to divide hyphae into compartments or cells. The septum is an annulus of cell wall material which vaginates the plasmalemma and grows from all round (in acropetal manner) inward completely or partially closing the passage of cytoplasm in the hypha. In lower fungi the septum, when formed, is completely closed and is called ADVENTITIOUS SEPTUM. In higher fungi the septum has one or more septal pores permitting continuity of cytoplasmic streaming between adjacent cells. These are called PRIMARY SEPTA. The adventitious septa usually occur in association with changes in the local concentration of cytoplasm as it streams from one part of the thallus to another. The primary septa are always associated with true mitotic or meiotic nuclear divisions, separating the daughter nuclei.

VII. Asexual reproduction

Fungi multiply by means of their propagules, the units capable of growing into new thallus. In most fungi these propagules are differentiated as SPORES which are basic units of typical reproduction. The simple or branched spore bearing hyphae are known as SPOROPHORES (Gr. *sporos* = seed, spore+ *phoreus* = bearer). The sporogenous organs have following characters in common.

(1) They are vertically oriented, perpendicular to the plane in which the mycelium spreads.

(2) Their growth is limited, and

(3) Endowed with specially differentiated cellular characters.

A. Types of Sporophores (Plate-IV)

1. Simple or Filamentous sporophores

These are branches of the somatic hyphae which give rise to sporogenous cells and spores on reaching a length which is roughly determinate. The spore bearing branches usually arise vertically and may be distinctly branched. When these branches bear sporangia, they are called SPORANGIOPHORES and when they bear conidia, they are called CONIDIOPHORES.

2. Compound Sporophores

The aggregation of hyphae form stromatic or semistromatic structures which grow into compound sporophores.

B. The Spores

The term spore is applied to any small propagative, reproductive or survival unit which separates from a hypha or sporogenous cell and can grow independently into a new individual. Thus, spore is a nucleate portion delimited from the thallus, characterized by cessation of cytoplasmic movement, small water content, and slow metabolism, lack of vacuoles, and specialized for dispersal, reproduction and survival. They may be grouped into two categories: ENDOGENOUS and EXOGENOUS spores.

(1) Endogenous spores: These are formed within a swollen sac like spore producing structure called **sporangium** (pl. sporangia Gr. *sporos* + *angeion* = vessel) which may be terminal or intercalary in position. The entire contents or a part of the sporangium is converted into spores known a SPORANGIOSPORES. The spores may be motile or nonmotile. Motile spores are called ZOOSPORES (Gr. *Zoon* = animal + sporos = seed, spore) and the nonmotile APLANOSPORES. The zoorpores are naked reniform or pyriform or amoeboid, uninucleated, haploid spores equipped with one or two flagella (Sing. *flagellum*; L. flagellum= whip). They are of two types in fungi, the whiplash and the tinsel. In whiplash flagellum, the structure resembles the whip. The basal part is rigid and the tip part elastic and flexible. In tinsel the entire flagellum is covered by hairy projections and look like comb. The zoospores may be uniflagellate or biflagellate. They may be posteriorly or anteriorly attached. Lateral attachment also occurs. After their differentiation in the zoosporangium, the released zoospores move in the film of water (movement), then encyst (encystment) and thereafter germinate by germtube. When zoospores move only once, they are **monoplanetic** (monoplanetism). If they move twice, they are **diplantic**, and if more than twice, they are referred as **Polyplantic**. The aplanospores are uninucleate or multinucleate, having two layered cell wall and formed in side the sporangium.

2. Exogenous spores : All asexually produced spores except sporangiospores (zoospores and aplanospores) are called CONIDIA (sing. conidium; Gr. *Konis* = dust+ *idion*-suffix). They are formed on branched or unbranched

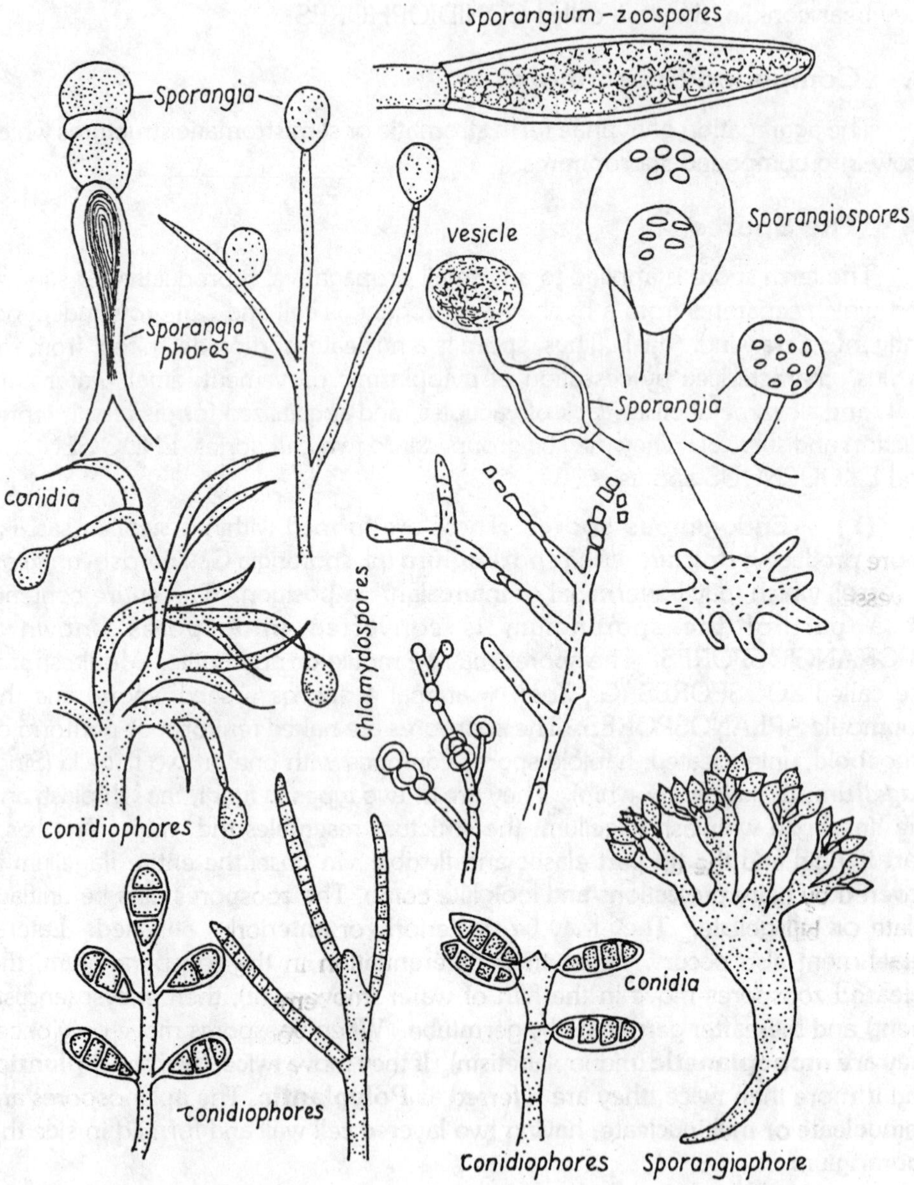
Sporangium-zoospores
Sporangia
Sporangiospores
vesicle
Sporangia
phores
Sporangia
conidia
Chlamydospores
Conidiophores
Conidia
conidiophores
Conidiophores
Sporangiaphore

Plate-IV: Asexual reproductive organs of fungi

hyphal tips termed as CONIDIOPHORES. The conidia may be formed singly or in chains or clusters, basipetally or acropetally. They may be unicellular, multicellular, uni or multi nucleate. In some fungi, conidia may be formed by transformation of existing cells of the thallus and are set free when the parent hyphae decay. They are referred as THALLOSPORES. They are of two kinds:

a. Arthrospores or oidia: (Gr. *arthron* = joint+ *sporos*= spore). Produced by fragmentation of hyphae from apex to base. Each cell thus formed rounds off and separates as a spore.

b. Chlamydospores: (Gr. *chlamys* = mantle + sporos = spore) formed by rounding off and enlargement of terminal or intercalary cells of hyphae. These can be single or formed in chains.

The CONIDIOSPORES or true CONIDIA are formed as new elements from the thallus and on maturity get cut-off from the conidiophores. They are either formed as buds from the somatic cells of a hypha called BLASTOSPORES or produced by the inflation of the apex of the conidiophores called **Aleuriospores** or produced as PHILOSPORES which get cut off from flask shapped, cylindrical phialides.

The shape, colour and septation of conidia (Plate-V & VI) serve as the basis for identification and description of the fungal species. The section proposed by Saccardo (1899) are the following:

Amerosporae : Conidia continuous, spherical, ovoid to elongated, or short cylindric.

Allantosporae : Conidia cylindric, curved (allantoid), hyaline to pale.

- **Hyalosporae** : Conidia hyaline.
- **Phaeosporae** : Conidia colored

Didymosporae : Conidia ovoid to oblong, one septate

- **Hyalodidymae** : Conidia hyaline
- **Phaeodidymae** : Conidia colored

Phragmosporae : Conidia oblong two to many-septate (transversely septate).

- **Hyalophragmiae** : Conidia hyaline.
- **Phaeophragmiae** : Conidia colored

Dictyosporae : Conidia ovoid to oblong; net septate (transversely and longitudinaly septate).

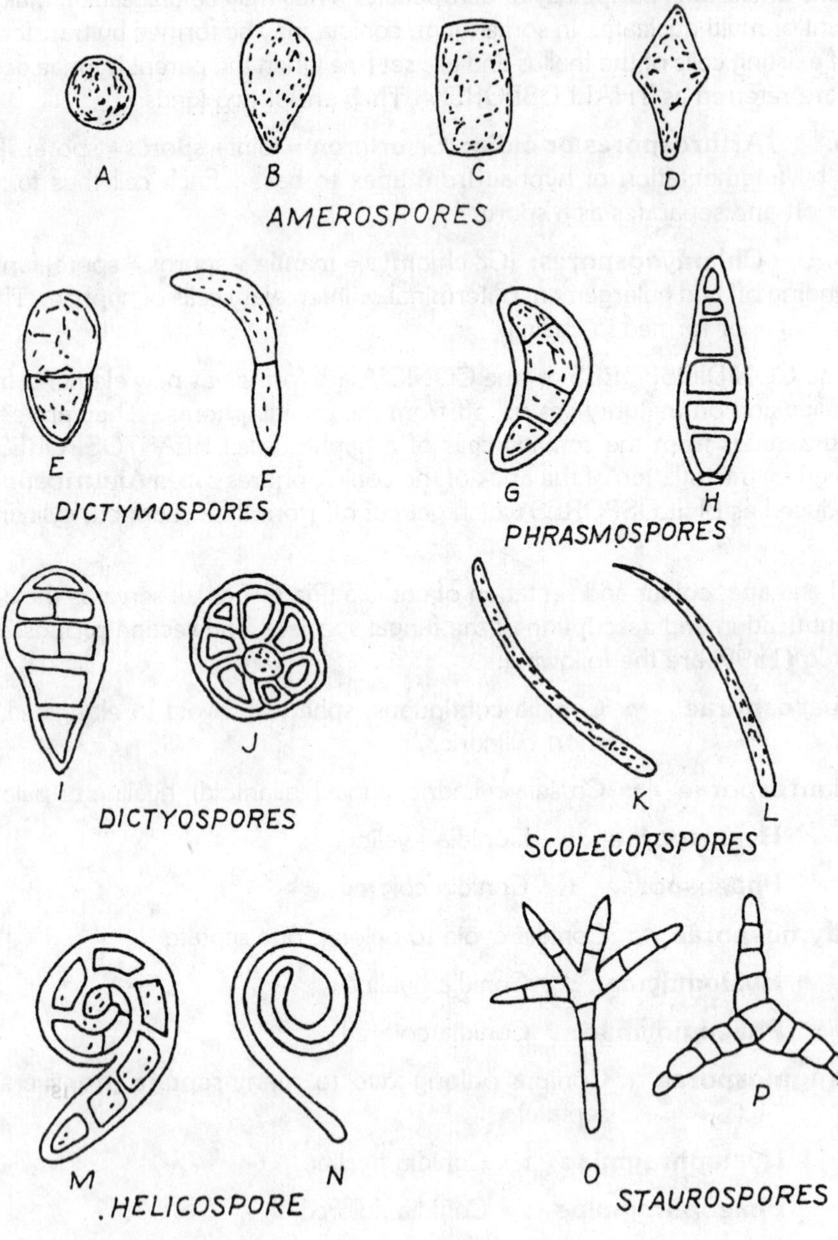
A
B
C
D
AMEROSPORES
E
F
DICTYMOSPORES
G
H
PHRASMOSPORES
I
J
DICTYOSPORES
K
L
SCOLECORSPORES
M
N
. HELICOSPORE
O
P
STAUROSPORES

Plate-V: Various types of fungal spores

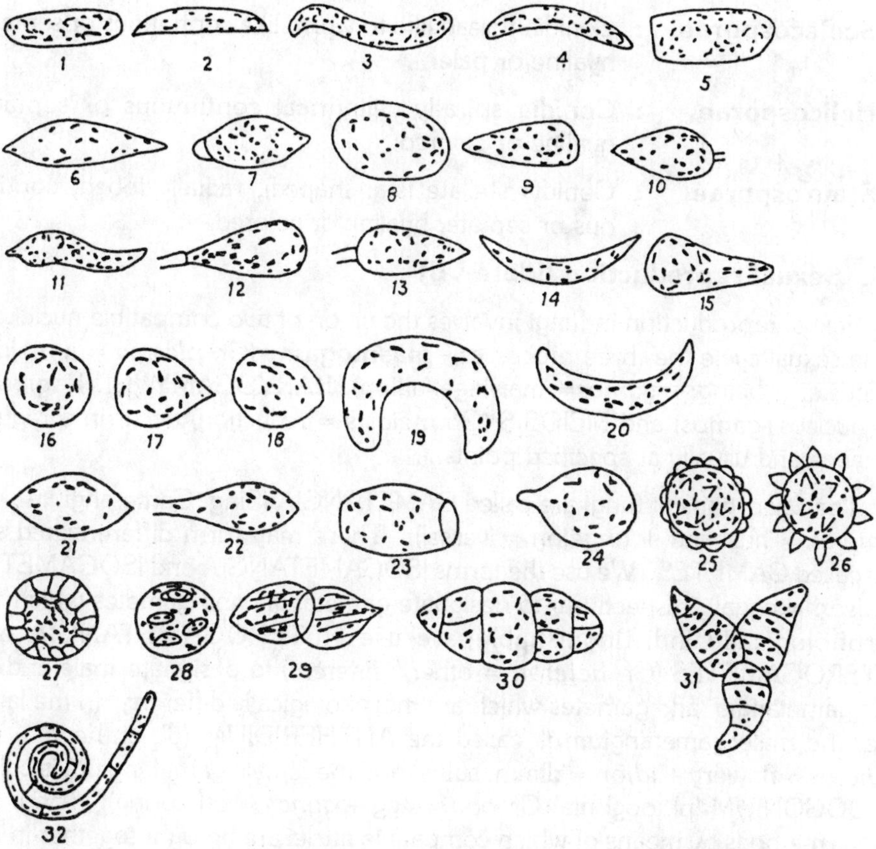
1
2
3
4
5
6
7
8
9
10
11
12
13
14
15
16
17
18
19
20
21
22
23
24
25
26
27
28
29
30
31
32

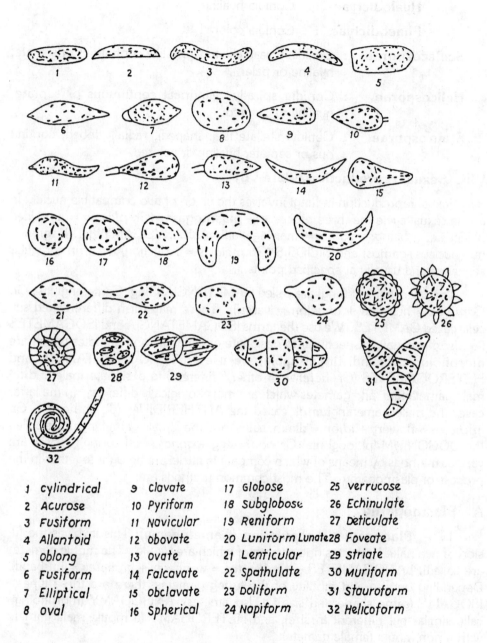

1 cylindrical	9 Clavate	17 Globose	25 verrucose
2 Acurose	10 Pyriform	18 Subglobose	26 Echinulate
3 Fusiform	11 Novicular	19 Reniform	27 Deticulate
4 Allantoid	12 obovate	20 Luniform Lunate	28 Foveate
5 oblong	13 ovate	21 Lenticular	29 Striate
6 Fusiform	14 Falcavate	22 Spathulate	30 Muriform
7 Elliptical	15 obclavate	23 Doliform	31 Stauroform
8 oval	16 Spherical	24 Napiform	32 Helicoform

Plate-VI: Various shapes of spores in fungi

Hyalodictyae : Conidia hyaline

Phaeodictyae : Conidia colored

Scolecosporae : Conidia thread like to worm like continuous or septate, hyaline or pale.

Helicosporae : Conidia spirally cylindrical continuous or septate, hyaline or colored.

Staurosporae : Conidia stellate (star-shaped), radially lobed; continu ous or septate; hyaline or colored.

VIII. Sexual reproduction (Plate-VII)

Sexual reproduction in fungi involves the union of two compatible nuclei. In a true sexual cycle the three processes - **plasmogamy** (Gr. *plasma* = a molded object i.e. a being + *gamos*= marriage/union), KARYOGAMY (Gr. Karyon = nut/nucleus+*gamos*) and MEIOSIS (Gr. *meiosis* = reduction) occur in a regular sequence and usually at specified points.

The sex organs of fungi are called GAMETANGIA (Sing. Gametangium. Gr. *Gametes* = husband + *angeion* = vessel). These may form differentiated sex cells called GAMETES. We use the terms ISOGAMETANGIA and ISOGAMETES (Gr. *ison* = equal), respectively to designate gametangia and gametes which are morphologically indistinguishable; we use HETEROGAMETANGIA and HETEROGAMETES (Gr. *heteron* = other/ different) to designate male and female gametangia and gametes which are morphologically different. In the latter case, the male gametangium is called the ANTHERIDIUM (pl. antheridia; Gr. *antheros* = flowery + *idion* = dimin, suffix) and the female gametangium is called the OOGONIUM (pl. oogonia; Gr. *oon* = egg + *gonos* = off spring). There are various methods by means of which compatible nuclei are brought together in the process of plasmogamy. The most common methods are:

A. Plasmogamy

1. Planogametic copulation (Gametogamy): This involves the fusion of two naked gametes one or both of which are motile. The motile gametes are called PLANOGAMETES (Gr. *planetes* = wanderer + *gametes* = husband). Depending on size and motility of the fusing gametes, there are three types - ISOGAMY (same shape and size i.e. Isogametes), ANISOGAMY (morphologically similar but different in size), and HETEROGAMY (a motile male gamete with a non motile female gamete).

2. Gametangial contact (Gametangiogamy) : Plasmogamy is brought about by the contact of male and female gametangia where both the

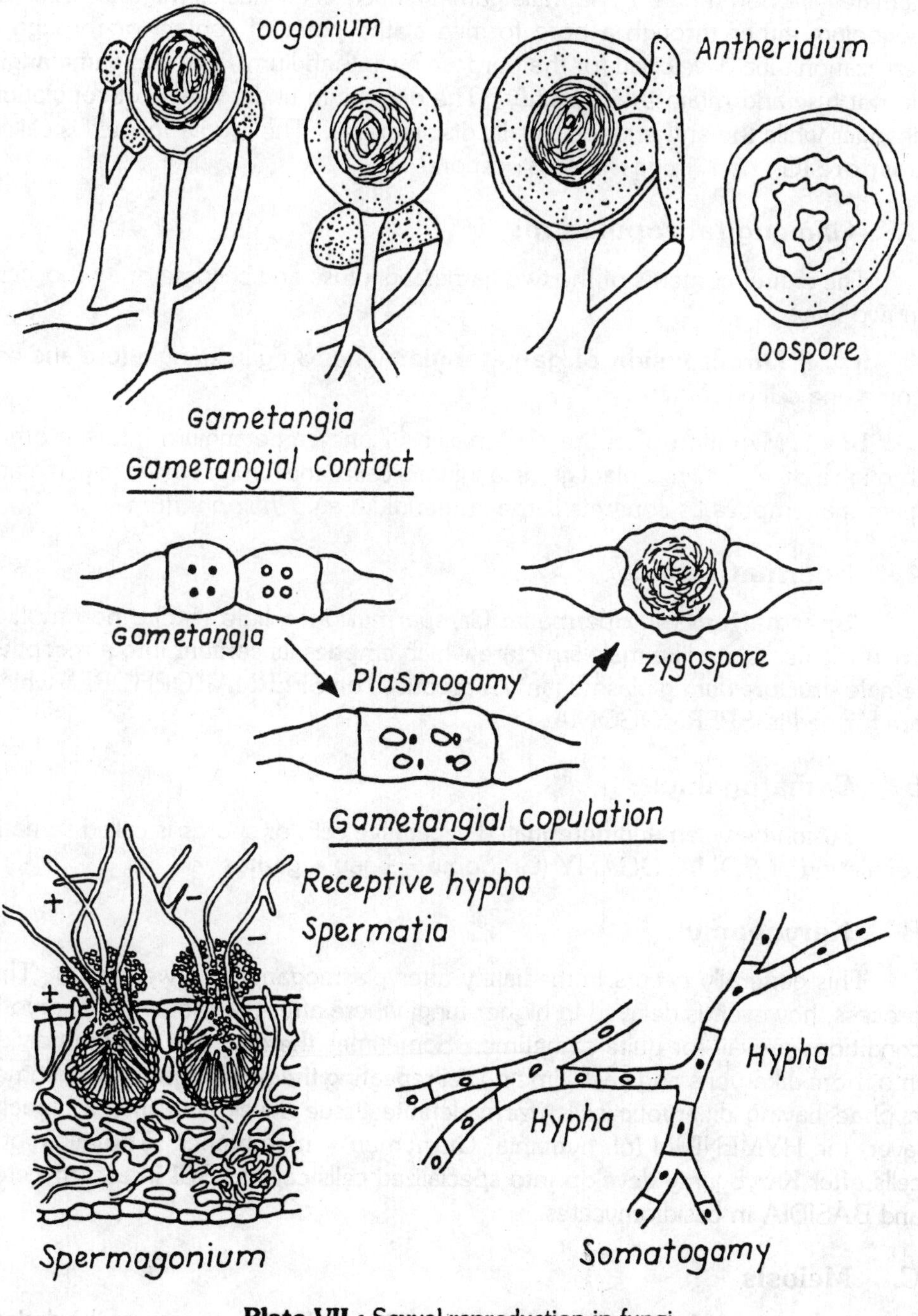
oogonium
Antheridium
oospore
Gametangia
Gametangial contact
Gametangia
Plasmogamy
zygospore
Gametangial copulation
Receptive hypha
Spermatia
+
-
Hypha
Hypha
Spermagonium
Somatogamy

Plate-VII : Sexual reproduction in fungi

gametes are non-motile. The male gametic nucleus (or nuclei) migrates into the oogonium either through a pore formed at the point of contact or through a fertilization tube developed for the purpose by antheridium. The two gametangia do not fuse and retain their identify. The oogonium undergoes post-copulation changes while the antheridium usually disintegrates. The zygote formed is called **oospore** (Gr. oon = egg+ sporos = spore).

3. Gamengital copulation:

The entire contents of the two gametangia fuse and become one. It occurs in two ways:

a. Direct fusion of gametangia: The two gametangia fuse and become one cell eg. *Mucor*.

b. Migration of entire protoplasm of one gametangium into the other through a pore. The recepient gametangium is called the female while the gametangium that empties its contents is the antheridium eg. *Rhizophidium*.

4. Spermatization:

Spermatium (pl. Spermatia; Gr. *spermation* = little seed) a non motile, uni nucleate, spore like male structure which empties its content into a receptive female structure during plasmogamy is produced on SPERMATIOPHORES which are formed in SPERMOGONIA.

5. Somatogamy:

Fusion between undifferentiated vegetative cells or spores is called somatic copulation or SOMATOGAMY (Gr. *Soma* = body + *gamos*).

B. Karyogamy

This generally occurs immediately after plasmogamy in lower fungi. This process, however, is delayed in higher fungi where after plasmogamy, dikaryotic conditions prevail for quite sometime. Sometimes the dikaryon cell may divide into more dikaryons and each time nuclei repeating the original pair. Sometimes hyphae having dikaryotic cells, form definite tissue developing into a, special layer, the HYMENIUM (pl. hymania, Gr. *hymen* = membrane). The dikaryotic cells after Karyogamy develop into specialized cells called ASCI in ascomycetes and BASIDIA in basidiomycetes.

C. Meiosis

Karyogamy is followed by a reduction division. During meiosis of the diploid nucleus, the chromosome do not split but separate out as a whole into two complete sets. Each of the sets forms the chromosome complement of the haploid daughter nucleus. In this process some of the chromosomes in each daughter nucleus are derived from one parent and others from the other parent. Subsequently, mitotic divsions take place resulting in an increase in the number of haploid nuclei within the zygote cell. In case of isogamy or anisogamy, the resulting zygote is a resting sporangium. This produces zoospores which encyst and germinate producing germ tube. Where oospore is the zygote, it may germinate by germtube or by zoospores.

D. Nuclear Cycle

Although the life-cycle of fungi vary greatly, the great majority of them go through a series of steps that are quite similar. Almost all fungi have a spore stage with a simple, haploid nucleus (IN). The spore germinates into a hypha containing haploid nuclei. The hypha may either produce simple, haploid spore again or it may fuse with another hypha to produce a fertilized hypha in which the nuclei unite to form a diploid nucleus called a zygote (2N). In oomycetous fungi, zygote divides to produce diploid mycelium and spores. The mycelium produces gametangia in which meiosis occurs, then fertilization and production of zygote. In most ascomycetes and generally in basidiomycetes, the two nuclei of fertilized hypha do not unite but remain separate within the cells in pairs (n+n) and divide simultaneously to produce more hyphal cells with pairs of nuclie. In the ascomycetes the dikaryotic hyphae are found only inside the fruiting body, in which they become ascogenous hyphae, since the two nuclei of one cell of each hypha unite into a zygote (2N), which divides meiotically to produce ascospores that contain haploid nuclei.

In basidomycetes, haploid spores produce only short haploid hyphae. On fertilization, dikaryotic mycelium (N+N) is produced, and this develops into the main body of the fungus. Such dikaryotic spores may produce asexually, dikaryotic spores that will grow again into a dikaryotic mycelium. Finally, however, the paired nuclei of the cells unite and form zygotes. The zygotes divide meiotically and produce basidiospores that contain haploid nuclei. In the imperfect fungi, of course, only the asexual-cycle (haploid spore-haploid mycelium-haploid spore) is found.

E. Heterokaryosis

All fungal cells, do not necessarily have the same number and kind of nuclei or even the same proportion of each kind in a mixture of nuclei. This phenomenon of the existence of different kinds of nuclei in the same individual is called HETEROKARYOSIS (Gr. *heteros* = different + *karyon* = nut, nucleus), and the individuals that exhibit it are HETEROKARYOTIC. Heterokaryons may originate in a fungus thallus in four ways:

1. by the germination of a heterokaryotic spore, which will give rise to heterokaryotic soma,
2. by the introduction of genetically different nuclei into a HOMOKARYON (Gr. *homo* = same + *Karyon*= nucleus, nut), a soma in which all nuclei are similar;
3. by mutation in a multinucleate, homokaryotic structure and the subsequent survival, multiplication, and spread of mutant nuclei among the wild type nuclei, and
4. by fusion of some nuclei in a haploid homokaryon, and the subsequent survival, multiplication, and spread of the diploid nuclei among the haploid.

F. Sexual compatibility

On the basis of sex, most fungi may be classified into 3 categories:

1. **Hermaphroditic (Monoecious):** Each thallus bears both male and female organs that may or may not be compatible.
2. **Dioecious:** Some thalli bear only male and some thalli bear only female organs; and
3. **Sexually undifferentiated:** Functional structures that are produced are morphologically indistinguishable as male or female.

Fungi in the categories outlined above belong to one or another of the following three groups on the basis of compatibility:

1. **Homothallic fungi:** Those in which every thallus is sexually self fertile and can, therefore, reproduce sexually by itself without the aid of another thallus.
2. **Heterothallic fungi:** Those in which every thallus is sexually self sterile, and requires another compatible thallus of a different mating type for sexual reproduction.
3. **Secondary Homothallic Fungi:** In some heterothallic fungi an interesting mechanism operates during spore formation whereby two nuclei of apposite mating type are incorporated regularly into each spore or at least some spores. Germlings arising from these spores are therefore self-fertile and behave as if they are homothallic when in reality they are actually heterothallic.

Heterothallic fungi belong to one or the other two general groups. One group includes species in which mating is controlled by one pair of *loci*. This type of heterothallism has been referred to as UNIFACTORIAL or BIPOLAR HETEROTHALLISM. In the other group, mating is controlled by more than one pair of *loci* located on different chromosomes. This is called BIFACTORIAL or TETRAPOLAR HETEROTHALLISM.

G. Parasexuality:

Some fungi do not go through a true sexual-cycle as described. They may derive the benefits of sexual recombination through a process known as PARASEXUALITY (Gr. *para* = besides + sex). In this process, plasmogamy, karyogamy and haploidization take place, but not at specified points in the thallus or the life-cycle. The parasexual cycle involves the following steps:

1. Formation of heterokaryotic mycelium;
2. Nuclear fusion and multiplication of the diploid nuclei
3. Mitotic crossing over during the division of the diploid cells
4. Sorting out of the diploid strains, and
5. Haploidization

H. Fructifications and fruit bodies (Plate-VIII)

Any fungal structure which contains or bears spores is called frustification (L. *fructus* = fruit). They may be asexual or sexual in nature. In lower fungi, the asexual spores are usually enclosed in a simple sac called *sporangia* or *zoosporangia*. In higher forms of fungi, however, there is a tendency to organise complex aggregates of spore bearing hyphae, frequently surrounded by supporting and protective tissues.

1. **Synnema** (pl. synnemata; Gr. syn = together + *nema = yarn).* A group of conidiophores cemented together and forming an elongated spore bearing structure
2. **Acervulus** (pl. acervuli; L. *acervus* = heap, dimin, form) A mat of hyphae giving rise to short conidiophores closely packed together forming a bed like mass.
3. **Sporodochium** (pl. sporodochia; Gr. *sporos* = seed/spore + *docheion* = container). A cushion shaped stroma covered with conidiophores.
4. **Pycnidium** (pl. pycnidia; Gr. *pyknon* = concentrated + *idion* = dimin / suffix). An asexual, hollow fruiting body, lined inside with conidiophores.
5. **Perithecium** (pl. perithecia; Gr. *Peri* = around + *theka* = a case). A closed ascocarp with a pore at the top, a true ostiole, on a wall of its own.
6. **Apothecium** (pl. apothecia; Gr. *apotheka* = store house). An open *ascocarp.* (Gr. *askos* = sac + *Karpos* = fruit), a fruiting body containing asci and ascospores.

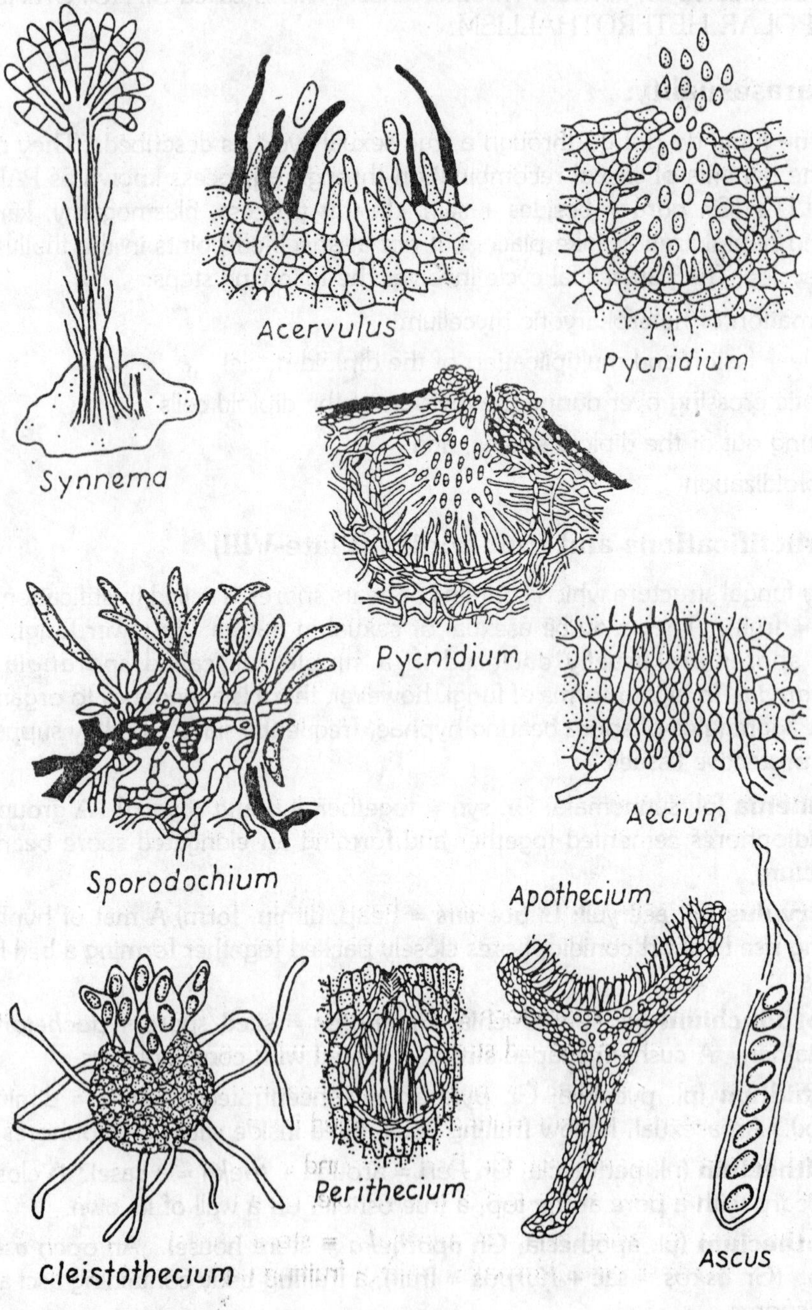
Synnema
Acervulus
Pycnidium
Pycnidium
Sporodochium
Aecium
Apothecium
Perithecium
Cleistothecium
Ascus

Plate-VIII: Spore fruits (Fructifications)

7. **Cleistothecium** (pl. cleistothecia; Gr. *kleistos* = closed + *theke* =case) a completely closed ascocarp.
8. **Spermagonium** (pl. spermagonia; Gr. *sperma* = seed, sperm + *gennao* = I give birth). A structure resembling a pycnidia and containing spermatiophores and spermatia.
9. **Aecium** (pl. aecia; Gr. *aikia* = injury). A structure consisting of binucleate hyphae cells, with or without peridium, which produce spore chains consisting of aeciospores alternating with disjunctor cells, by the successive conjugate division of nuclei.

IX. CLASSIFICATION OF FUNGI

1. INTRODUCTION

Since the begining of the study of fungi, many systems of classification have been proposed but still there is no satisfactory and final classification. The system satisfactory today may become obsolete tomorrow and may be replaced by a new one.

2. THE EVER CHANGING ATTEMPTS

The first attempt to classify fungi was made by Persoon (1801). He classified fungi into classes, orders and families and published his work in "Synopsis Methodica Fungorum". Fries (1821-1929), tried to classify fungi and published his work in "Systema Mycologium". Eicher (1866) divided division THALLOPHYTA into two classes *Algae* and *Fungi*. The fungi included Schizomycetes, Eumycetes, and Lichens. Saccardo (1822-1931) wrote "Sylloge Fungorum": in 25 volumes and classified fungi into 6 classes. Schroeter (1897) classified fungi into Phycomycetes, Eumycetes and Fungi-imperfectii. Gaumann and Dodge (1928) created five classes while Fitzpatrick (1930) divided Thallophyta into two groups - Myxothallophyta and Euthallophyta. Tippo (1942) divided Thallophyta into 3 phyla - Schizomycophyta (bacteria), Myxomycophyta (slime molds) and Eumycophyta (Fungi).

Martin (1957) classified fungi into four classes - Phycomycetes (two subclasses and 12 orders), Ascomycetes (two sub-classes and 16 orders), Basidiomycetes (2 subclasses and 11 orders), and fungi-imperfectii (four orders). Smith (1955) divided fungi into two phyla Myxomycophyta (3 classes) and Eumycophyta with classes, sub classes and orders.

3. ATTEMPTS FOLLOWED AND STUDIED

Alexopoulos (1962) created division MYCOTA with two sub-divisions

MYXOMYCOTINA and EUMYCOTINA. Myxomycotina included class Myxomycetes with two sub-classes and six orders. Sub-division Eumycotina included nine classes, sub-classes, series and orders. Ainsworth (1966) classified fungi as follows:

KINGDOM - MYCOTA

DIVISIONS

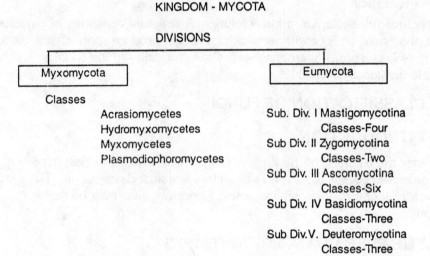

Myxomycota	Eumycota
Classes	
Acrasiomycetes	Sub. Div. I Mastigomycotina
Hydromyxomycetes	Classes-Four
Myxomycetes	Sub Div. II Zygomycotina
Plasmodiophoromycetes	Classes-Two
	Sub Div. III Ascomycotina
	Classes-Six
	Sub Div. IV Basidiomycotina
	Classes-Three
	Sub Div.V. Deuteromycotina
	Classes-Three

Alexopoulos(1979) further modified his previous classification. The classification is as follows;

DIVISION-I		Gymnomycota
With two Sub-	-	Acrasiogymnomycotina
divisions	-	Plasmodiogymnomycotina
DIVISION-II		Mastigomycota
Sub Divisions	-	*Haplomastigomycotina*
	-	Chytridiomycetes
	-	Hypochytridiomycetes
	-	Plasmodiophoromycetes
	-	*Diplomastigomycotina* **(Sub-Div. II)**
	-	Oomycetes **(Class)**
Division-III		Amastigomycota
Sub Division-I	-	*Zygomycotina*
Classes	-	Zygomycetes
	-	Trichomycetes
Sub Division II	-	*Ascomycotina*
	-	5 sub classes

Sub Division III - *Basidiomycotina*
- 3 sub classes

Sub Division IV - *Deuteromycotina*
- One Form-class and
3 form sub-class

According to Ainsworth and Bisby's dictionary of fungi (1983) the kingdom fungi has been classified as follows:

MYXOMYCOTA

Classes

1. Protosteliomycetes
2. Ceratiomyxomycetes
3. Dictyosteliomycetes
4. Acrasiomycetes
5. Plasmodiophoromycetes
6. Myxomycetes
7. Labyrinthulomycetes

EUMYCOTA

A. Mastigomycotina
- Chytridiomycetes (8)
- Hyphochytridiomycetes (9)
- Oomycetes (10)

B. Zygomycotina
- Zygomycetes (11)
- Trichomycetes (12)

C. Ascomycotina
- No classes recognized

D. Basidiomycotina
- Hymenomycetes (13)
- Gasteromycetes (14)
- Urediniomycetes (15)
- Ustilaginomycetes (16)

E. Deuteromycotina
- Coelomycetes (17)
- Hyphomycetes (18)

4. RECENT DEVELOPMENTS

Agrios (1996) after having taken into consideration the work of Alexopoulos et.al. (1996), Arc (1987), Barr (1992), Carlile and Warkinson (1994), Hanlin (1990), Samules and Seifert (1995) and a few others, has proposed the classification of plant pathogenic fungi. A brief out-line of the classification is reproduced below:

Kingdom - Fungi **Phylum - I** - Chytridiomycota
Class - Chytridiomycetes

Orders -
1. Spizellomycetales (*Rozella, Olpidium)*
2: Neocallimasticales (*Neocallimastix)*
3. *Chytridiales (Synchytrium, Rhizophydium, Chytriomyces, Nowakowskiella)*
4. Blastocladiales (*Allomyces, Blastocladiella, Blastocladia, Coelomomyces, Catenaria, Physoderma)*
5. Monoblepharidales (*Monoblepharis, Monoblepharella, Gonapodya, Oedogonimyces)*

Phylum-II Zygomycota

Class: 1. Zygomycetes **Order :** Mucorales **Families:** 1. Mucoraceae (*Mucor, Rhizomucor*) 2. Gilber

Pseudofungi (Fungal like organisms)

Kingdom: Protozoa: Unicellular, plasmodial, colonial, Phagotrophic

Phylum: Myxomycota: Plasmodium, Pseudoplasmodium

Class: Myxomycetes : Naked,amorphous plasmodium, produce zoospores

Order: Physarales: Crusty frutification, biflagellate zoospores

Genus: *Fuligo*: *Mucilago, Physarum - slime molds*

Phylum: Plasmodiophoromycota

Class: Plasmodiophoromycetes: Endoparasitic slime molds.

Order: Plasmodiophorales: Obligate parasites

Genus: *Plasmodiophora brassicae:* - Club root of crucifers, *Polymyxa-gramimis* - Parasitic on wheat, *Spongospora subterranea* - Powdery scab of potato

Kingdom: Chromista-Uni or multicellular, filamentious or colonial, phototrophic

Phylum: Oomycota: Biflagellate zoospores (whiplash posterior, tinsel anterior). sexual spore (zygote) oospores.

Class: Oomycetes Water molds, white rusts, downy mildews

Order-I: Saprolegniales : Long cylindrical zoosporangia, several oospores in an oogonium

Genus: *Aphanomyces euteiches*: root rot of peas

Order-II: Peronosporales: Zoosporangia, oospores

Family-I: Pythiaceae: Little or no difference between hyphae and sporophores, if different then sporophores of inderminate growth.

Genus: *Pythium* spp., *Phytophthora spp.*

Family-II: Peronosporacae :Downy mildews, Sporangiophores determinate, branched, obligate parasite.

Genus: *Peronospora, Sceleraspora, Bremia, Plasmopara, Pseudoperonospora,* Peronosclerospora

Family-III: Albuginaceae: White-rusts, sporangiophores club shaped, sporangia borne in chain

Genus: *Albugo candida* : White rust of crucifers

The True Fungi

Kingdom : FUNGI

Mycelial thallus, wall contains glucans and chitin, lack chloroplast

↕

Phylum-I : Chytridiomycota

Posteriorly uniflagellate, flagellum whiplash

↕

Genera : *Olpidium brassicae* : Parasitic on cabbage roots

Physoderma maydis : Brown spot of corn

Synchytrium endobioticum : Potato wart

Urophlyctis alfalfae : Crown wart of alfalfa

↕

Phylum II : Zygomycota

No zoospores, aplanospores in sporangia, zygozpore is the zygote

↕

Class : Zygomycetes : The bread molds

↕

Order-I : Mucorales

Non-motile asexual spores formed in terminal sporangia

↕

Genera : *Rhizopus : Mucor, Choanephora*

↕

Order-II : Glomales : VAM Fungi

↕

Genera : Glomus, Acaulospora, Gigaspora

↕

Phylum-III : Ascomycota : Asci and ascorpores produced

↕

Class-I : Archiascomycetes : A group of diverse fungi

↕

Order : Taphrinales
Asci arising from binucleate ascogenous cells

↕

Genus : *Taphrina maculans, T. deformans*

↕

Class II : Saccharomycetes
Yeasts, asci naked, no ascocarp

↕

Genus : *Galactomyces* – Citrus sour rot, *Saccharomyces cerevisiae*-bread yeast

FILAMENTOUS FUNGI-III

↕

Order : Erysiphales : powdery mildew fungi

↕

Genera : *Erysiphe, Blumeria, Microsphaera, Podosphaera, Sphaerotheca, Uncinula*

A. Pyrenomycetes : Perithecial ascomycetes

Order-I : Hypocreales

↕

Genera : *Hypocrea, Melanoslpora, Ophiostoma, Nectria, Gibberella, Claviceps*

↕

Order-II : Microascales, Genus-*Ceratocystis*

↕

Order-III : Phyllachorales, Genus-*Glomerella, Phyllachora*

↕

Order-IV : Ophiostomatales-Genus *Ophiostoma,*

Order V : Diaporthalles-Genera *Diporthe, Gnomonia, Gaeumannomyces, Magnaporthe*

↕

Order-VI : Xylariales,

↕

Genera : *Hypoxylon, Roselinia, xylaria*

B. Loculoascomycetes : Ascomycetes without ascostromata

Order-I : Dothidiales, Genera-*Mycorsphaerella,* Elsinoe

↕

Order-II: Capnodiales-Genus-*Capnodium*

↕

Order-III : Pleosporales-Genera-*Cochliobolus, Pyrenophora, Pleospora, Leptosphaeria, Venturia,* Guinardia

C. Discomycetes- Ascomycetes with apothecia

Order-I : Rhytismales,

Genera : *Hypoderma, Lophodermium, Rhytisma*

↕

Order-II : Helotiales

Genera : *Monilina, Sclerotinia, Stromatinia, Diplocarpon*

D. Denteromycetes : Imperfect or asexual fungi

Phylum-IV : BASIDIOMYCOTA : Basidium and basidiospores produced

↕

Order-I : Ustilaginales : The smut fungi

↕

Genera : *Ustilago, Tilletia, Neovossia, Tolyposporium, Sphacelotheca*

↕

Order-II : Uredinals : The rust fungi

↕

Genera : *Puccinia, Uromyces, Melampsora, Phragmedium, Hemileia, Cronartium*

↕

Order-III : Exobasidiales, Genus : *Exobasidium*

↕

Order-IV : Ceratobasidiales,
Genera : *Thanatephorus, Typhula*

↕

Order-V : Agaricales
Genera : *Armillaria, Marasmiues, Pleurotus, Pholiota*

↕

Order-IV : Aphyllophorales
Genera : *Aethalium, Corticium, Heterobasidion, Gonoderma, Postia, Peniophora, Polyporus*

5. In the present text book classification proposed by Alexopoulos *et al.* (1996) is being followed. Alexopoulos *et. al.* (1996) have classified fungi that reflects hypothesized evolutionary relationships of the organisms. Such classification is known as PHYLOGENETIC CLASSIFICATION and texa, the names of the groups of organisms all correspond to monophyletic lineages. The classification proposed by them recognises the fact that the organism that have been called "fungi" are not all closely related. Though these organisms do not share a common evolutionary history, they do form a closely knit group on the basis of their morphology, nutritional modes, and ecology.

The classification proposed by them is outlined with emphasis on the parasitic fungi as follows :

Kingdom : Fungi

Phylum-Chytridiomycota
Phylum-Zygomycota
Phylum-Ascomycota
Phylum-Basidiomycota

Kingdom : Stramenopila

Phylum-Oomycota
Phylum-Hyphochytriomycota
Phylum-Labyrinthulomycota

Protists

Phylum-Plasmodiophoromycota
Phylum-Dictyosteliomycota
Phylum-Acrasiomycota
Phylum-Myxomycota

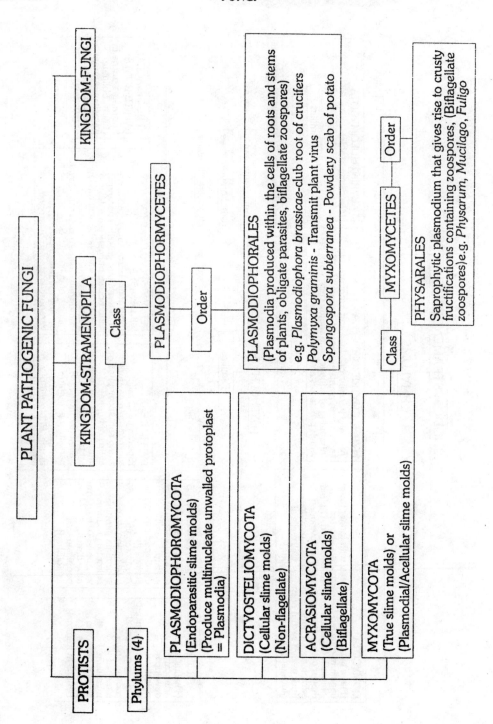

PLANT PATHOGENIC FUNGI

- **PROTISTS**
  - Phylums (4)
    - PLASMODIOPHOROMYCOTA (Endoparasitic slime molds) (Produce multinucleate unwalled protoplast = Plasmodia)
      - Class: PLASMODIOPHORMYCETES
        - Order: PLASMODIOPHORALES (Plasmodia produced within the cells of roots and stems of plants, obligate parasites, biflagellate zoospores)
          - e.g. *Plasmodiophora brassicae*-club root of crucifers
          - *Polymyxa graminis* - Transmit plant virus
          - *Spongospora subterranea* - Powdery scab of potato
    - DICTYOSTELIOMYCOTA (Cellular slime molds) (Non-flagellate)
    - ACRASIOMYCOTA (Cellular slime molds) (Biflagellate)
    - MYXOMYCOTA (True slime molds) or (Plasmodial/Acellular slime molds)
      - Class: MYXOMYCETES
        - Order: PHYSARALES — Saprophytic plasmodium that gives rise to crusty fructifications containing zoospores, (Biflagellate zoospores)e.g. *Physarum, Mucilago, Fuligo*
- **KINGDOM-STRAMENOPILA**
- **KINGDOM-FUNGI**

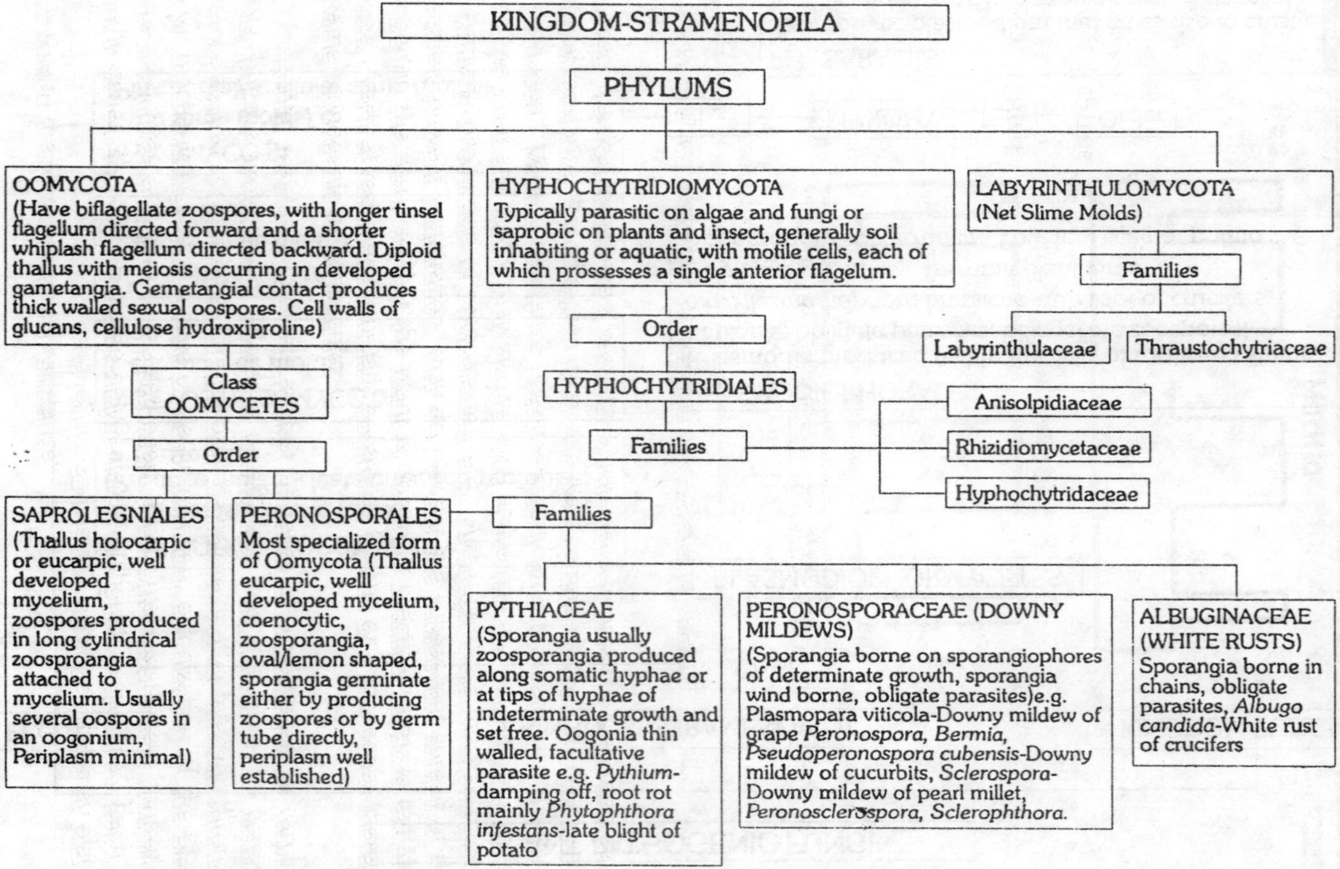
KINGDOM-STRAMENOPILA
PHYLUMS
OOMYCOTA
(Have biflagellate zoospores, with longer tinsel flagellum directed forward and a shorter whiplash flagellum directed backward. Diploid thallus with meiosis occurring in developed gametangia. Gemetangial contact produces thick walled sexual oospores. Cell walls of glucans, cellulose hydroxiproline)
Class OOMYCETES
Order
SAPROLEGNIALES
(Thallus holocarpic or eucarpic, well developed mycelium, zoospores produced in long cylindrical zoosproangia attached to mycelium. Usually several oospores in an oogonium, Periplasm minimal)
PERONOSPORALES
Most specialized form of Oomycota (Thallus eucarpic, welll developed mycelium, coenocytic, zoosporangia, oval/lemon shaped, sporangia germinate either by producing zoospores or by germ tube directly, periplasm well established)
Families
PYTHIACEAE
(Sporangia usually zoosporangia produced along somatic hyphae or at tips of hyphae of indeterminate growth and set free. Oogonia thin walled, facultative parasite e.g. *Pythium*-damping off, root rot mainly *Phytophthora infestans*-late blight of potato
PERONOSPORACEAE (DOWNY MILDEWS)
(Sporangia borne on sporangiophores of determinate growth, sporangia wind borne, obligate parasites)e.g. Plasmopara viticola-Downy mildew of grape *Peronospora, Bermia, Pseudoperonospora cubensis*-Downy mildew of cucurbits, *Sclerospora*-Downy mildew of pearl millet, *Peronosclerospora, Sclerophthora.*
ALBUGINACEAE (WHITE RUSTS)
Sporangia borne in chains, obligate parasites, *Albugo candida*-White rust of crucifers
HYPHOCHYTRIDIOMYCOTA
Typically parasitic on algae and fungi or saprobic on plants and insect, generally soil inhabiting or aquatic, with motile cells, each of which prossesses a single anterior flagelum.
Order
HYPHOCHYTRIDIALES
Families
Anisolpidiaceae
Rhizidiomycetaceae
Hyphochytridaceae
LABYRINTHULOMYCOTA
(Net Slime Molds)
Families
Labyrinthulaceae
Thraustochytriaceae

FUNGI

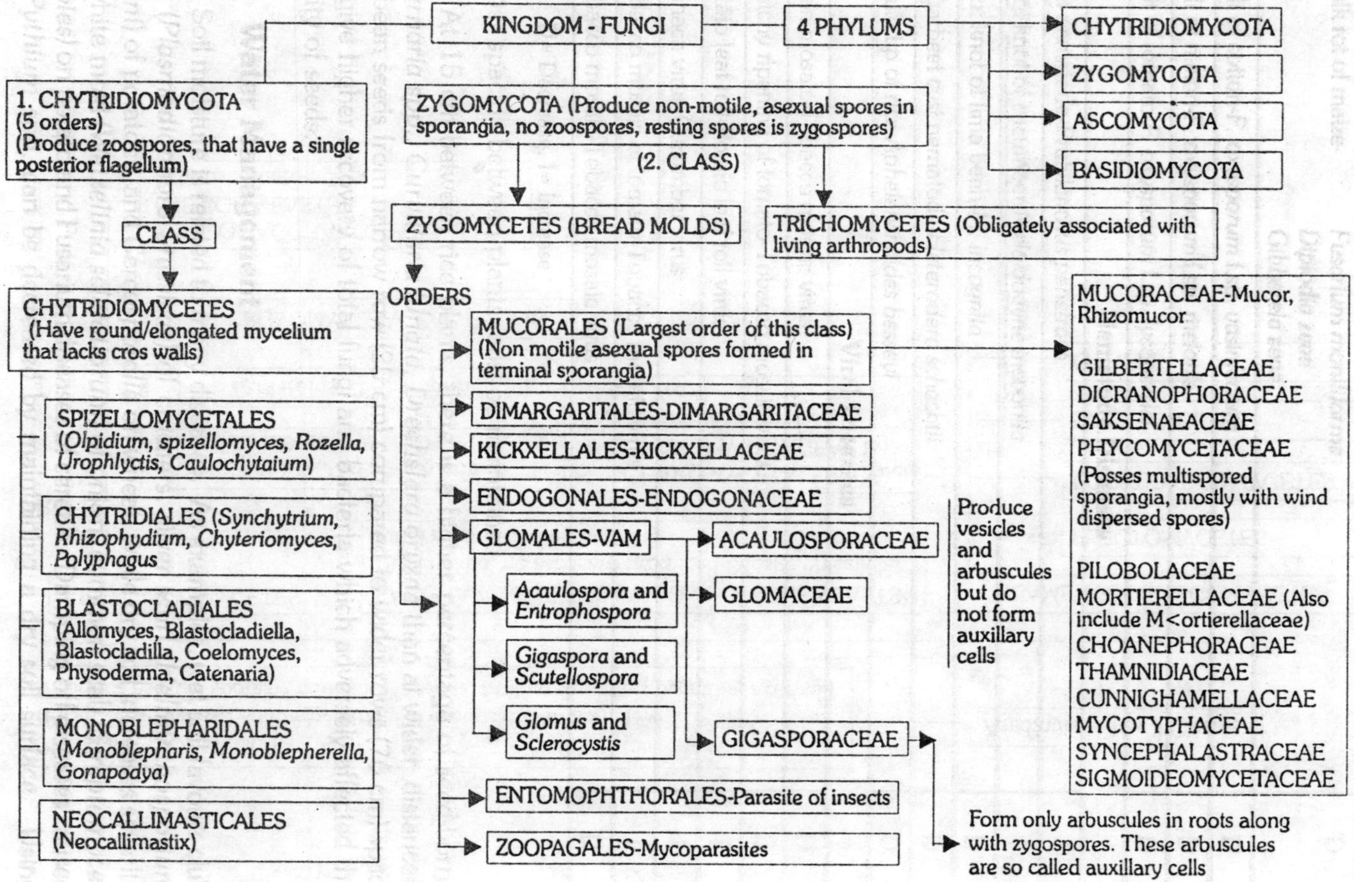
KINGDOM - FUNGI
4 PHYLUMS
CHYTRIDIOMYCOTA
ZYGOMYCOTA
ASCOMYCOTA
BASIDIOMYCOTA
1. CHYTRIDIOMYCOTA (5 orders) (Produce zoospores, that have a single posterior flagellum)
CLASS
CHYTRIDIOMYCETES (Have round/elongated mycelium that lacks cros walls)
SPIZELLOMYCETALES (Olpidium, spizellomyces, Rozella, Urophlyctis, Caulochytaium)
CHYTRIDIALES (Synchytrium, Rhizophydium, Chyteriomyces, Polyphagus
BLASTOCLADIALES (Allomyces, Blastocladiella, Blastocladilla, Coelomyces, Physoderma, Catenaria)
MONOBLEPHARIDALES (Monoblepharis, Monoblepherella, Gonapodya)
NEOCALLIMASTICALES (Neocallimastix)
ZYGOMYCOTA (Produce non-motile, asexual spores in sporangia, no zoospores, resting spores is zygospores)
(2. CLASS)
ZYGOMYCETES (BREAD MOLDS)
TRICHOMYCETES (Obligately associated with living arthropods)
ORDERS
MUCORALES (Largest order of this class) (Non motile asexual spores formed in terminal sporangia)
DIMARGARITALES-DIMARGARITACEAE
KICKXELLALES-KICKXELLACEAE
ENDOGONALES-ENDOGONACEAE
GLOMALES-VAM
Acaulospora and Entrophospora
Gigaspora and Scutellospora
Glomus and Sclerocystis
ACAULOSPORACEAE
GLOMACEAE
GIGASPORACEAE
Produce vesicles and arbuscules but do not form auxillary cells
Form only arbuscules in roots along with zygospores. These arbuscules are so called auxillary cells
ENTOMOPHTHORALES-Parasite of insects
ZOOPAGALES-Mycoparasites
MUCORACEAE-Mucor, Rhizomucor
GILBERTELLACEAE
DICRANOPHORACEAE
SAKSENAEACEAE
PHYCOMYCETACEAE
(Posses multispored sporangia, mostly with wind dispersed spores)
PILOBOLACEAE
MORTIERELLACEAE (Also include M<ortierellaceae)
CHOANEPHORACEAE
THAMNIDIACEAE
CUNNIGHAMELLACEAE
MYCOTYPHACEAE
SYNCEPHALASTRACEAE
SIGMOIDEOMYCETACEAE

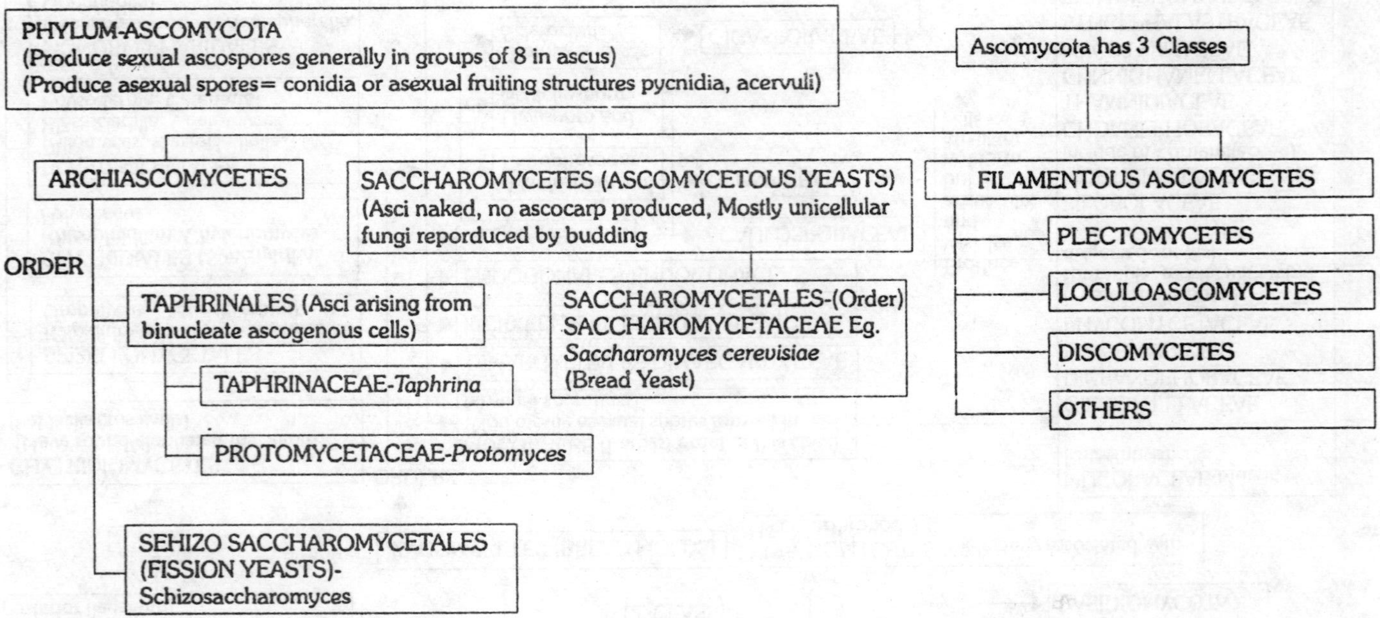
PHYLUM-ASCOMYCOTA
(Produce sexual ascospores generally in groups of 8 in ascus)
(Produce asexual spores= conidia or asexual fruiting structures pycnidia, acervuli)
Ascomycota has 3 Classes
ARCHIASCOMYCETES
SACCHAROMYCETES (ASCOMYCETOUS YEASTS)
(Asci naked, no ascocarp produced, Mostly unicellular fungi reporduced by budding
FILAMENTOUS ASCOMYCETES
PLECTOMYCETES
LOCULOASCOMYCETES
DISCOMYCETES
OTHERS
ORDER
TAPHRINALES (Asci arising from binucleate ascogenous cells)
TAPHRINACEAE-Taphrina
PROTOMYCETACEAE-Protomyces
SEHIZO SACCHAROMYCETALES
(FISSION YEASTS)-
Schizosaccharomyces
SACCHAROMYCETALES-(Order)
SACCHAROMYCETACEAE Eg.
Saccharomyces cerevisiae
(Bread Yeast)

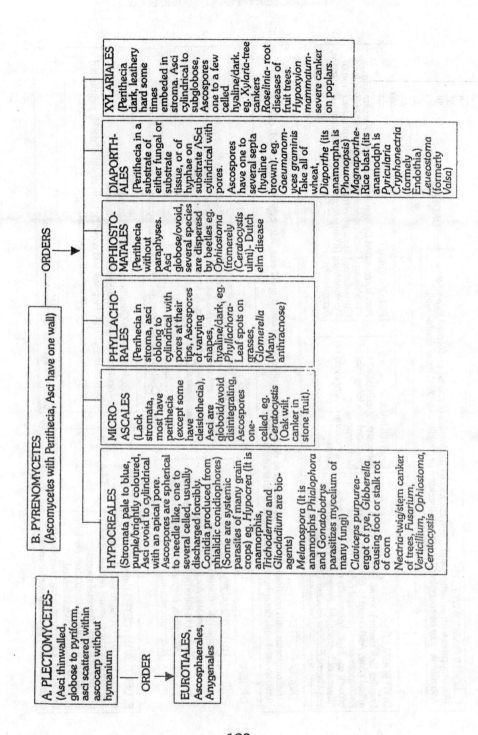

A. PLECTOMYCETES-
(Asci thinwalled, globose to pyriform, asci scattered within ascocarp without hymanium

ORDER →

EUROTIALES, Ascosphaerales, Anygenales

B. PYRENOMYCETES
(Ascomycetes with Perithecia, Asci have one wall) — ORDERS →

HYPOCREALES
(Stromata pale to blue, purple/brightly coloured, Asci ovoid to cylindrical with an apical pore. Ascospores are spherical to needle like, one to several celled, usually discharged forcibly. Conidia produced from phialidic conidiophores) (Some are systemic parasites of many grain crops) eg. *Hypocrea* (It is anamorphis, *Trichoderma* and *Gliocladium* are bio-agents)
Melanospora (It is anamorlphs *Phialophora* and *Gonatobotrys* parasitizes mycelium of many fungi)
Claviceps purpurea- ergot of rye, *Gibberella* causing foot or stalk rot of corn
Nectria-twig/stem canker of trees, *Fusarium*, *Verticillium*, *Ophiostoma*, *Ceratocystis*

MICRO-ASCALES
(Lack stromata, most have perithecia (except some have cleistothecia), Asci are globoid/avoid disintegrating, Ascospores one-celled. eg. *Ceratocystis* (Oak wilt, canker in stone fruit).

PHYLLACHO-RALES
(Perihecia in stroma, asci oblong to cylindrical with pores at their tips, Ascospores of varying shapes, hyaline/dark, eg. *Phyllachora*- Leaf spots on grasses, *Glomerella* (Many anthracnose)

OPHIOSTO-MATALES
(Perithecia without paraphyses. Asci globose/ovoid, several species are disperesd by beetles eg. *Ophiostoma* (fromerely *(Ceratocystis* ulmi)- Dutch elm disease

DIAPORTH-ALES
(Perithecia in a substrate of either fungal or substrate tissue, or of hyphae on substrate ASci cylindrical with pores.
Ascospores have one to several septa (hyaline to brown). eg. *Gaeumanomyces graminis* Take all of wheat, *Diaporthe* (its anamorpha is *Phomopsis*) *Magnaporthe*- Rice blast (its anamodph is *Pyricularia* *Cryphonectria* (formely Endothia) *Leueostoma* (formerly *Valsa*)

XYLARIALES
(Perithecia dark, leathery hard some times embeded in stroma. Asci cylindrical to subglobose, Ascospores one to a few celled hyaline/dark. eg. *Xylaria*-tree cankers *Roselinia*- root diseases of fruit trees. *Hypoxylon mammatum*- severe canker on poplars.

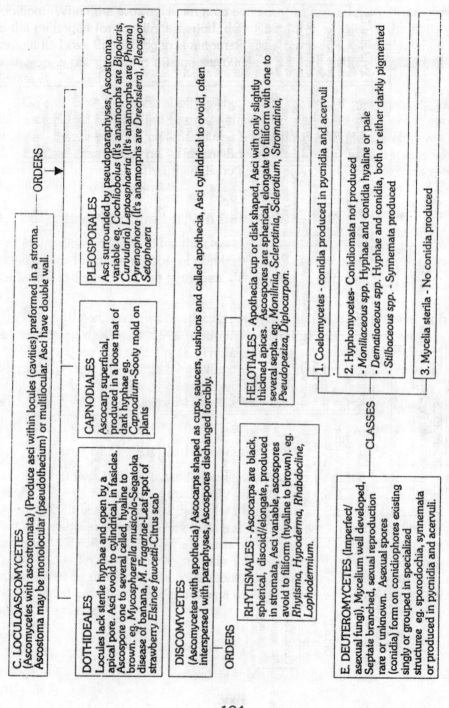

C. LOCULOASCOMYCETES
(Ascomycetes with ascostromata) (Produce asci within locules (cavities) preformed in a stroma. Ascostoma may be monolocular (pseudothecium) or multilocular. Asci have double wall.

ORDERS

- **DOTHIDEALES**
  Locules lack sterile hyphae and open by a apical pore. Asci ovoid to cylindrical, in fasicles. Ascospore one to several celled, hyaline to brown. eg. *Mycosphaerella musicola*-Segatoka disease of banana, *M. Fragariae*-Leaf spot of strawberry *Elsinoe fawcetti*-Citrus scab
- **CAPNODIALES**
  Ascocarp superficial, produced in a loose mat of dark hyphae eg. *Capnodium*-Sooty mold on plants
- **PLEOSPORALES**
  Asci surrounded by pseudoparaphyses, Ascostroma variable eg. *Cochliobolus* (It's anamorphs are *Bipolaris, Curvularia*) *Leptosphaeria* (It's anamorphs are *Phoma*) *Pyrenophora* (It's anamorphs are *Drechslera*), *Pleospora, Setophaera*

DISCOMYCETES
(Ascomycetes with apothecia) Ascocarps shaped as cups, saucers, cushions and called apothecia, Asci cylindrical to ovoid, often interspersed with paraphyses, Ascospores dischanged forcibly.

ORDERS

- **RHYTISMALES** - Ascocarps are black, spherical, discoid//elongate, produced in stromata, Asci variable, ascospores avoid to filiform (hyaline to brown). eg. *Rhytisma, Hypoderma, Rhabdocline, Lophodermium.*
- **HELOTIALES** - Apothecia cup or disk shaped, Asci with only slightly thickned apices. Ascospores are spherical, elongate to filiform with one to several septa. eg. *Monilinia, Sclerotinia, Sclerotium, Stromatinia, Pseudopeziza, Diplocarpon.*

E. DEUTEROMYCETES (Imperfect/ asexual fungi), Mycelium well developed, Septate branched, sexual reproduction rare or unknown. Asexual spores (conidia) form on conidiophores existing singly or grouped in specialized structuree eg. sporodochia, synnemata or produced in pycnidia and acervuli.

CLASSES

1. Coelomycetes - conidia produced in pycnidia and acervuli
2. Hyphomycetes- Conidiomata not produced
   - *Moniliaceous spp.* Hyphae and conidia hyaline or pale
   - *Dematiaceous spp.* Hyphae and conidia, both or either darkly pigmented
   - *Stilbaceous spp.* - Synnemata produced
3. Mycelia sterila - No conidia produced

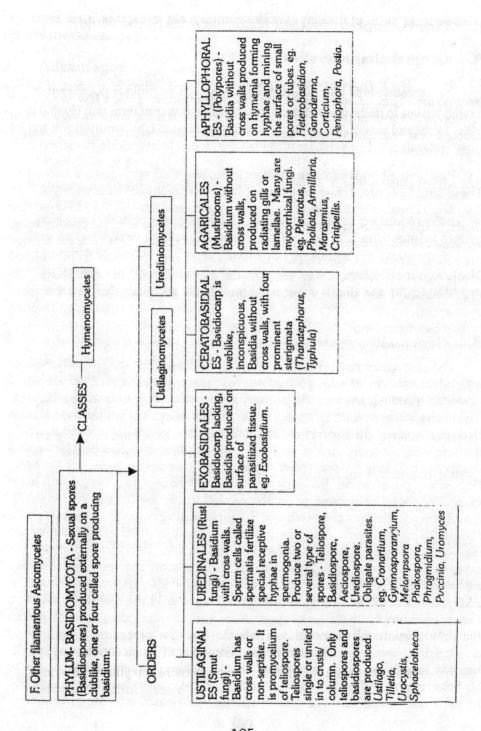

F. Other filamentous Ascomycetes

PHYLUM- BASIDIOMYCOTA - Sexual spores (Basidiospores) produced externally on a clublike, one or four celled spore producing basidium.

→ CLASSES — Hymenomycetes

Ustilaginomycetes

Urediniomycetes

ORDERS

USTILAGINALES (Smut fungi) - Basidium has cross walls or non-septate. It is promycelium of teliospore. Teliospores single or united in to crusts/ column. Only teliospores and basidiospores are produced. *Ustilago, Tilletia, Urocystis, Sphacelotheca*

UREDINALES (Rust fungi) - Basidium with cross walls. Sperm cells called spermatia fertilize special receptive hyphae in spermogonia. Produce two or several type of spores - Teliospore, Basidiospore, Aeciospore, Urediospore. Obligate parasites. eg. *Cronartium, Gymnosporangium, Melampsora Phakospora, Phragmidium, Puccinia, Uromyces*

EXOBASIDIALES - Basidiocarp lacking, Basidia produced on surface of parasitiized tissue. eg. *Exobasidium.*

CERATOBASIDIALES - Basidiocarp is weblike, inconspicuous, Basidia without cross walls, with four prominent sterigmata (*Thanatephorus, Typhula*)

AGARICALES (Mushrooms) - Basidium without cross walls, produced on radiating gills or lamellae. Many are mycorrhizal fungi. eg. *Pleurotus, Pholiota, Armillaria, Marasmius, Crinipellis.*

APHYLLOPHORALES - (Polypores) - Basidia without cross walls produced on hymenia forming hyphae and mining the surface of small pores or tubes. eg. *Heterobasidion, Ganoderma, Corticium, Peniophora, Postia.*

X. Plasmodiophoromycota (Endoparasite Slime Molds)

A. General characterstics

(I) The members of this phylum are necrotrophic endoparasites of vascular plants. They cause *Hypertrophy* and *Hyperplasia*. Only a few species are of economic importance. *Plasmodiophora brassical* is the cause of club root of cabbage and related crucifers, and *Spongospora subterranea* causes powdery scab of potatoes.

Plasmodiophorids produce multinucleate, unwalled protoplasts (Plasmodia). These structures are, however, different from the plasmodia of Myxomycota as they are incapable of translocational movement, lack the ability to phagocytize food materials, and exist wholly within the cells. The life-cycle of plasmodiophorid involves the production of two different plasmodial phase. One phase is termed the PRIMARY or SPORANGIAL PLASMODIUM and gives rise to thin walled zoosporangia, while the other is called the SECONDARY or SPOROGENIC PLASMODIUM and produce resting spores which eventually germinate to produce zoospores.

B. General life- cycle

The resting spores formed in the sporosorus are deposited in soil or water after disintegration of infected host tissues. Each resting spore is capable of germination usually to form one *primary zoospore* capable of infecting the host. The primary zoospore attaches to the wall of root hairs. Thereafter, the flafellum becomes inactive, the axonemes are retracted, and the zoospore encysts within the encysted zoospore *Rohr*(Gen.tube) and with the rohr, *Stachel*(Ger.=spike) are formed. Finally the protoplast of the encysted zoospore enters the host cell. Following peretration by encysted primary zoospore, the uninucleate plasmodiophorid protoplast is carried around inside the host cell by cyclosis. Cruciform mitotic divisions occur and the protoplast enlarges to form primary or sporangial plasmodium. After the primary plasmodium reaches a certain size, it cleaves into segments that develop into zoosporangia. Such zoosporangia may occur singly or they may be loosely aggregated. In some species they are organized into sori. *Secondary zoospores* are caleaved within and released from the zoosporangium either directly into other host cells or to the outside of the root. Following contact with tissues of appropriate host, the secondary zoosposes from the soil penetrate the host. Once inside the host, the protoplast is reported to develop into secondary or sporogenic plasmodium. As secondary plasmodia become established in the cortex and vascular tissues, the host cells often under go hypertrophy and hyperplasia. Eventually these plasmodia undergo cleavage to form resting spores. The resting spores are released into the environment following disintegration of the host cells, where some are thought to lie dormant for up to 8 years before germinating to form primary zoosporse.

C. Classification

Plasmodiophoromycota is well defined as a monophyletic group. The group is related to ciliate protists. This phylum has the single class *Plamodio phoromycetes*, with the single order - *Plasmodiophorales* and the single family *Plasmodiophoraceae.* A total of 10 genera and 29 species are recognized. Important genera are *Plasmodiophora, Sorosphaera, Spongospora, Sorodiscus, Octomyxa, Polymyxa,* and *woronina.*

D. Plasmodiophora brassicae (Figure-1 A-Q)

The species causes club root or finger and toe disease of brassicas. The resting spores possess a dark wall which contains chitin but no cellulose. On germination usually a single anteriorly biflagellate primary zoospore emerges. The larger flagellum points forward and the shorter one is directed almost at a right angle. Both the flagella are of the truncated whiplash type, no tinsel structure being seen on any one of them. Before infection of the host the zoospores become amoeboid. In contact with host root hair or epidermal cells of the root the zoospore discards its flagella and bodily enters the root through a hole dissolved on the cell wall. The hole is then closed by host action. At first a small uninucleate stage is seen in the root hair or epidermal cell. Later, with enlargement of the amoeboid structrue, mitotic divisions of the nucleus occur until a small plasmodium is formed containing 30 to 100 nuclei. The plasmodium cleaves into a variable number of uninucleate, roughly spherical zoosporangia or gametangia lying packed together in the host cell. This stage can be seen within 4 days of infection. The chemical composition of the wall of sporangia is not known. The nucleus in the zoosporangium divides mitotically 2-3 times and 4 to 8 uninucleate anteriorly flagellate zoospores are formed within the sporangium. These are swarm cells or secondary zoospores, smaller than the primary zoospores. The mature zoosporangia become attached to the host cell wall and a pore is developed at this point through which the swarm cells escape. Occasionally they are released within the host cells.

It is believed that the swarm cells function as gametes and fuse in pairs. Quadriflagellate or even six flagellated swarm cells have also been reported. However, their origin is not clear. It is believed that fusion of swarmers in pairs results in the formation of amoeboid zygote. The zygote is the origin of large plasmodium which gives rise to resting spores. The zygote and young plasmodia which arise from them may unite to form larger plasmodia. It has been claimed by some workers that the plasmodium has the power of penetrating the cell walls

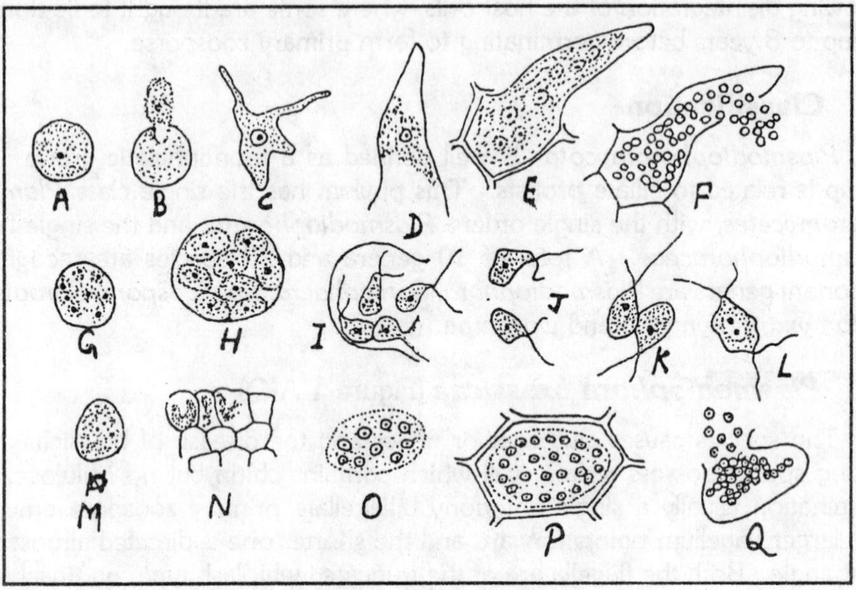
A
B
C
D
E
F
G
H
I
J
K
L
M
N
O
P
Q

Figure-1: *Plasmodiophora brassicae*. Stages in life cycle.

of the host and thus pass from cell to cell but others claim that as the host cells divide, the plasmodium is passively distributed to daughter cells. Possibly both mechanisms are involved (Webster, 1970). The plasmodium has no specialized feeding structrures such as haustoria. It is immersed in the cell cytoplasm surrounded by a thin plasmodial envelop. The plasmodium enlarges, undergoes division following nuclear divisions.

The cells containing plasmodia become hypertrophied. The host nucleus remains functional during this period. Hypertrophy of host cells (abnormal enlargement) is brought about by blocking of the mechanism for cell division and is accompanied by enhanced DNA synthesis. Starch also accumulates in infected cells(William, 1966). Eventually the cell contents of the infected cells are almost completely exhausted and the cell is prectically killed by the plasmodium. The nuclei of this plasmodium undergo two rapid successive divisions, probably meiotic. Then the protoplasrn rounds up into uninucleate spores surrounded by dark chitinous wall. Thus the plasmodiurn is transformed into a mass of resting spores which lies free in the host cell. With decay of the root tissues these spores are released into the soil.

E. *Spongospora subterranea* (Plate-IX)

The main characters of *Spongospora* are: spores in a hollow sphere with several openings; zoosporangia may be formed, zoospores are anteriorly biflagellate and heterokont (unequal flagella), similar in size whether from sporangia or from resting spores. The genus *Spongosr ora* differs from *Plasmodiophora* mainly in two respects, viz, in *Plasmodiophora* the secondary zoospores are smaller than the primary zoospores while in *Spongospora* their size is not different, and while the spores in *Plasmodiophora* are free, not in balls or discs, in *Spongospora* they are in spongy balls. The best known species is *Spongospora subterranea* which causes powdery scab disease of potato tubers and also attacks the stems and roots of this host and of related plants.

The life-cycle of *Spongospora subterranea* is possibly similar to that of *Plasmodiophora brassicae.* On germination the resting spore forms a single uninucleate amoeba. The plasmodial mass causes infection of tubers at or near the "eyes" or root hairs of potato and tomato. Lenticels and wounds are other avenues of entry. The young uninucleate plasmodium in the root hair or potato tuber develops into a mulinucleate plasmodium after nuclear divisions while the proof of sexuality is lacking there is evidence that nuclei fuse and their reduction division occurs just before formation of spores. At the same time aggregation of plasmodia also occurs. The plasmodium then may develop into zoosporangium or may form balls of resting spores. The zoosporangia release anteriorly biflafellate zoospores which can cause fresh infections and produce zoosporangia. The resting spores are formed from the plasmodium in spore balls or clusters. These balls are spongy with irregular internal channels. They appear as yellow brown dust in the mature sori. Each individual cell of the ball represents an individual uninucleate spore.

XI. Phylum CHYTRIDIOMYCOTA

A. GENERAL CHARACTERSTICS

These are the only members of the kingdom fungi that produce motile cells at some stge in their life-cycle history. There are about 100 genera consisting of about 1000 species, most of which are saprobes. The motile cells (Zoospores and motile cells) of these possess a single, posteriorly directed whiplash flagellum Other characters include globose or oval coenocytic thallus, well developed mycelium, zygote being converted into resting spore/resting sporangia, even into a diploid thallus. Cell walls are chitinous. Thalli of varied structures and characters are formed in chytrids. They may be endobiotic or epibiotic, holocarpic, monocentric or polycentric. A number of chytrids are important plant parasites, for example, *Synchytrium endobioticum* (wart of potatoes), *Physoderma maydis* (brown spot of corn) and *olpidium brassicae.*

B. Somatic Structures — The most primitive chytrids are unicellular and holocarpic. Such organisms have no mycelium and, in the early stages of their development, may lack cell walls. In some what more advanced species, a few *rhizoids* (Gr. *rhiza* =root +oeides =like) or *rhizomycelium* (*rhiza* + mycelium) are produced. In still more advanced forms, a scanty mycelium, represented only by a few short hyphal branches, is produced. The most advanced have a true mycelial thallus. Although the hyphae of such species are typically coenocytic, a septum is formed at the base of each reproductive organ. Such septa are solid plates. In addition, the mycelium of higher chytrids may form *pseudosepta* (Gr. pseudo = false). These are septum like partitions or plugs of a chemical composition different from that of the hyphal walls, which are deposited at intervals in the hyphae.

C. A Sexual Reproduction — The sporangium is the asexual reproductive structure of the chytrids. In the young stage, sporangia are full of protoplasm containing nuclei. As the sporangium develops, the entire protoplast undergoes cleavage into numerous minute sections each of which develops into a uninucleate zoospore. After discharge, the zoospore swims for a time, encysts, withdrawing or loosing its flafellum in the process, and then germinates, usually after a short rest period.

D. Sexual Reproduction — Sexual reproduction in chytrds is accomplished by one of the following methods :

1 Planogametic copulation — The two gametes that are morphologically similar, but physiologically different, unite in water to form a motile zygote (conjugation of isogamous planogametes). In some species, gametes originating in the same gametangium will not copulate. Examples of fungi which produce isogamous planogametes are *olpidium brassical and S. endobioticum*. In some species, copulation takes place between planogametes in which one is larger than the other (congugation of anisogamous planogametes). This is known to occur only in some species of *Blastocladiales*. Still in some chytrids, fertilization of a non-motile female gamete(egg) by a motile male gamete (antherozoid) takes place. The male motile gametes are released from antheridia into water and swim away. Some of them reach oogonia, where upon one antherozoid enters each oogonium and unites with the egg within. This method is common in species belonging to *Monoblepharidales*.

2. Gametangial copulation — In the chytrids this is accomplished by the transfer of the entire protoplast of one gametangium into the other.

3. Somatogamy — Fusion between rhizomycelial filaments is reported to precede resting spore formation in some chytrids, but this has not been confermed.

E. Classification —Traditionally, the class *chytridiomycetes* has been divided into orders on the basis of morphological features of somatic and reproductive structures. The validity of most characters emphasized in older classification schemes is now in doubt. A new approach emphasizing zoospore ultrastructure has been proposed (Barr, 1990). Using this approach Barr (1990) recognized the orders *Spizellornycelates (Rozella,olpidium), Chytridiales (Synchytrium, Rhizophydium, Chytriomyces), Blastocladiales (Allomyces, Blastocladiella, Blastocladia, Coelomomyces, Catenaria, Physoderma) and Monoblepharidales (Monoblepharis)* Li and Heath (1993) proposed another order *Neocallimas Ficales (Neocalli mastix).*

F. Order - Chytridiales

Members of this order are water or soil inhabiting fungi. They are parasitic on algae, water molds and vascular plants. A few parasitize animal eggs and protozoa while others are saprophytic on the decaying remains of dead plants. This order is defined on the basis of zoospore ultrastructure. The zoospore is characterized by a castellation of characters. One or more mitochondria are included in the MLC of the zoospore, and the nucleus appears to occupy the space not taken by the MLC and the ribosomes. The nucleus is not connected to the kinetosome. The robosomes do not occur dispersed throughout the cell, but rather, are packaged by a double membrane in the central part of the cell. Rootlet microtubules typically extend from the side of the kinetosome into the cytoplasm. Some of the better known genera are *Chytridium, Chytriomyces, Polyphagus, Rhizophydium, Endochytrium, Synchytrium, Cladochytrium* and *Nowakowskiella.* Only a few species in the entire order are economically important. *S.endobioticum*, is perhaps the most destructive plant parasite. Its life-cycle is discussed.

G. Genus - *Synchytrium(Plate-IX)*

This genus belongs to phylum - Chytridiomycota, class- *Chytridiomycetes*, order - *Chytridiales* and family - *synchytriaceae*. The members of the family are endobiotic, holocarpic fungi with inoperculate sporangia. At the time of reproduction it may become converted directly into a group of *sorus* of sporangia or may form a *prosorus* which later gives rise to a sorus of sporangia. Some times the thallus may be converted into a resting sporangium which can either function directly as a sporangium or give rise to zoospores. It can also function as a prosorus producing a vesicle, the contents of which divide to form a sorus of sporangia. Asexual reproduction takes place by means of zoospores (possessing a posterior whiplash flagellum) formed within the zoosporangium. Sexual reproduction is accomplished by the union of two isogametes. The zygote secretes a thick wall and behaves as *resting Sporangium.*

Genus *Synchytrium* consists of more than 100 species that are parasitic on plants. The fungus is an abligate parasite and produces hypertrophy. *S.endobioticum* causes the "black wart of potato". In India, the disease is restricted to Darjeeling hills of West Bengal.

H. Asexual Reproduction

S.endobioticum absorbs food material from the host, enlarges considerably and after attaining a certain size develops two layers around the thallus, the outer wall is thick and golden yellow while the inner one thin and hyaline. This structure is termed as *summer sporangium* and behaves as *prosorus* (Fig. 2). The uninucleate protoplasm migrates in a vesicle formed by the parasite in the host cell. Mitosis takes place in the nucleus of the prosorus forming 32 nuclei. The segmentation of the protoplast occurs and many multinucleate segments are formed. Each segment develops into a sporangium. The protoplast of each sporangium produces a large number of zoospores by cleavage. The zoospores after liberation swim for sometime in the film of water and when any one of them comes in contact with potato tubers, it withdraws its flagellum and penetrates the host epidermis by forming a penetration tube. The uninucleate naked protoplasm enlarges in the host tissue, becomes round and secretes two cell layers. This structrue is called *Prosorus.* Due to presence of prosorus, the infected host cells undergo repeated divisions forming a warty outgrowth on the surface of the tubers. This process is repeated several times.

The behaviour of zoospores is variable ones. If they are released soon after their formation and there is plenty of water available, they function as zoospores and complete the asexual phase of the life-cycle as usual and if the release of zoospores is delayed and the conditions are comperatively dry, they are smaller in size and behave as *planogametes* and fuse in pairs (fig. 2).

I. Sexual Reproduction

During unfavourable conditions, the swarmers behave as gametes. It is belived that gametic fusion between the gametes produced from the same gametangium does not take place but the planogametes produced from different gametangia of the same prosorus may fuse. It is clear from the behaviour of these gametes that those produced from the different gametangia, although morphologically similar, are physiologically different from one another. The fusion of the gametes takes place out side the host and fuse to form usually a biflagellate planozygote. Sometimes it has been seen that one of the gametes loses its flagellum before fusion, thus forming a uniflagellate zygote. Each zygote swins for a while and the planozygote withdraws its flagellum and secretes a thin wall around itself (fig. 2)

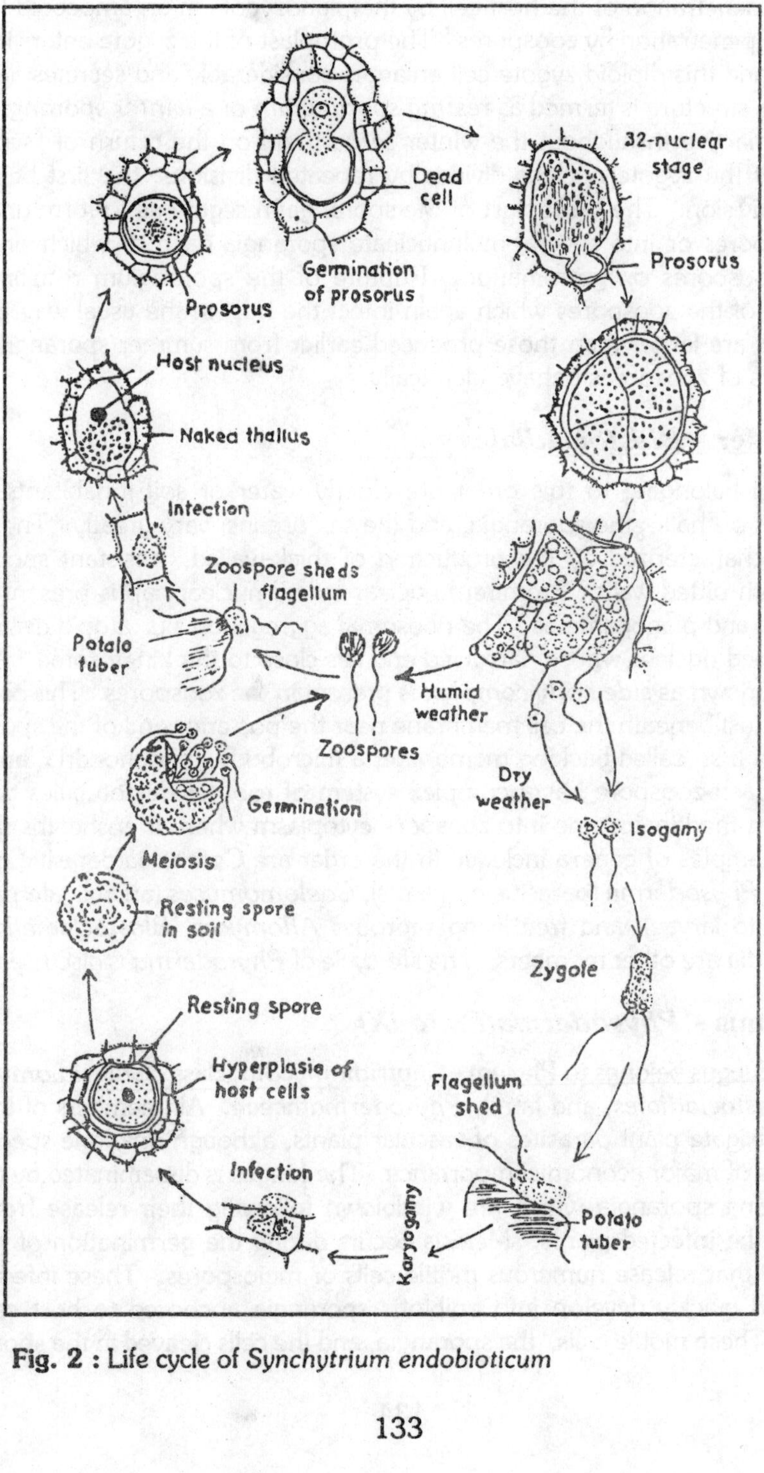

Dead cell
Germination of prosorus
32-nuclear stage
Prosorus
Prosorus
Host nucleus
Naked thallus
Intection
Zoospore sheds flagellum
Potalo tuber
Humid weather
Zoospores
Germination
Dry weather
Isogamy
Meiosis
Resting spore in soil
Zygote
Resting spore
Hyperplasia of host cells
Flagellum shed
Infection
Potato tuber
Karyogamy

Fig. 2 : Life cycle of *Synchytrium endobioticum*

The penetration of the host cell by the planozygote in an amoeboid fashion is like the penetration by zoospores. The protoplast of the zygote enters into the host cell and this diploid zygote cell enlarges considerably and secretes a double wall. This structure is termed as *resting sporangium* or a *winter sporangium*. It remains inactive throughout the winter season and on the return of favourable conditions the zygote nucleus divides by repeated divisions, the first being the reduced division. The protoplast of the sporangium segments to form uniflagellate zoospores or into several multinucleate sporangia each of which produces several zoospores on germination. Rupture of the sporangium results in the liberation of the zoospores which again infect the host in the usual way. These zoospores are larger than those produced earlier from summer sporangium but both types of zoospores behave identically.

J. Order - *Blastocladiales*

Fungi belonging to this order are chiefly water or soil inhabitants. The characters of thalli, the sporangia, and the sex organs, vary greatly. The members are characterized by the production of thick-walled, resistant sporangia, usually with pitted walls. Promirent nuclear-bound nuclear cap is present in the zoospores and planogametes. The ribosomal aggregation sits atop a distinctive, cone shaped nucleus whose narrowed end lies close to the kinetosome. A form of MLC, known as *side body complex* is present in the zoospores. This complex is located just beneath the cell membrane near the posterior end of the spore and consists of a so called backing membrane, a microbody, mitochondria, and lipid bodies. Each zoospore has a complex system of root let mictotubules that extends from the kinetosome into zoospore cytoplasm where it ensheaths the nucleus. Examples of genera included in the order are *Catenaria*(parasitic on animals) and *Physoderma* (parasitic on plants), *Coelomomyces* (an obligate parasite of mosquito larvae), and free living saprobes *Allomyces*, *Blastocladiella*, and *Blastocladia* are other members. The life-cycle of *Phyroderma* is discussed here.

K. Genus - *Physoderma(Plate-IX)*

The fungus belongs to Phylum - *Chytridiomycota*, class - *Chytridiomycetes*, order- *Blastocladiales*, and family *Physodermataceae*. All members of this genus are obligate plant parasites of vascular plants, although only one species, *P. maydis* is of major economic importance. The fungus is disseminated by means of its resting sporangia which are windblown following their release from dry leaves of the infected plants. Meiosis occurs during the germination of resting sporangia that release numerous motile cells or meiospores. These infect corn leaves and quickly develop into epibiotic sporangia anchored to host cells by rhizoids. These motile cells, the sporangia, and the cells cleaved in the sporangia are thought to represent haploid phases of the life-cycle. Ephermal and thin walled, the sporangia from motile cells that Sparrow (1947) suggested were gametes. He was of the opinion that these cells fused in pairs to form zygote that reinfects host, giving rise to the diploid phase consisting of extensive, endobiotic, polycentric thalli. This has, however, not been confirmend. These thalli consist of rhizoids bearing terminal and intercalary swellings called TURBINATE CELLS. Resting sporangia develop from the ends of tubular outgrowths arising from the tips of turbinate cells.

L. Physoderma zea - maydis

1. Life-cycle (Fig. 3)

The fungus couses 'corn pox' or 'brown spot' disease of maize. The resting spores perennating in the soil or plant debris germinate duruing host crop season. They absorb water and swell. The circular slit appears in the wall and the inner wall protrudes out as a finger like structure called " endosporangium. The contents divide and form 20-25 posteriorly uniflagellate zoospores called "resting spore zoospores". Of the several resting spore zoospores that settle on the epidermal cells of the host leaves, some give rise to epibiotic zoosporangia while others to the endobiotic stage.

2. The Epibiotic Phase (Fig. 3)

(Monocentric, epibiotic or zoosporangial phase). The resting-spore zoospores present on the leaf epidermis, become flat and develop a wall. The wall on the dorsal surface is thicker than on the ventral side. Rhizoids are sent into the epidermal cells and the cyst develops into a sporangium. Its contents split and form zoospores. The zoospores are liberated by the dissolution of papilla. These are much smaller than the resting spore zoospores which is considered to be gametes by some mycologists.

After discharge of the zoospores, a basal sterile portion of the sporangium (at the point of origin of the rhizoids) enlarges and forms a new sporangium within the old one. This is called sporangial proliferation which may occur at least three times.

3. Endobiotic Phase

Some of the resting spore zoospores that settle down on the host epidermis after retracting the flagellum dissolve a pore and enter the epidumal cells as an amoeboid body. The first element established withih the host epidermal cells is the "Primary turbinate cells". It is broadly ovate to spinde shaped cell which by repeated transverse divisions becom many - celled. These cells are nucleated and

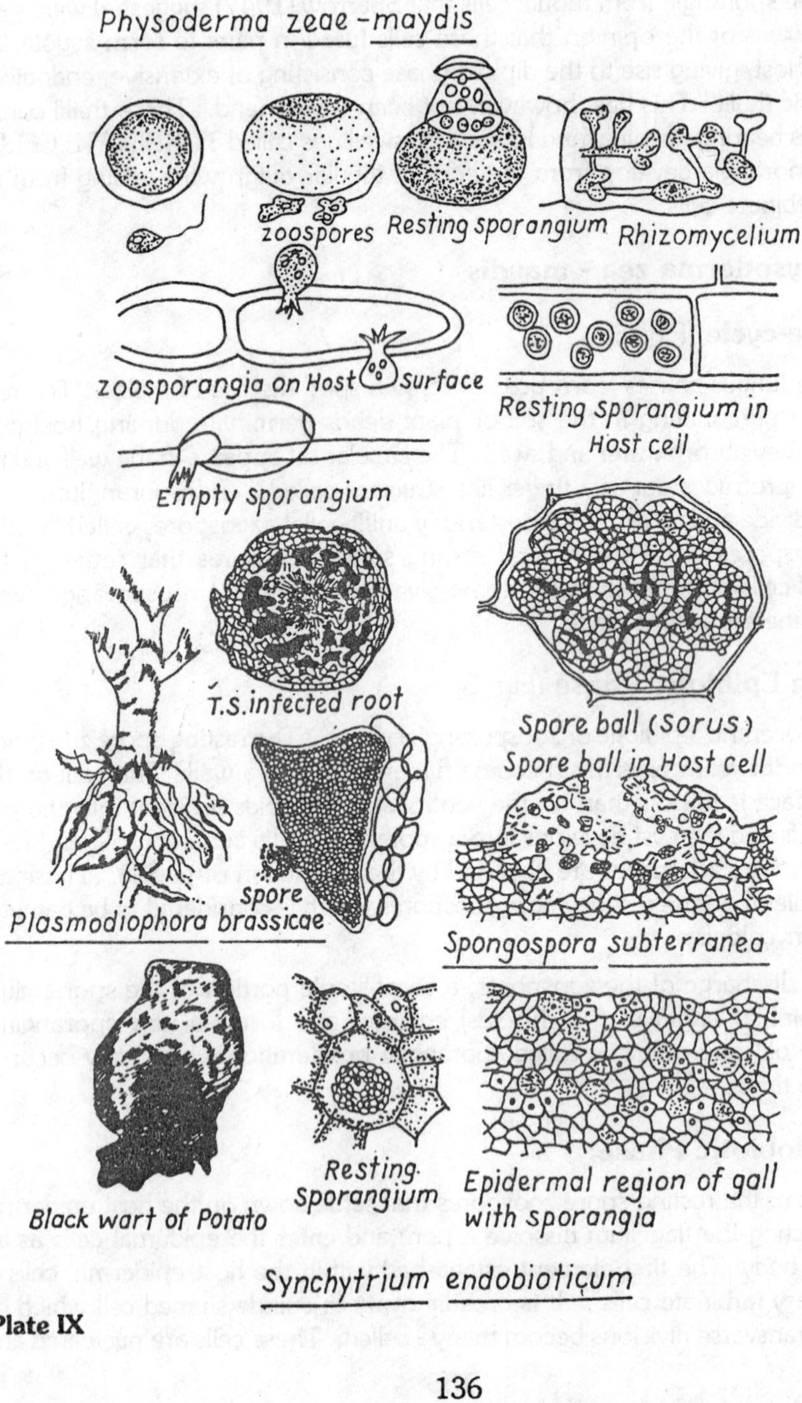
Physoderma zeae-maydis
zoospores
Resting sporangium
Rhizomycelium
zoosporangia on Host surface
Resting sporangium in Host cell
Empty sporangium
T.S. infected root
Spore ball (Sorus)
Spore ball in Host cell
spore
Plasmodiophora brassicae
Spongospora subterranea
Black wart of Potato
Resting Sporangium
Epidermal region of gall with sporangia
Synchytrium endobioticum

Plate IX

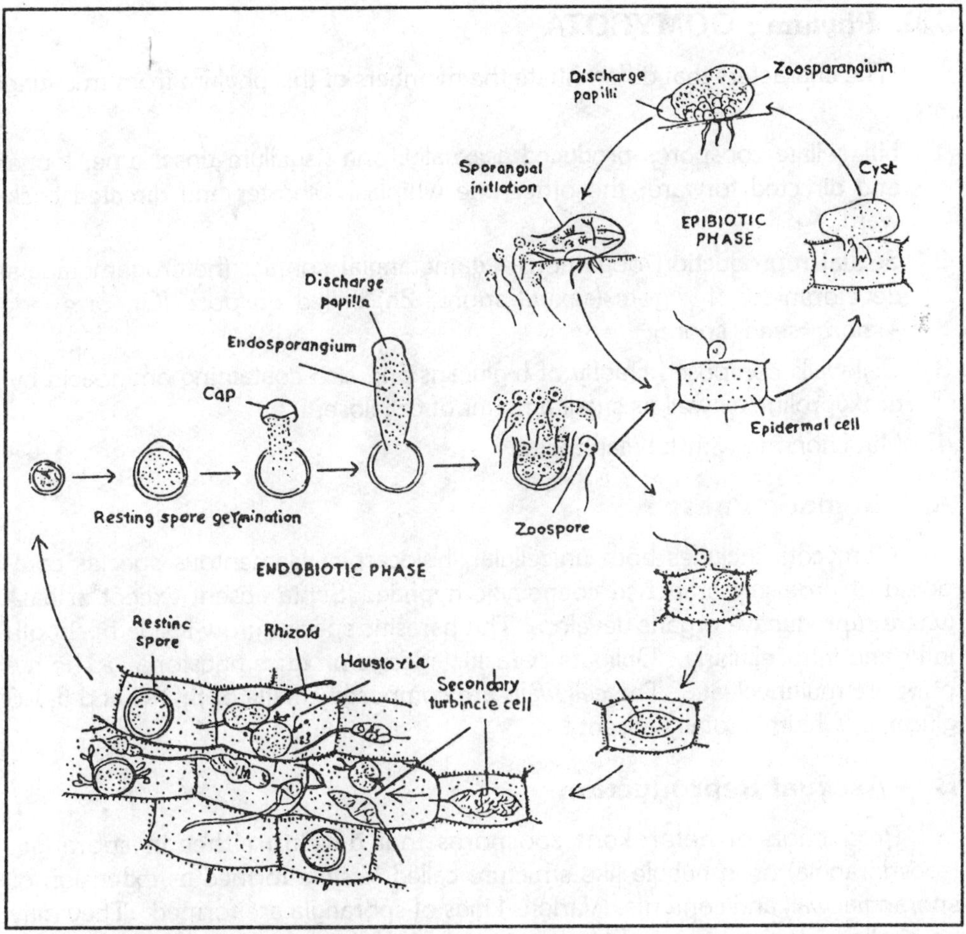
Discharge papilli
Zoosporangium
Sporangial initiation
Cyst
EPIBIOTIC PHASE
Discharge papilla
Endosporangium
Cap
Epidermal cell
Resting spore germination
Zoospore
ENDOBIOTIC PHASE
Restinc spore
Rhizoid
Haustoria
Secondary turbincie cell

Fig. 3 : Life cycle of *P. zea-maydis.*

mitotic divisions occur in them. Rhizoids emerge from these cells and spread to adjoining cells. Secondary turbinate cells are formed in many host cells by swelling of the rhizoids. Thus, the infection spreads to healthy cells. A short lateral outgrowth, the rudiment of resting spore, emerges from one of the turbinate cells where lip develops into a thin walled spherical body. It absorbs nutrients, and finally develops into a resting spore. The rhizoids act primarily to spread infection from cell to cell.

After the resting spore has attained full growth, a pigmented wall is laid down. Fully mature resting spores are broadly ellipsoidal and distinctly flattened at one face. Later the rhizoids disintegrate and wither, leaving the resting spores free inside the host cells. Planogametic fusion has not been reported.

XII. Phylum : OOMYCOTA

The characters that differentiate the members of this phylum from true fungi are

1. biflagellate zoospores produced asexually; one flagellum tinsel type, longer and directed forward; the other one whiplash, shorter and directed backward,
2. Sexual reproduction oogamous - gametangial contact (heterogametangia) development of zygote (sexual spore, 2n) called *oospore* (Gr. *oon*=egg +*spora*=seed, spore);
3. Cell walls composed chiefly of b-glucans and also coataining aminoacid hydroxyproline as well as small amount of cellulose;
4. Mitochondria with tubular cristae

A. Somatic Phase :

Oomycota includes both unicellular, holocarpic, filamentous species composed of profusely branched coenocytic hyphae. Septa absent except at base where reproductive organs develop. The parasitic species grow inside host both inter and intracellularly. Obligate parasitic species produce haustoria. The hyphae are multinucleate. The cell walls are composed chiefly of β, 1-3 and β, 1-6 glucans. Chitin is totally absent.

B. Asexual Reproduction

Production of heterokont zoospores that develop either in sporangia (zoosporangia) or in bubble like structure called *Vesicle* formed as extension of sporangial wall and contents. Various kinds of sporangia are formed. They may be swallen lobed portions of hyphae to spherical, oval, or lemonshaped. After zoospores release an other sporangium develops in some species and grow through the first sporangium. In the obligate parasites the sporangia resemble conidia and are produced at the tips of specialized, branched sporangiophores or in chains at tips of short clubshaped sporangiophores.

Two morphologically distinct types of biflagellate zoospores are produced in oomycetes. The one type is termed the *Primary zoospore* and is pyriform (pear shaped) with flagella attached at the anterior end of the spore. *Primary zoospores* are poor swimmers and thus considered primitive. The second type is termed *Secondary zoospores*. It is reniform (Kidney or bean shaped) with flagella inserted laterally in a groove on the spore surface. The flafella are directed away from one another at an angle of about 130^0, with the longer tinsel flagellum having flagellar hairs direated forward and the shorter whiplash flagellum trailing behind. Regardless of shape, each zoospore typically contains a single-pear shaped nucleus. Near the tapered end of the nucleus lies a pair of kinetosomes (or basal bodies) from which the flagella arise. The function of zoospores in the oomycetes life-cycle is to swim short distances through water, find substantial substrates or hosts, eneyst, and ultimately germinate by forming germtubes that subsequently give rise to new thalli. Various patterns of swimming and encystment are exhibited by different species and are important characters in the identification of certain species.

Not all oomycetes produce zoospores. Zoospores appear to have been eliminated from the life - cycle of some species. In addition, some species that are obligate parasites of angiosperms tend to produce zoosporangia that conidium like in appearance and tend to germinate by germ tubes rather than zoospores.

C. Sexual Reproduction

Sexual reproduction in oomycetes is almost always heterogametangic. In most species, the gametangia typically are differentiated into small hyphal-like male structrues termed *antheridia* and larger, globase female structures called oogonia. Both the structures may develop from the same thallus or from different thalli. Gametangia produced on the same thallus may or may not be compatible. Following meiosis one or more nonmotile eggs or *oospheres* (Gr. *oon*=egg + *spora* = spore/seed), develop within each oogonium. Depending upon the species, oospheres are differentaiated either by a centrifugal cleavage process or by the centripental aggregation of peripheral oogonial cytoplasm known as *periplasm* (Gr. *peri*=around +*plasma*=protoplasm), cytoplasm that is excluded from the eggs. At maturity each oosphere contains a prominent storage vacuole termed an *ooplast*, and one or more nuclei. Developing antheridia are attracted to oogonia by hormones and give rise to *fertilization tube*. A haploid nucleus resulting from meiosis in the antheridium is introduced into the oosphere via the fertilization tube and fuses with the nucleus of the oosphere, following fertilization and oosphere develops into an oospore that matures in the oogonium. oospores are thickwalled, resistant structures capable of surviving unfavourable environmental conditions. The mature oospore wall consists of three layers. These include an exospore layer, an epispore layer or zygote wall, and an endospore layer.

D. Classification: The phylum *oomycota* contains the single class *oomycetes*. Dick (1990) has recognized six orders. They are *Leptomitales, Rhipidiales, Sclerosporales, Pythiales, Peronosporales* and *Saproleganiales*. Alexopoulos etal (1996) have however, followed troditional approach and described five orders. They are *Saprolegniales*, *Rhipidiales*, *Leptomitales*, *Lagenidiales*, and *Pesonosporales*.

E. Order - PERONOSPORALES

This large order includes aquatic, amphibious, and terrestrial species, culminating in a group of highly specialized obligate parasites(Plate-X). Many species of this order are destructive parasites of economic plants.

1. Somatic Structures

The mycelium of Peronosporales is well developed, consisting of coenocytic slender hyphae that branch freely. The hyphae of parasitic species are intercellular or intracellular, those of most specialized parasites growing between host cells and producing houstoria of various shapes.

2. Asexual Reproduction

In some species of *Peronosporales* the sporangia are borne on ordinary somatic hyphae and remain attached even after the zoospores have been released. In the more derived types the sporangia are borne on sporangiophores and are deciduous upon maturity, depending on the wind for dissemination. In this respect, the whole sporangium acts as a spore and, in most specialized species, actually germinates by germ tube instead of producing zoospore. The majority of species, however, produce zoospores, which are reniform and biflagellate. The zoospores are formed as a result of the cleavage of the contents of the zoosporangium into uninucleate segments. Upon their release the zoospores swarm for sometime, come to rest, encyst and germinate by germ tube that develops into mycelium. *Diplanetism* (Gr. *dis* = twice + *planetes* = wanderer) and *polyplanelism* (Gr. *poly* = much/many + *planetes*), however, have been reported in a few species.

3. Sexual Reproduction

Peronosporales reproduce sexually by means of well differentiatted oogonia and antheridia borne on the same or on different hyphae. Meiosis takes place in the gametangia resulting in the formation of haploid oospheres and antheridial nuclei. The globose oogonia contain, with few exceptions, a single uninucleate or multinucleate, depending on the species. When gametangial contact is affected, a fertilization tube is formed by the antheridium, pushes through the oogonial wall and the periplasm, and reaches the oospeher. The male nucleus or nuclei then pass through the fertilization tube and are shed into the oosphere. If the oosphere is uninucleate, a single male nucleus fuses with the female nucleus and forms the zygote. It the oosphere is multinucleate, one or more of its nuclei may be functional, and the number of male nuclei that will affect fertilization is regulated accordingly. Thus, either a single zygote nucleus or a number of nuclei may result from a simple or multiple fertilization.

After fertilization, the oosphere develops a thick wall and becomes a oospore. The periplasm serves as nourishment for the developing oospore. It is aiso responsible for the deposition of external thickenings and ornamentations on the oospore walls of some species. The oospore wall consists of an outer, a middle, and an innerlayer. The outer layer may be smooth or variously sculpted or ornamented; it may be spiny, warty, wavy ridged or otherwise marked. The mature oospore generally lies free within the oogonial wall, filling it (plerotic, Gr. *Pleres* = full) or not (aplerotic, Gr. *a* = not + *pleres*), but in many species adhereing to it so closely that the walls appear to be united. Only in the genus *Sclerospora* is the oogonial wall actually fused with the oospore wall. After over wintering, the oospores germinate either by giving rise to zoospores, thus behaving as zoosporangia, or by putting out germ tubes that soon afterward produce sporangia. This type of germination varies with species.

4. Classification

The classification of *Peronosporales* is based mainly on the characters of sporangia and sporangiophores. The order is divided into three well defined families — *Pythiaceae*, *Peronosoraceae* and *Albuginaceae*.

5. Family Pythiaceae

Members of this family generally bear their sporangia directly on the somatic hyphae. In some species, the fertile hyphae are no different than the somatic hyphae. The more specialized species produce well defined and recognizable sporangiophores that are of indeterminate growth. This means that sporangiophares contunue to grow indefinitely, producing sporangia as they grow.

This family includes aquatic, amphibious, and terrestrial fungi, most of the last causing deseases of economic significance. In the simplest forms the sporangia remain attached to the hyphae that bear them. Upon maturity they produce and liberate zoospores. In specialized species the sporangia are deciduous and often germinate by germtube instead of producing zoospores. The type of germination, whether by zoospores or by germ tube, appears to be governed to a great extent by environmental conditions, especially by temperature. Sexually, members of *Pythiaceae* conform to the general pattern described for *Peronosporales* as a whole. Dice (1990) has included seven genera in the family. The most common genera are *Pythium* and *Phytophthora*.

Genus - *Pythium*

Members of the genus *Pythium* are soil-inhabitants and occur worldwide. They cause seed rots, damping-off, root rots and fruit rots. Important species are

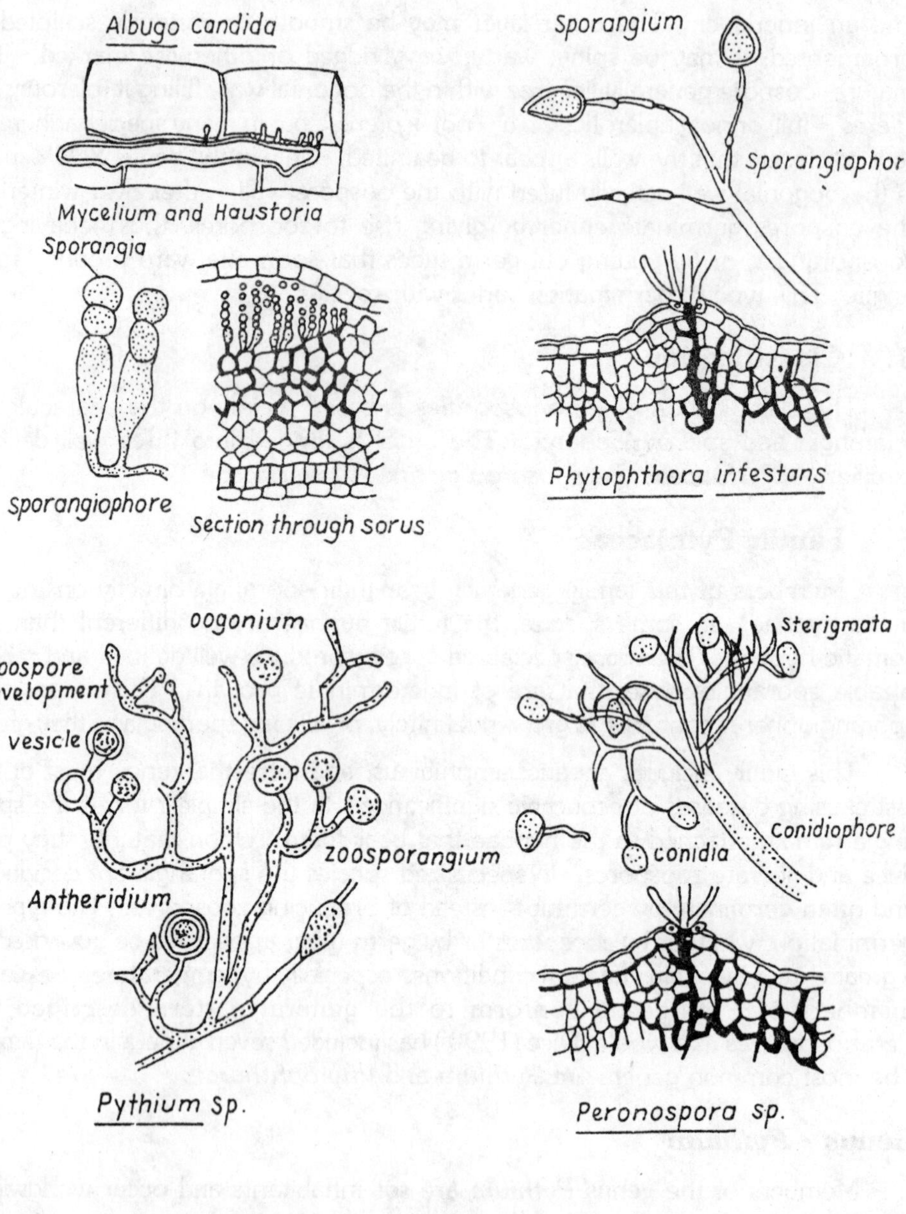
Albugo candida
Mycelium and Haustoria
Sporangia
sporangiophore
Section through sorus
Sporangium
Sporangiophor
Phytophthora infestans
oogonium
zoospore Development
vesicle
Antheridium
zoosporangium
Pythium Sp.
Sterigmata
Conidiophore
conidia
Peronospora sp.

Plate-X : *Oomycetous fungi*

P aphanidermatum, P debaryanum, P insidiosum, P graminicolum, P irregularae etc.

Life-cycle (Plate-XI) (Fig. 4)

The mycelium of most species consists of slender, coenocytic hyphae. It survives in the soil saprophytically on crop refuge or parasitically on the young seedlings of several seed plant species. In host tissues the hyphae are both inter and intracellular. The fungus reprodues repeatedly asexually by forming globose to oval sporangia on the somatic hyphae. They may either be terminal or intercalary. Sporangia remain attached to the hyphae and germinate at place. Germination is either zoospores or by a germtube. Production of zoospores is preceded by the formation of a bubble like *vesicle* at the tip of a tube that issues from the sporangium.

The sporangial protoplast flows into the vesicle through the tube, and differentiation of zoospores takes place in the vesicle. Eventually the vesicle breaks and the zoospores swim away. After a period of swimming, the zoospores encyst, and germinate by germ tubes. Polyplanetism, however, has been reported

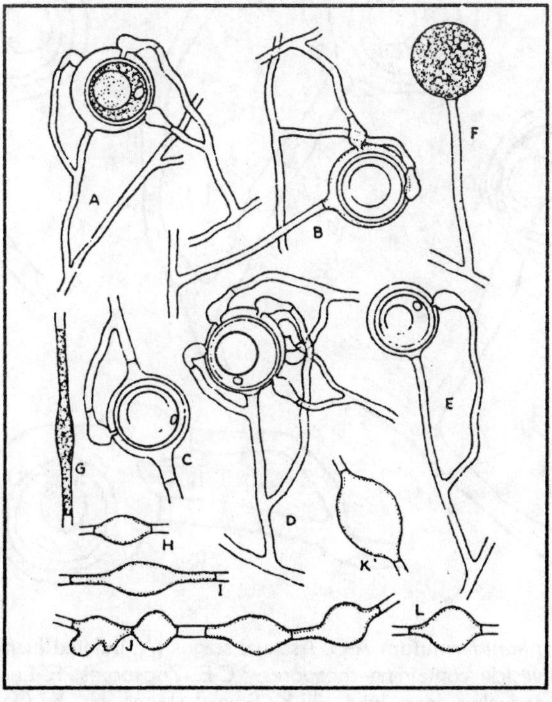
A
B
C
D
E
F
G
H
I
J
K
L

Fig. 4 : Pythium debaryanum. A-E, Typical sexual stage, F-L, Sporangial types

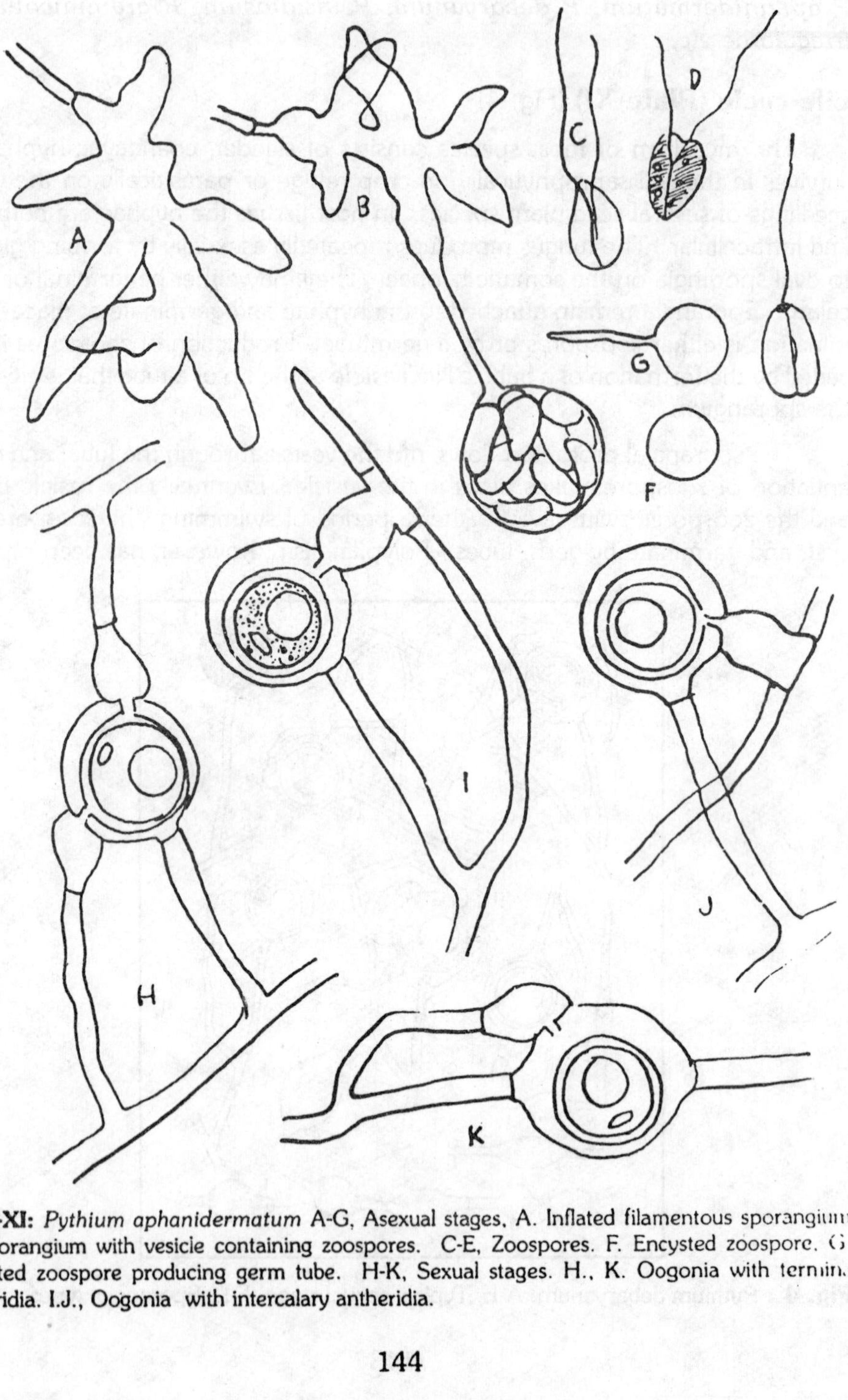
A
B
C
D
G
F
H
I
J
K

Plate-XI: *Pythium aphanidermatum* A-G, Asexual stages. A. Inflated filamentous sporangium B, Sporangium with vesicle containing zoospores. C-E, Zoospores. F. Encysted zoospore. G. Encysted zoospore producing germ tube. H-K, Sexual stages. H., K. Oogonia with terminal antheridia. I.J., Oogonia with intercalary antheridia.

in some members of the genus. Most species of *Pythium* are homothallic, several heterothallic are also known.

The sexual reproduction is oogamous. Antheridia and oogoria are formed in close proximity, often on the same hyphae with the antheridium (small, elongated/clubshaped) just below the oogonium which is globose. The oogonium contains multinucleate oosphere surrounded by periplasm. Upon gametangial contact, a fertilization tube develops and penetrates the oogonial wall and the periplasm. In the mean time meiosis has taken place in both gametanigia, and all but one functional nucleus in each have disintegrated. The male nucleus now passes through the tube into the oosphere, approaches the female nucleus, unites with it, and forms the zygote. The oosphere develops into a thick-walled, more or less spherical,unornamented oospore which germinates after a rest period. At high temperatures (28°c) the oospore germinates by a germ tube, which develops into mycelium. However, at lower temperature (10-17°C) , the germ tube stops growing after it has attained a length of 5-20 µm, and the protoplast of the oospore migrates through the tube, pushes out through the tip, and forms a vesicle in which zoospores develop.

Genus : *Phytophthora*

Phytophthora species cause a wide variety of diseases on a large number of hosts. *P. infestans* which causes late blight of potato, contributed directly to the "Great Irish Famine" in Ireland in 1845 and 1846. This resulted in the deaths of perhaps as many as 1.0 million Irish people and the emigration of up to 1.5 million people principally to North America.

The chief distinctions between *Pythium* and *Phytophthora* are the formation of indeterminate sporangiophores and sporangial germination. In general, no vesicle is formed in *Phytophthora*, or if one is formed, the zoospores differentiate in the sporangium proper and pass into the vesicle as mature zoospores and subsequently released. Important species and diseases caused by *Phytophthora* include *P. infastans* (late blight of potato, tomato), P. *nicotianae* var *nicotiane* (black shank of tobacco), P. *nicotianae* Var. *parasitica* (damping off of several crops/plants), *P parasitica* var *sesami* (leaf blight of sesamum), *P. citrophthora* (citrus gummosis), *P. colocasiae* (colocasia blight), *P cinnamomi* (avocado root rot), *P syringae* (collar and fruit rot of temperate trees) *P. mesasperma* var *megasperma* and var. *sojae* (root and stalk rot of soybean), *P. drechsleri* var *cajni* (stem blight of pigeonpea) etc.

Somatic structure

The mycelium is non-septate producing branches often constricted hyaline and coenocytic. Septa are formed at the time of sex organ formation. Sometimes old hyhae also develop septa. The mycelium may be **intercellular** or **intracellular**. The **haustoria** are formed by the intercellular mycelium and absorb food material from the host tissue. In *P. infestans* within potato tubers the haustoria are finger-like protuberances which may be in part surrounded by thickenings of host wall material. Electron micrographs of haustoria in potato leaves, however, show that the haustoria are not surrounded by host cell wall. They are surrounded by an encapsulation whose origin and function are not certain. In *P. colocasiae* the haustoria remain slender and unbranched. The walls of the hyphae show a similar constitution as found in *Pythium* which lack chitin and are composed mostly of glucans.

Asexual Reproduction

It takes place by means of chlamydospore and sporangia.

(i.) **Chlamydospores** They are produced as additional asexual bodies. In shape, they are spherical, ovoid or slightly irregular. They may be hyaline, deep brown but usually **lemon-coloured**. They are thick-walled bodies produced either **acrogenously** on short lateral hyphae or **intercalarily**. Each chlamydospore germinates to form 3-11 germ tubes which bear sporangia on their tips.

(ii.) **Sporangia** The sporangiophores (Plate-XII) in *Phytophthora* may be of following types:

(a) compound with ultimate sporangium bearing branch a monochasial sympodium (Gr. *monos*-one, *khasis*-separation, *syn*-with, *pod* a foot). This type of sporangiophore is seen in *Phytophthora infestans* and *P. phaseoli*.
(b) simple consisting of a single monochasial sympodium such as in *P.cactorum*,
(c) Proliferating such as in P. megasperma.

In the plant parasitic forms the sporangiophores emerge through the epidermis of the host, piercing it or passing through stomata. The zoosporangia of the soil inhabiting types may be submerged and then remain persitent but in association with the host the zoosporangia form in air on the sporangiophore and are detached probably due to hygroscopic twisting of the drying sporangiophore.

The sporangia are colourless, usually terminal but sometimes intercalary especially in species with a pipillate sporangium. The shape varies from ovoid to pyriform (pear shaped). The papilla may be inconspicuous or very prominent. Presence of apical papilla on the sporangium is a characteristic feature of *Phytophthora*. The mode of germination of sporangia in *Phytophthora* varies not only within the species but also in the same individual depending on environmental conditions. The main types of germination are (i) by germ tube and (ii) indirect by zoospores. When the zoospporangium falls in water (film of water on

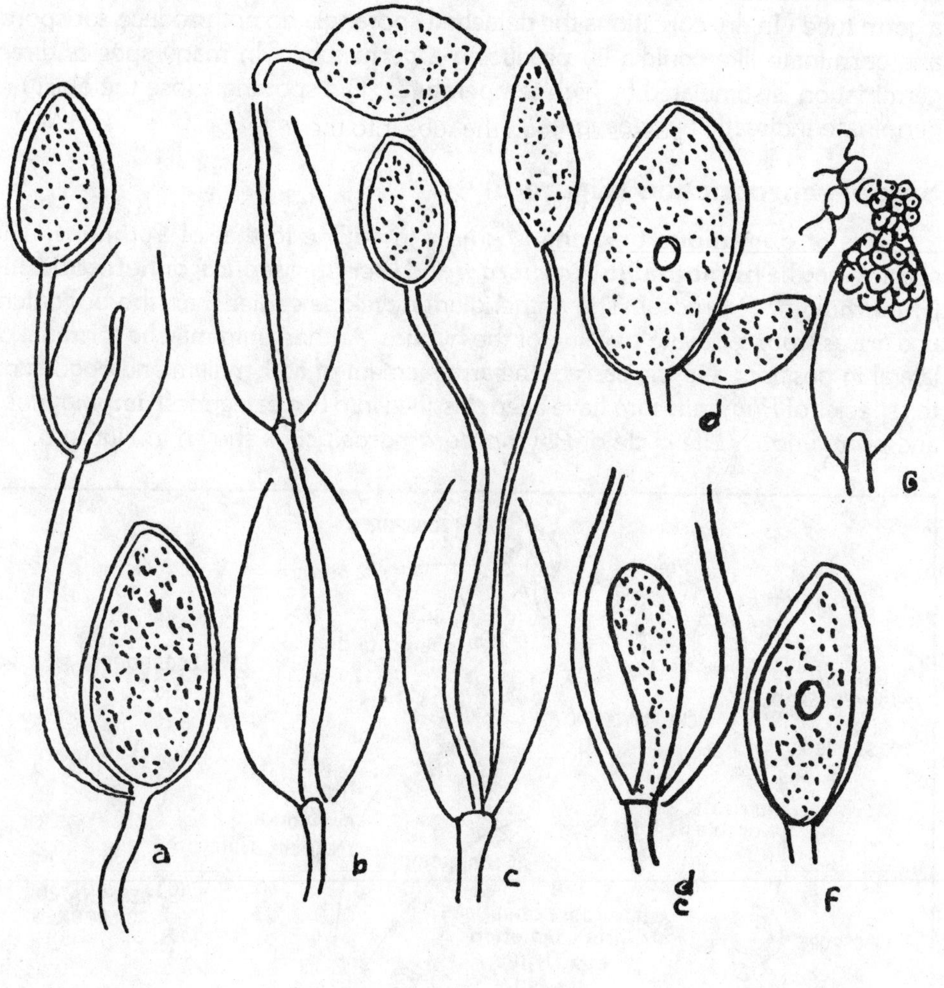
a
b
c
d
d
c
e
f

Plate-XII: Different stages of sporangial development of *Phytophthora cinnamomi.*

host surface) it germinates by producing zoosppores. The zoospores are formed within the zoosporangium by cleavage of its contents. They escape through the apical papilla. In rare cases they pass into a vesicle and then escape by rupturing its wall. Of the two flagella the shorter anteriorly directed one is tinsel type and the longer posteriorly directed one is whiplash type. The zoospores are usually uninucleate. After swimming for a while the zoospores encyst and germinate by a germ tube. In dry conditions the detached sporangia do not produce zoospores and germinate like conidia by producing a germ tube. In many species direct germination is stimulated by high temperature. Old sporangia lose the ability to germinate indirectly by zoospores.

Sexual reproduction (Figure-5)

It is of **oogamous** type and is similar in outline to that of *Pythium*. The species may be **homothallic** (*P.cactorum* and *P. erythroseptica*) or **heterothallic** (*P. palmivora, P. arecae)*. The antheridium develops earlier than the oogonium and arises from a clavate swelling of the hypha. Antheridium may be terminal or lateral in position. On the basis of the arrangement of antheridium and oogonium. the species of *Phytophthora* have been classified into two categories: *amphigynous* and *paragynous*. Life cycle of *Phytophtora parasitica* is shown in Figure 5.

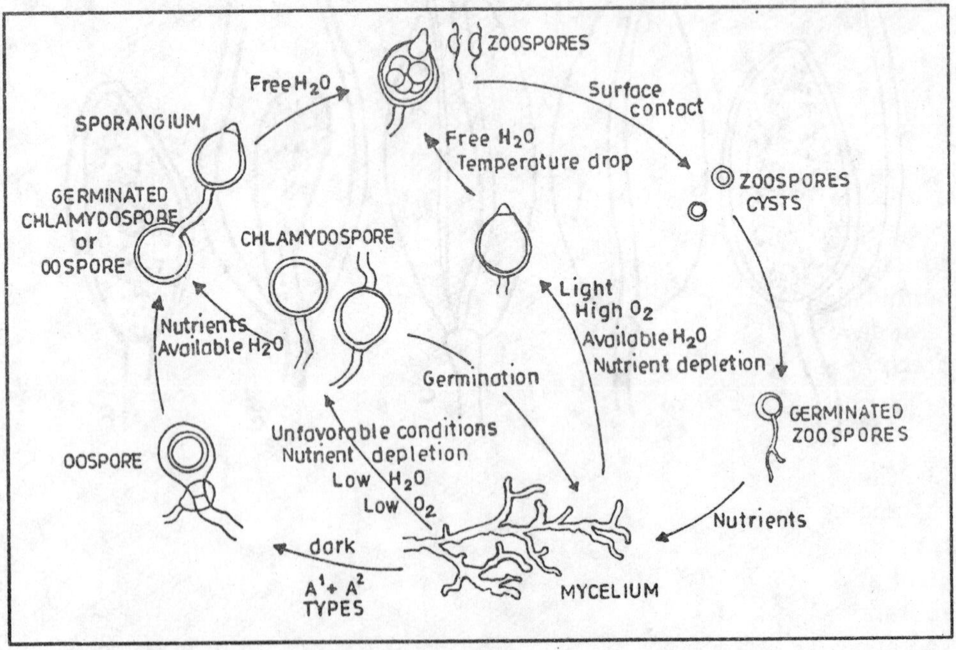
ZOOSPORES
Free H_2O
Surface contact
SPORANGIUM
Free H_2O
Temperature drop
ZOOSPORES CYSTS
GERMINATED CHLAMYDOSPORE or OOSPORE
CHLAMYDOSPORE
Light
High O_2
Available H_2O
Nutrient depletion
Nutrients
Available H_2O
Germination
GERMINATED ZOOSPORES
Unfavorable conditions
Nutrient depletion
Low H_2O
Low O_2
OOSPORE
Nutrients
dark
$A^1 + A^2$
TYPES
MYCELIUM

Fig. 5 : Life cycle of *Phytophthora parasitica*

A fully developed antheridium includes one or two nuclei. In the early stages of development the antheridium is multinucleate structure but later on one or two nuclei persist and the rest degenerate. The oogonium (in amphigynous species) develops from a lateral branch of hypha piercing through antheridium, swells and becomes large globular structure. The oogonium is a multinucleate structrue and contains dense protoplasm. The oogonial cytoplasm differentiates into an outer vacuolated periplasm and an inner dense ooplasm. The nuclei of the peripheral region gradually degenerate while the nucleus of the ooplasm divides to form two nuclei. One of the nucleus thus formed degenerates and the other functions as an egg nucleus.

At the time of fertilization, mature antheridium forms a number of papillae but only one papila grows further and the tip becomes swollen which comes in close contact with the ooginium . The wall at the point of contact dissolves and a pore is formed. The male nucleus which has since migrated from the antheridium to the papilla is delivered into the oogonium. Both male and female nuclei fuse and form an oospore. In paragynous species the fertilization takes place in the same manner as described for amphigynous species. The oospore in majority of the species is aplerotic as it does not fill the oogonium cavity entirely. A mature oospore has an outer exospore and an inner endospore composed of cellulose, protiens and other reserve substances and is thick.

The oospores of *Phytophthora infestans* are rare and may not play a significant role in the survival or propagation of the fungus. In Mexico , they have been reported on potato leaves. In India the fungus probably over winters in the form of mycelium in infected tubers. Germination takes place after the oospores undergo a resting period and may survive in soil for long periods. The oospores germinate by the cracking of the outer wall and inner wall forms a tube which bears sporangium at the tip. In *P.cactorum* oospores exihibit a long dormacy period and upon germination form a new mycelium. It is believed that before germination, nuclear fusion might have occured and the fused nucleus might have divided several times. There is now evidence that in *P. cactorum* and *P. drechsleri* meiosis occours in the process of gamets formation .

6. Family: Peronosporaceae

Members of Peronosporaceae have sporangia borne on unmistakable sporangiophores that are characterstically branched and of determinate growth. No sporangia are produced until the sporangiophores completes the development and matures. Then a single crop of sporangia is produced, with all the sporangia approximately the same age. After the sporangia fall off, the sporangiophores wither and die. The type of branching exhibited by the sporangiophores of the species of Peronosporaceae serves as the chief distingwishing feature of the genera in this family. Priplasm in the oospores is quite conspicuous. All the members of the family are considered to be obligate parasites of plants. Another taxonomic character is the mode of germination of spores. In *Peronospora* and *Bremia* and certain species of *Sclerospora* the germination is typicaliy by germtube while in others it is mostly by zoospores.

Sexual reproducion is like that in Albuginaceae. The oogonia are larger and the antheridia broader in comparison with the Pythiaceae. The ogonial wall is thin or not much thickened and unornamented except few species. Mature oogonium contains a single uninucleate egg. The residual protoplasm (periplasm) envelops the oospore, forming a thick looking dark, often folded layer of wall. The oospore germinates by a germtube or by the formation of a sporophore terminated by a single, large conidium or sporangium. Sometimes they may directly form zoospores.

Genera of Peronosporaceae (Figure-6 - 9)

1. ***Sclerophthora*** : Sporophores determinate, short hypha like, unbranched or sympodially branched, sporangia citriform or obpyriform, not maturing synchronously, germinating by zoospores: antheridia always paraynous; oogonial wall thick and confluent with that of the oospores; oospores germinate by germ tube or a sporophore terminated by a sporongium. Important species are *S. macrospora* (occurs on 46 species belonging to 31 genera of the graminae; S. rayssiae var. zea (brown stripe downy mildew of maize).

2. ***Basidiophora*** : Sporophores determinate, unbranched, apex swollen and with short sterigmata bearing papillate sporangia germinating by zoospores, and oospores aplerotic.

3. ***Sclerospora(Figure-8(b))*** : Sporophores determinate, repeatedly branched in the upper portion, dichotomous, spores mature synchronosusly, oogonial wall thick, oospores plerotic, sporangia germinate by zoospores or germtube, oospores germinate by germtube.

4. ***Peronospora*** : Sporophores determinate, narrow, dichotomously branched at the apex, conidia non-poroid, germinate typically by germtube, oogonial wall unornamented.

5. ***Plasmopara(Figure-8(a))*** : Sporophores determinate, branched at right angles, tips of branches blunt, spore wall poroid, germination by zoospores or plasma emerging through an apical pore with or without papilla.

6. ***Pseudoperonospora*** : Sporophore branching dichotomous at acute angle and tips of branches acute.

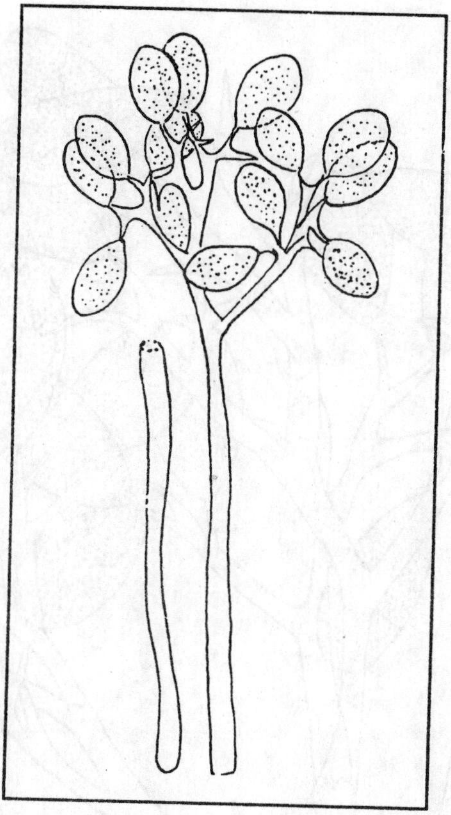

Fig. 6 : Sporangiophore of *Schlerophthora sp*

7. ***Bremia* :** Tips of the branches much enlarged and bearing 3-4 peripheral sterigmata, oogonial wall and oospore wall thin and unornamented.

8. ***Bremiella* :** Tips of branches blunt, and slightly enlarged, oogonial wall thick and ornamented.

Genus : *Sclerospora*

The genus *Sclerospora* now contains only those species in which macronemous sporophores (distinct from vegetative mycelium) with determinate growth are formed. If the method of sporangial germination is taken as criterion for classification, the genus can be broken into 2 sub-genera (Shaw, 1970), viz, *Eusclerospora* and *Peronosclerospora* with gemination of sporangia typically by zoospores in the former and by germ tube in the latter.

All species of *Sclerospora* attack only Gramineae. Some are host specific

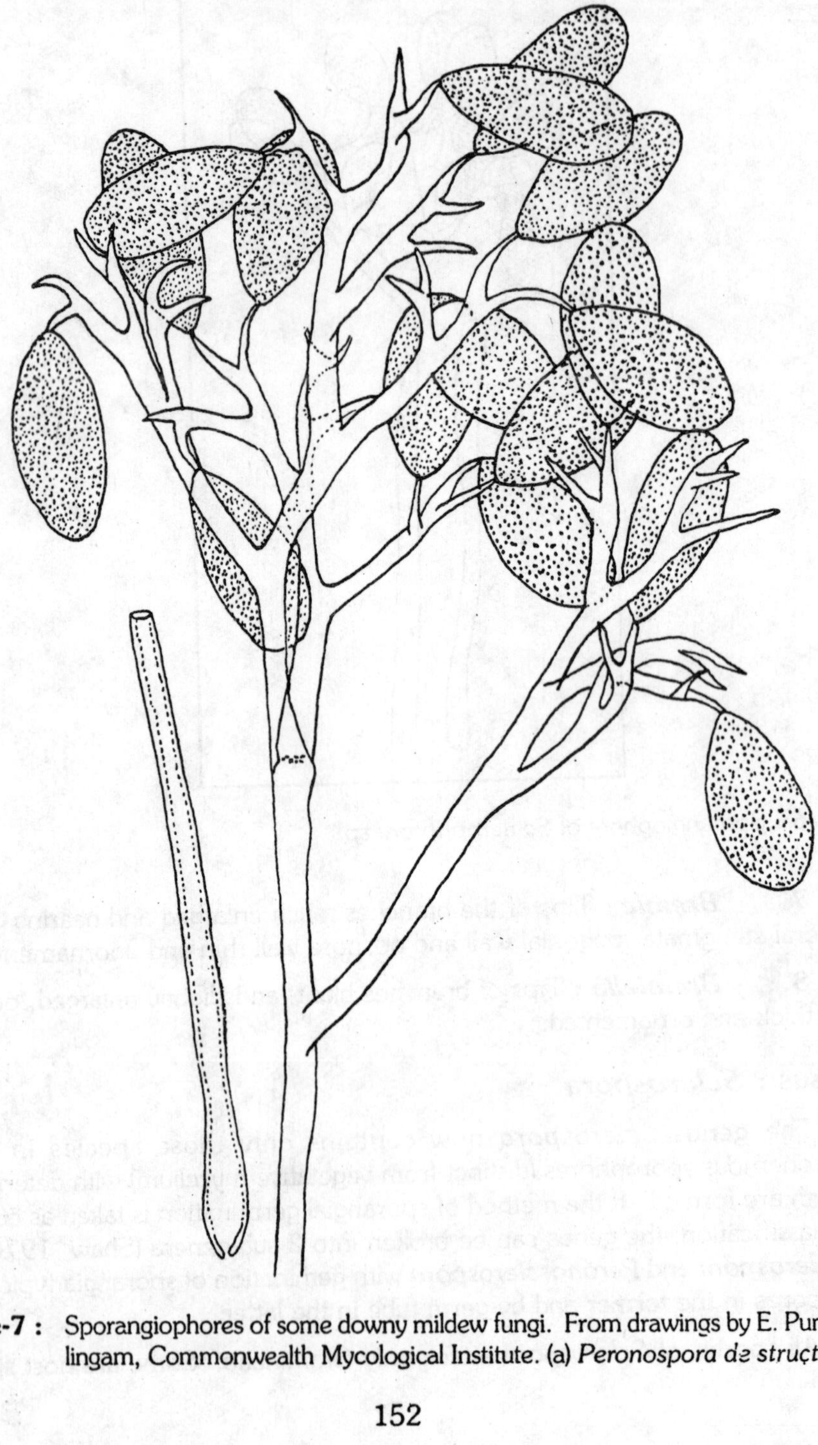

Figure-7 : Sporangiophores of some downy mildew fungi. From drawings by E. Punithalingam, Commonwealth Mycological Institute. (a) *Peronospora destructor.*

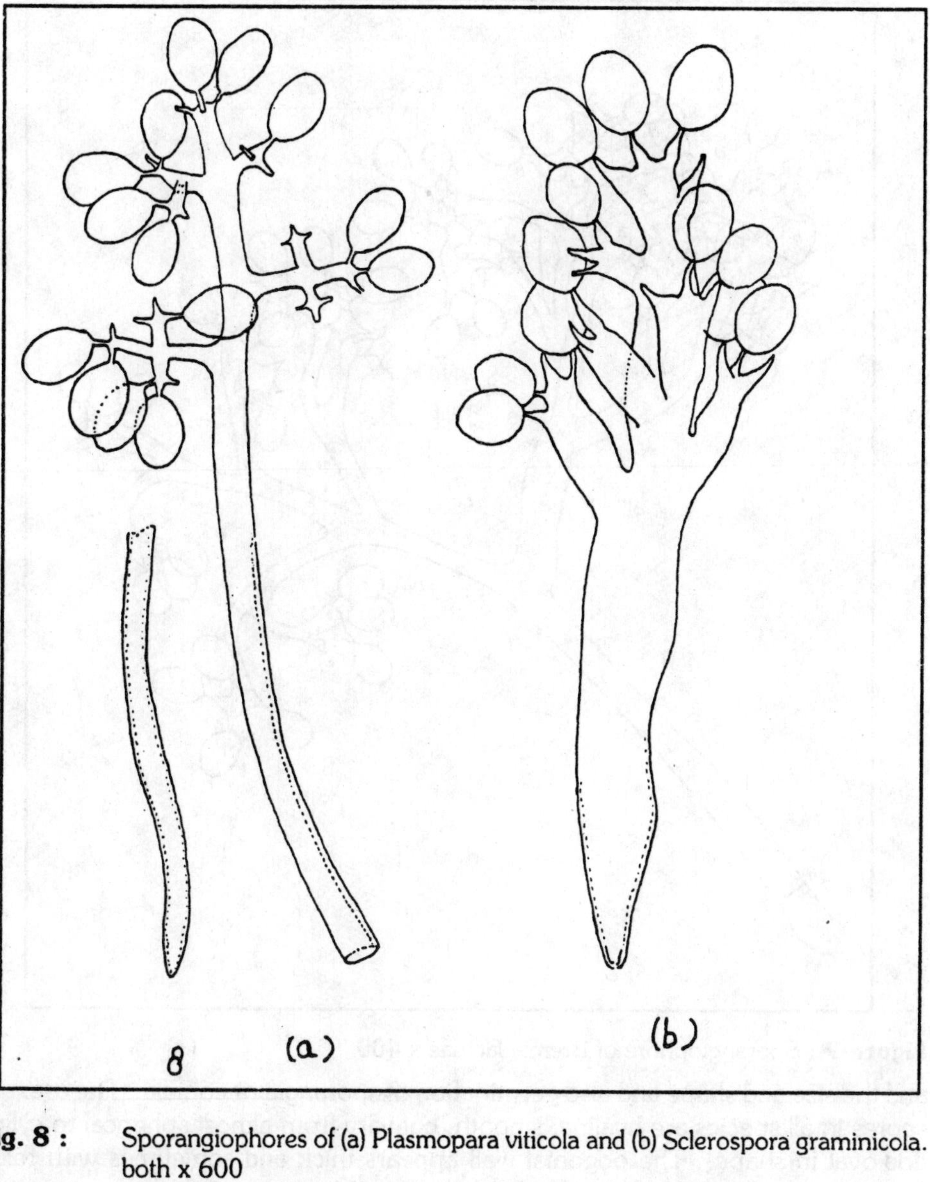
8
(a)
(b)

Fig. 8 : Sporangiophores of (a) Plasmopara viticola and (b) Sclerospora graminicola. both x 600

but those restricted to maize-sorghum-sugarcane-millet group exhibit cross infectivity, although morphologically distinct on the basis of conidial and oogonial characters.

The species of *Sclerospora* can be distinguished by their conidiophores which have a basal cell differing in length and bulbosity of the base, length of sterigmata

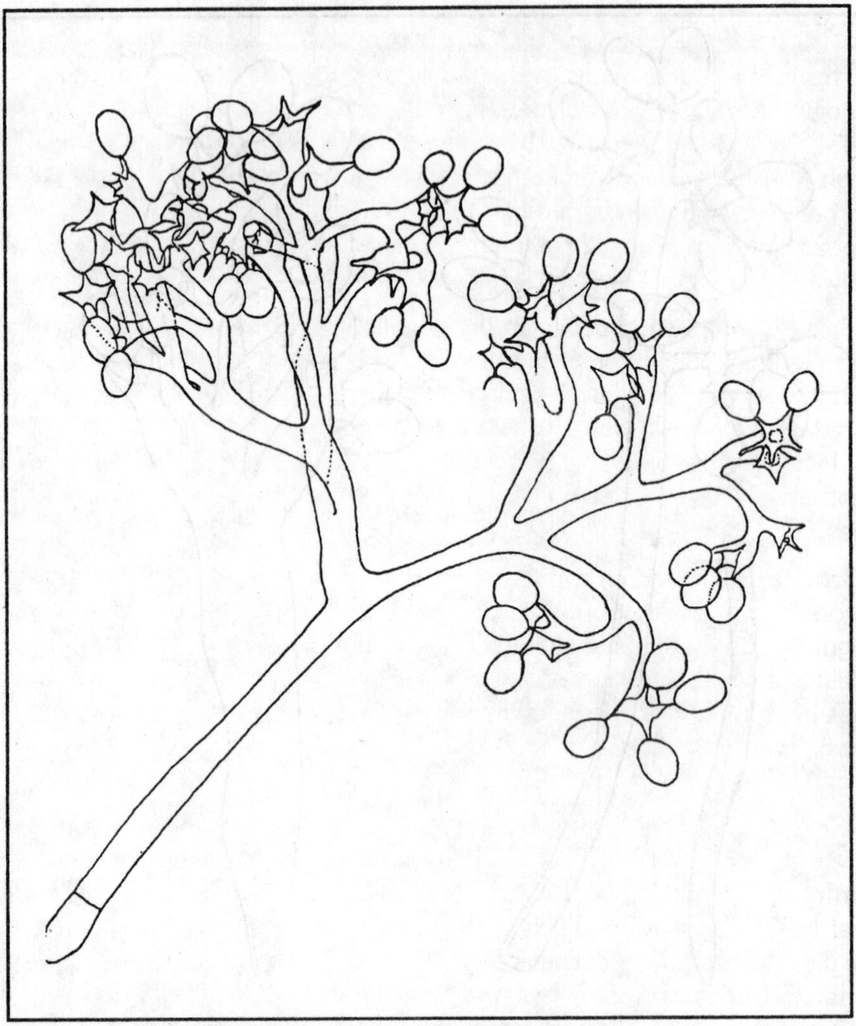

Figure-9 : Sporangiophore of Bremia lactuae x 400

and the size and shape and also germination of sporangia or conidia. The asexual spores in all species are hyaline, smooth, but vary from almost spherical to cylindric-oval in shape. The oogonial wall appears thick and sometimes with folds (ornamentation). The oospores are produced in deeper tissues of the host, mostly mesophyll, and germinate, with difficulty, by a germtube.

Important species and the diseases caused by them include *S. graminicola* (downy mildew of pearl millet), *S. maydis* (downy mildew of maize), *S. sacchari* (downy mildew of sugarcane and maize), *S. phillippinensis* (downy mildew of maize and sorghum) and *S. sorghi* (downy mildew of sorghum).

Genus : *Peronospora*

This is the largest genus in the family, recorded on many dicotyledons and few monocotyledons. Species of the genus are typically conidial members of the family. They are characterized by long slender conidiophores which emerge through the stomata and remain unbranched for a considerable length then dichotomously branched several time. The final branches (Sterigmata) taper to a thin point and bear a single conidium. The conidia are oval or nearly spherical and always germinate by a lateral germtube.

Sexual reproduction occurs in the host tissues. Antheridia and oogonia are at first plurinucleate but only one male nucleus and one female nucleus participate in fertilization. The nuclei lie in pair and fuse only after the oospore wall has matured. The oospore wall is usually hard and germination by a germtube is the rule. However, in some species, such as *P. tabacina*, zoospores are reported. Homothallism, heterothallism and intermediate conditions are reported in many species.

Taxonomy of *Peronospora* has gone through many changes. Conidiophores and conidia are of taxonomic value. At least 15 types of conidiophores are distinguished on the basis of stiffness or curvature of branches, density of branching, features of the sterigmata, and mode of germination of conidia. Important species include *P. parasitica* (downy mildew of crucifers), *P. pisi* (downy mildew of peas), *P. destructor* (downy mildew of onion), and *P. effusa* (downy mildew of spinach).

Life - cycle (*Peronospora tabacina*) (fig. 10)

Infection is initiated by the air borne, thin walled conidia, which germinate on wet leaf surfaces. The germtube may grow either through stomata or penetrate directly into the epidermal cells. The coenocytic hyphae then grow within the inter cellular spaces of the leaf, producing branched haustoria within the cells. After penetration two developmental phases may be distinguished : 1. the expansion phase, during the first 48 hrs. after penetration; and 2. the growth phase, between 3 to 7 days after infection (Figure-10)

In the growth phase, the parasite greatly enlages its surface in contact with the host which culminates in the formation of a large number of conidia (as many as 1.5 million/sq cm of leaf area) produced on conidiophores emerging from the end of first cycle. The conidia are generally produced at night and are disseminated during morning. They are the primary means of spreading the disease. The asexual cycle takes 5 to 10 days and may be repeated several times during the growing season.

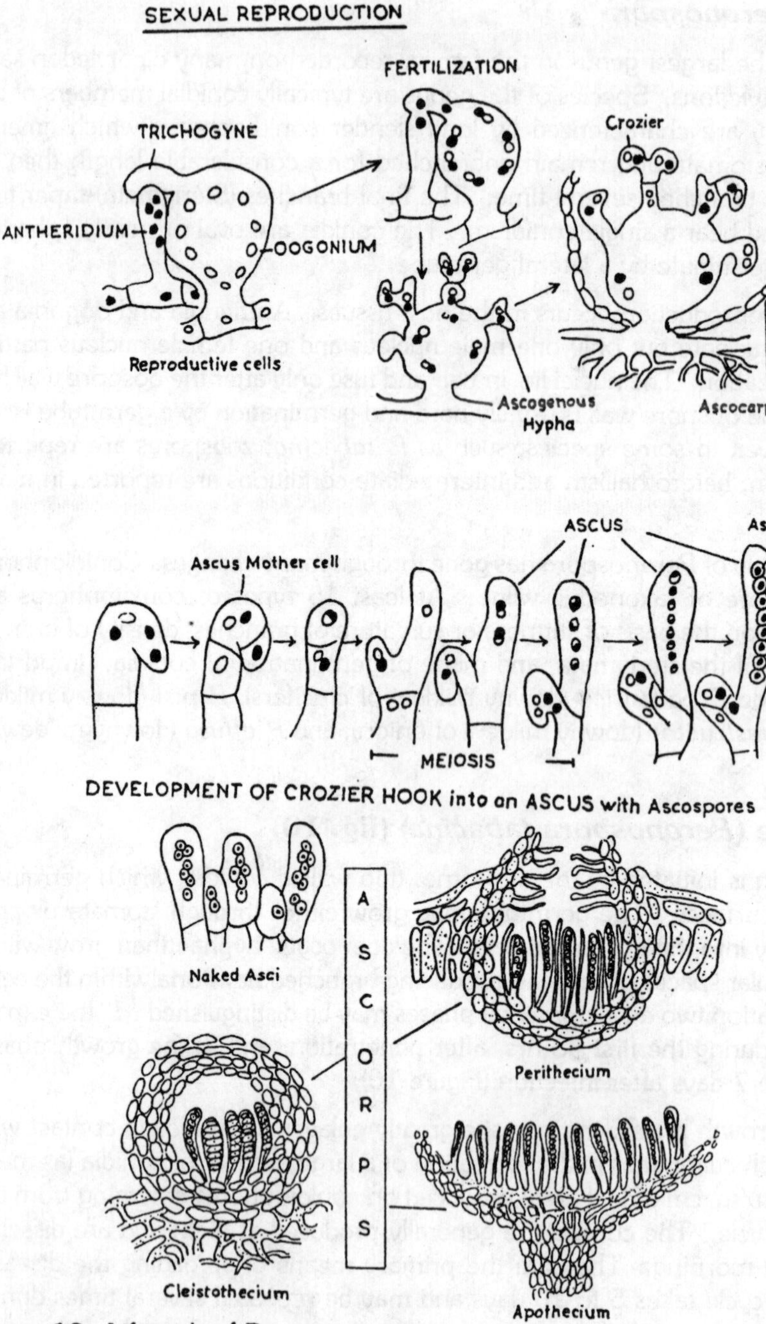
SEXUAL REPRODUCTION
FERTILIZATION
TRICHOGYNE
Crozier
ANTHERIDIUM
OOGONIUM
Reproductive cells
Ascogenous Hypha
Ascocarp Initial
ASCUS
Ascospores
Ascus Mother cell
MEIOSIS
DEVELOPMENT OF CROZIER HOOK into an ASCUS with Ascospores
Naked Asci
ASCOCARPS
Perithecium
Cleistothecium
Apothecium

Figure-10: Life cycle of *Peronospora spp.*

At a later stage sexual reproduction results in the formation of oospores (Fig-10) within the host tissues and may represent the resting or over wintering stage of the fungus.

Pseudoperonospora

The spores of *Pseudoperonospora* are true sporangia with a poroid apex. While this character of the genus is near to that of some species of *Sclerospora* or *Plasmopara* the branching of the sporophore is somewhat similar to that of some species of *Peronospora*. However, the sporophore is more delicate, branching is not always dichotomous and is much less profuse and the branches are straight. Because of the nature of sporangial germination and monopodial branching the species of this genus for some time were placed under *Plasmopara*. But while in the latter the branches arise nearly at right angles, in *Pseudoperonospora* they are at acute angles. The sporangia are ellipsoid, poroid and papillate, and nearly always germinate by zoospores. They are faint brownish or mauvish gray in colour. The oogonium, if formed, has a thin wall and the aplerotic oospore has a thick, dark, slightly rough wall. Haustoria are knob-like to digitate. *Pseudoperonospora cubensis, P. humuli, P. urticae, P. connabina and P. celtidis* are described species. The last four species are morphologically close and occur on closely related host families (Urticaceae, Cannabinaceae and Ulmaceae) and might prove to be forms of one species.

Family : Albuginaceae

Fungi of this family produce short, club shaped and indetermate sporangiophores which bear chains of globose sporangia at their tips. The members are obligate parasites and cause the disease termed " white rust/white blister". There are several species of *Albugo. A candida*, which attacks crucifers, is the only one causing diseases of economic importance. Some of the other species commonly found are *A ipomoeae*-pandurance on sweet potato and morning glory, *A. portulacae* on portulaca, A *accidentalis* on spinach, and *A bliti* on various members of *Amaranthaceae.*

Life - Cycle : In *A. candida* the mycelium is intercellular and feeds by means of globose knobshaped haustoria. The hyphae colonize host cells and attain maturity. Subsequently, short clubshaped sporangiophores develop. These sporangiophores produce chain of sporangia the oldest being at the tip and youngest at the base. As sporangia mature, they become detached and accumulate beneath epidermis. The epidermis ruptures facilitating spore release which form a white crust/blister on the host surface. Later on wind, water (rain splash) or some others disseminate the sporangia which under suitable environmental and nutritional conditions germinate either by producing zoospores or directly by germtube. Germination is basically temperature dependent. During zoospore production, the sporangia extrude 4-12 zoospores into a sessile vesicle. Subsequent details of asexual-cycle follow the pattern typical of *Peronosporaceae*.

In sexual reproduction oogonia and antheridia are formed within the host tissues. Both organs are multinucleate initially but only one nucleus in each is functional. Meiosis is beleived to occur in the gametangia. Gametangia are formed terminally near each other on somatic hyphae. They soon establish contact, with the antheridium contacting the oogonium at the side. The mature oogonium contains a single oosphere surrounded by periplasm. The functional nucleus moves to the centre of oosphere while the other nuclei-migrate to the periplasm. The antheridium then forms a fertilization tube which conducts movement of nucleus with cytoplasm into the oogonium. Subsequently both male and female nuclei fuse leading to formation of disploid(2n) zygote which becomes oospore. During development, oospore develops a thick ornamental wall and simultaneously zygote nucleus divides several times mitotically. The oospores germinate by forming zoospores which swim, encyst, and then germinate by forming germtube.

XIII. Phylum - Ascomycota

A. Genral Characteristics

The primary morphological characters that distinguishes fungi of ascomycota from all other fungi is the ascus (pl asci, Gr. *askos*=goat skin, sac), a sac like cell containing ascorpores (Gr. *askos +spora* =spore) formed after karyogamy and meiosis. Eight ascospores are typically formed within the ascus, but this number may vary from one to over a thousand according to the spoecies.

B. Somatic Structures

The somatic stages of ascomycetes may be single called, mycelial, or dimorphic. A large proportion of the cell walls of filamentous ascomyeetes is chitin. Hyphae are divided into compartments by septa that form from the hyphal periphery and advance toward the centre, thus invaginating the plasmamembrane. In most ascomycetes a small circular opening or pore is left near the centre of the septum through which the plasma membrane and cytoplasm extend from one hyphal compartment to the next. 'Woronin bodies' which are spherical, hexagonal, or rectangular membrane bound structures with a crytalline protein matrix usually associated with a septum, are also known to occur. In addition to woromin bodies that may plug septal pores, a more complex structure occures in some filamentous ascomycetes. These ascomycetes have membrane-bound "septal Pore organcells" often shaped like pully wheels, that are distributed in parts of mycelium so that structures involved in sexual reproduction are isolated from other regions of the mycelium. It has been suggested that these pore plugs may be derived from woronin bodies. Another unique organelle that occures primarily in lichenized ascomycetes is the "concetric bodies"

Hyphal compartments often are uninucleate, but mycelia consisting of multinucleate cells are well known. The perforations in the hyphal septa permit nuclei to migrate from one cell to another. Such migration of nucli throughout the mycelium is important in the phenomeman of "heterokaryosis". Ascomycetes may produce specialized mycelial structures, some of which are associated with host infection, In addition to appresoria and houstoria, HYPHOPODIA (sing. hyphopodium; Gr *hyphe* = web+*pous*=foot) which are small appendages on a hypha in which MUCRONATE hyphopodia function as conidiogenous cells and CAPITATE hyphopodia give rise to haoustoria.

C. Asexual Reproduction

Asexual reproduction in ascomycetes may be carried out by 1. fission 2. fragmentation, 3. thallospores, or 4. conidia accoding to species and factors concerning nutrition and environment. Fission and budding are common in yeasts and dimosphic fungi. Conidial production is the major method of inoculum. Conidia may either directly develop from somatic hyphae or from specialized conidiogenous cells. Various types of conidiophore branching are found. They may develop independently or in complex structures the asexual spore fruits Acervuli, Pycnidia, Sporodochia, or Synnemata.

D. Sexual Reproduction

In sexual reproduction two unlike but compatible nuclei are brough together in the same cell by one of several methods—

1. Isogametangiogamy

Two morphologically similar gametangia come together, touch at their tips or coil around each other and fuse. Karyogamy follows and zygote develops. In yeasts, which are mostly unicellular, two cells of similar morphology fuse and produce zygote which develops in ascus.

2. Hetetogametangiogamy (Gametangial Contact)

Antheridia and ascogonia develop. Plasmogamy takes place through transfer of cytoplasm and nuclei from antheridia to ascogonia through trichogyne.

(Gr. *thrix*=hair, *gyne*-woman). Later asci and ascosposes develop. The entire process is discussed later.

3. Spermatization

In some ascomycetes this method creates binucleate (heterokaryotic) cells. It is already discussed.

4. Somatogamy

This method involves fusion of two rather unspecialized but compalible hyphal cells, theryby creating heterokaryotic condition. In due course karyogamy and meiosis take place.

E. Life-cycle Pattern (Figure-11)

Ascomycetes (the sac fungi) produce sexual spores (ascospores) within a sac called ascus. They also produce asexual spores (conidia). The sexual stage is called TELEOMORPH or perfect stage and the conidial stage is the ANAMORPH or asxual stage.

The ascus in most ascomycetes is formed as a result of fertilization of the female sex cell, called an ASCOGONIUM, by either an ANTHERIDIUM or a minute male sex spore called spermatium (figure-11). Upon contact between the two sex organs, the nuclei pass from antheridium to ascogonium via a tubular hair like structure called TRICHOGYNE. After the entry of the antheridial cytoplasm and nuclei, the compatible nuclei from both organs get paired, but no fusion, and move toward the periphery of ascogonium. The ascogonial wall becomes elastic and ultimately assumes hyphal structures particulary in areas where paired nuclei are located. The many hyphal structures that are formed are termed as ASCOGENOUS HYPHAE. In these structures, the nuclei move in pairs. Later septa are formed dividing ascogenous hypha into compartments. Each compartment contains two (paired) compatible nuclei. The tip cell is invariably binucleate (one nucleus being male and the other female). This cell may directly become the "ascus mother cell" or it may elongate and bend to form a hook or crozier (figure-11). The two nuclei in this bent portion divide in such a way that their spindles lie parallel to each other. Two septa divide the hook into 3 cells. The tip and the basal cells which are lying side by side contain one nucleous each, one being the male and the other female. These cells unite to form a single cell and then contain two nuclei. Meanwhile, the middle cell which is at the tip of hook now and contains two nuclei (one from male and other from female) become the ascus mother cell. The united tip and basal cells, now forming a single cell with two nuclei repeat the process. In this way, a large number of ascus

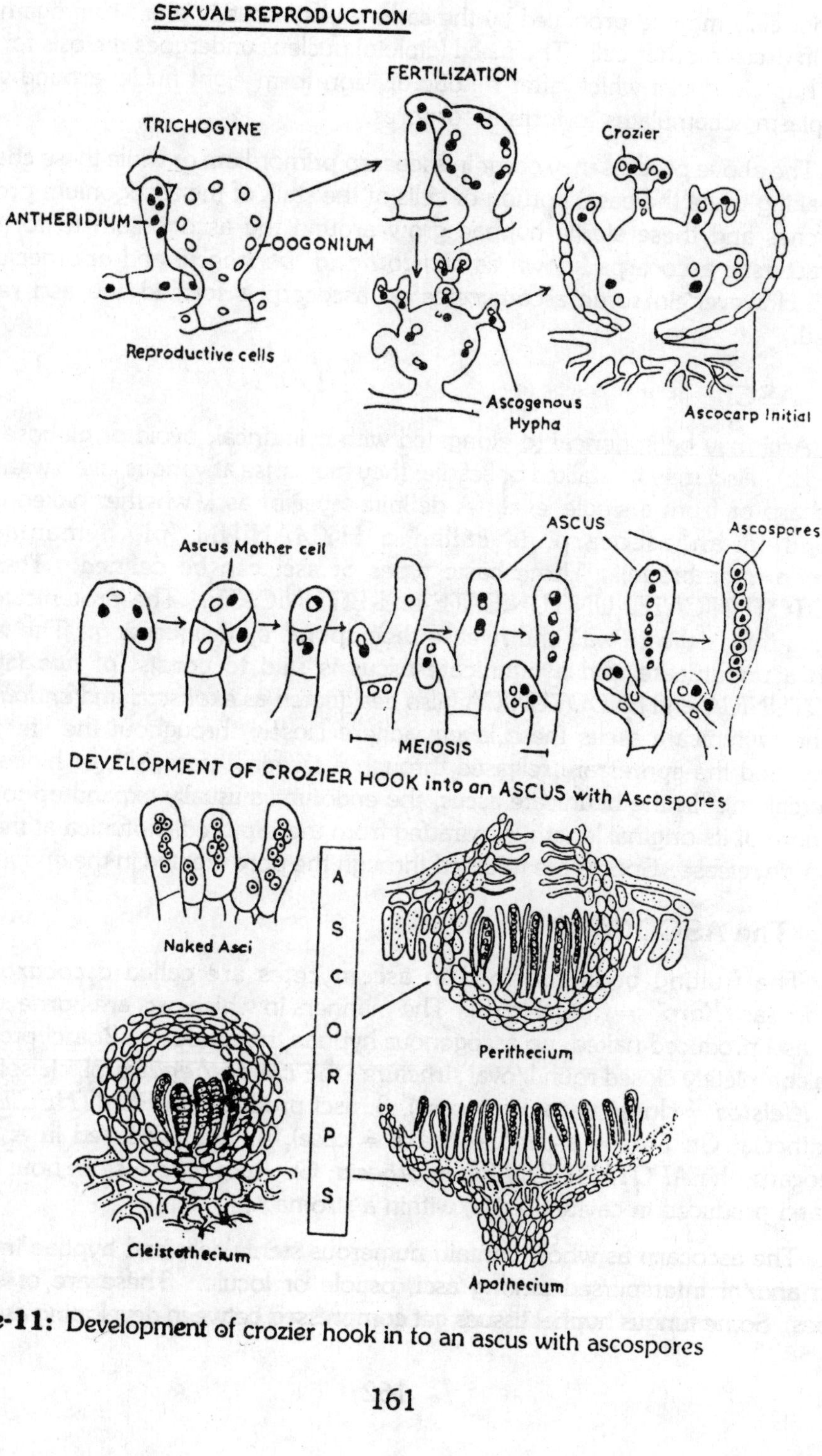
SEXUAL REPRODUCTION
TRICHOGYNE
ANTHERIDIUM
OOGONIUM
Reproductive cells
FERTILIZATION
Ascogenous Hypha
Crozier
Ascocarp Initial
Ascus Mother cell
ASCUS
Ascospores
MEIOSIS
DEVELOPMENT OF CROZIER HOOK into an ASCUS with Ascospores
Naked Asci
ASCOCARPS
Perithecium
Cleistothecium
Apothecium

Figure-11: Development of crozier hook in to an ascus with ascospores mother cells may be produced by the same ascogenous hypha. Karyogamy occurs in ascus mother cell. The fused (diploid) nucleus undergoes meiosis forming four haploid nuclei which after mitotic division form eight nuclei around whcih cytoplasm accumulates to form ascospores.

The above process may occur in ascocarp primordium or while these changes are taking place the basal portion or cells of the stalk of the ascogonium produce branches and these sterile hyphae grow around the ascogonium to form the characterstic ascocarps known as *cleistothecia, perithecia* and *apothecia* (fig-11). However, in some ascomycetes no ascocarp is formed and asci remain naked.

F. ASCI

Asci may be spherical to elongated with cylindrical, ovoid or globose form (Fig-12). Asci may be stalked or sessile; they may arise at various levels within the ascocarp or from a single level. A definite layer of asci, whether naked or enclosed in an ascocarp, is called a HYMANIUM (pl. hymania, Gr. *Hymen*=membrane). Three basic types of asci can be defined. They are PROTOTUNICATE, UNITUNICATE and BITUNICATE. The protunicate asci have a thin, delicate wall and release their spores by deliquescing. The wall in both a unitunicate and a bitunicate ascus is said to consist of two layers - EXOTUNICA and ENDOTUNICA (also designated as *exoascus* and *endoascus*). In the unitunicate ascus these layers adhere closely throughout the life of the ascus, and the spores are released through a terminal pore, slit, or hinged cap (operculum). In the bitunicate ascus, the endotunica usually expand up to twice or more of its original length, separating from the raptured exotunica at the time of spore release. Spores are released through the pore created in the endotunica.

G. The ASCOCARP

The fruiting bodies formed in ascomycetes are called *ascocarps* (Gr. askos=sac+*Karpos*= fruit) (Fig -8). The manners in which asci are borne include : 1. asci produced naked, no ascogenous hyphae, no ascocarp, 2. asci produced in a completely closed round/oval structure - the ***cleisothecium*** (pl. cleisothecia, Gr. *Kleistos* = closed + *theke* = case), 3. asci produced in *PERITHECIUM* (pl Perithecia; Gr. *Peri* = around + *theke* = case), 4. asci produced in an open ascocarp, the *APOTHECIUM* (pl. *apothecia*; Gr. *Apotheke* = store house), and 5. asci produced in cavity (*locule*) within a stroma.

The ascocarp as whole contain numerous sterile cells and hyphae interwoven and/or interspersed among asci, ostiole or locule. These are of several types. Some fungus hyphal tissues get compressed between developing asci. It is

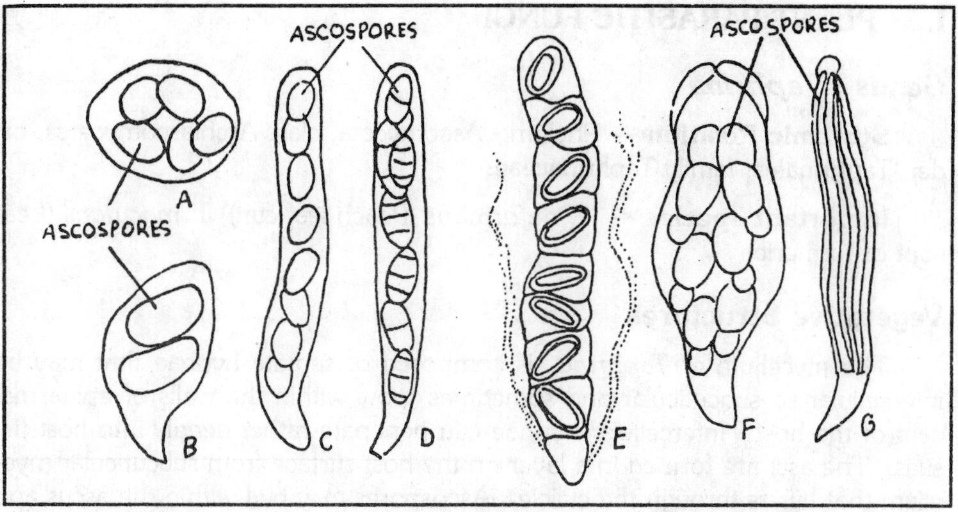
ASCOSPORES
ASCOSPORES
A
ASCOSPORES
B
C
D
E
F
G

Figure-12: Different forms of asci. A. Globose ascus. B. Broadly ovate ascus with a stalk. C. Cylindrical ascus with ascospores in a single row. D. Cylindrical ascus with multicellular ascospores. E. Septate ascus. F. Clavate ascus with ascospores in two rows. G. Tubular ascus containing thread like ascospores, in a bundle.

refeered as INTERASCAL PSEUDOPARENCHYMA. PARAPHYSES are alongated, cylindrical, or club shaped hyphae originating at the base of an ascocarp. They usually are unbranched and may be septate or aseptate. Paraphyses often absorb water and expand, and their movement may shake the asci in the hymanium to aid in ascospores release. PERIPHYSES (Sing. periphysis; Gr. *peri* = around + *Physis*) are short unbranched hyphae in the ostiolar canal of an ascocarp. They apparently serve to direct the asci toward the tip of ostiole prior to ascospore discharge.

H. Classification

The classification of ascomycetous fungi has always been controversial. Based on presence or absence of ascocarp, the type of ascocarp, type of asci and their arrangement in respective ascocarp has been the major basis for classification of ascomycetes. Ainsworth (1966) classified ascomycetes under subdivision ascomycotina with six classes. Several contemporary attemps followed almost the same criteria. Alexopoulos *et.al.* (1996) based on phylogenetic connections, DNA sequence analysis and several other additional characters have classified these fungi as " *filamentous ascomycetes*", *archiascomycetes,* and *saccharo mycetales*. In the present text book, the plant parasitic fungi discribed have been classified as per scheme proposed by Alexopoulos *etal* (1996).

I. PLANT PARASITIC FUNGI

Genus : *Taphrina*

Systemic Position — Phylum - Ascomycota, class-Archiascomycetes, order -Taphrinales; family Taphrinaceae.

Important species — *T. deformans* (Peach leaf curl), *T. maculans* (Leaf sopt of turmeric).

Vegetative Structures

The mycelium of *Taphrina* is composed of septate hyphae that may be intercellular or subcuticular and sometimes grow within the walls of epidermal cells of the host. Intercellular hyphae can penepate rather deeply into host tissues. The asci are formed in a layer on the host surface from subcuticular mycelium that bursts through the culicle. Ascospores rnay bud within the ascus and cells continue to bud after ascospores release to comprise the saprophytic yeast stage.

Life - cycle (Fig-13)

Several workers have investigated the life-cycle of *Taphrina* spp. but with somewhat contradictory observations. In *T. deformans*, after their formation the ascospores produce small, round or ovoid blastospores by budding (Fig-13). these sprout cells, like the ascospores, are uninucleate and haploid. On the host suface, the cells continue to bud. At the same point they initiate the mycelial stage, perhaps by fusing in pairs, after which they form the infective mycelium. The mycelium often has been said to be dikaryotic but several others have observed multinucleate hyphal compartments with the nuclei arranged in pairs. As the hypha alongates, conjugate division of the nuclei perpetuates the paired arrangement of the nuclei within the hyphal compartments. The mycelium grows and branches, spreading between the cells and penetrating into the host tissues. Hyphal strands eventually become massed in the subcuticular region, and some of the hyphal cells enlarge to form "ascogenous cells". The ascogenous cells may contain several paired nuclei. Karyogamy occurs, apparently between the paired nuclei present within each ascogenous cell. After Karyogamy, the cells alongate. During elongation, the diploid nucleus may divide mitotically, and the daughter nuclei move toward opposite ends of the cell. Later, a septum develops between these two nuclei, separating the ascogenous cell into a basal stalk cell and the upper ascus cell. The protoplast of the stalk cell soon disintegrates, leaving the cell empty. Meiosis and a subsequent mitotic division result in the formation of eight nuclei. Ascosporogenesis in *T. deformans* takes place by the invagination

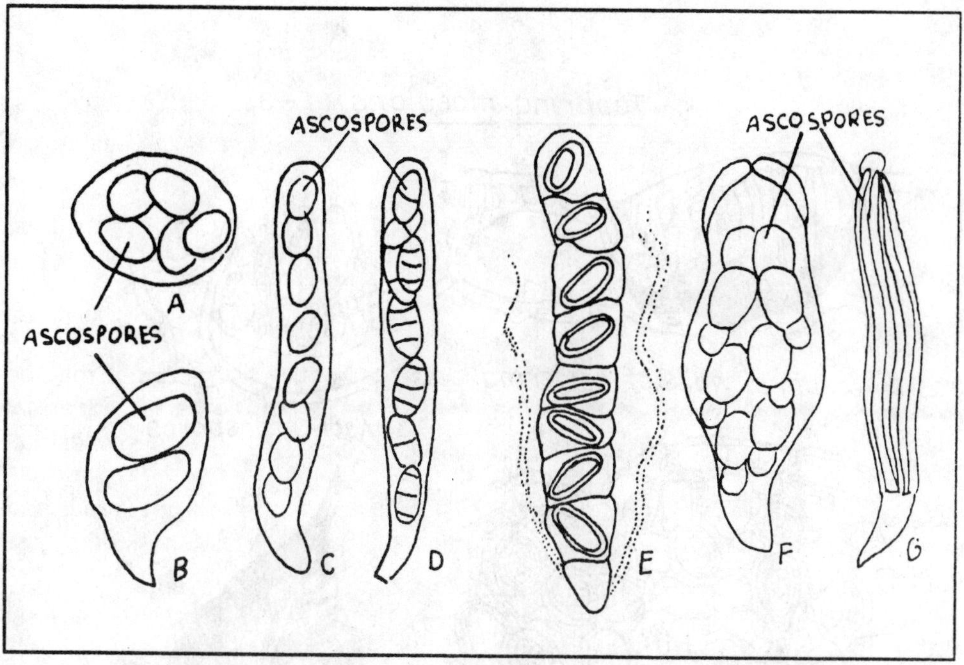
ASCOSPORES
ASCOSPORES
ASCOSPORES
A
B
C
D
E
F
G

Figure-13: Life cycle of *Taphrina deformans*

of the ascus plasmamembrane to form an EMS near the spindle pole body of each nucleus, and eventually ascosores are delimited. As the ascogenous cells enlarge and alongate, they exert pressure on the host cuticle from below and finally break through to form a compact layer of asci the hymanium on the epidermis of the host. The asci usually discharge the ascospores into the air.

The ascus of *Taphrina* is unitunicate and without a diffrentiated apical apparatus for release of ascospores. The tip of the ascus simply bursts at its thinnest point due to build up of pressure within the ascus. The ascospores after their release, bud to form numerous cells, a process that may have begun in the ascus, and the life cycle is ready to begin a new.

Genus : Protomyces (Plate-XIII)

Systemic Position - Phylum - Ascomycota, class Archiascomycetes; order Taphrinales; family Protomycetaceae

Important species - *P. macrosporus* (stem gall of coriander)

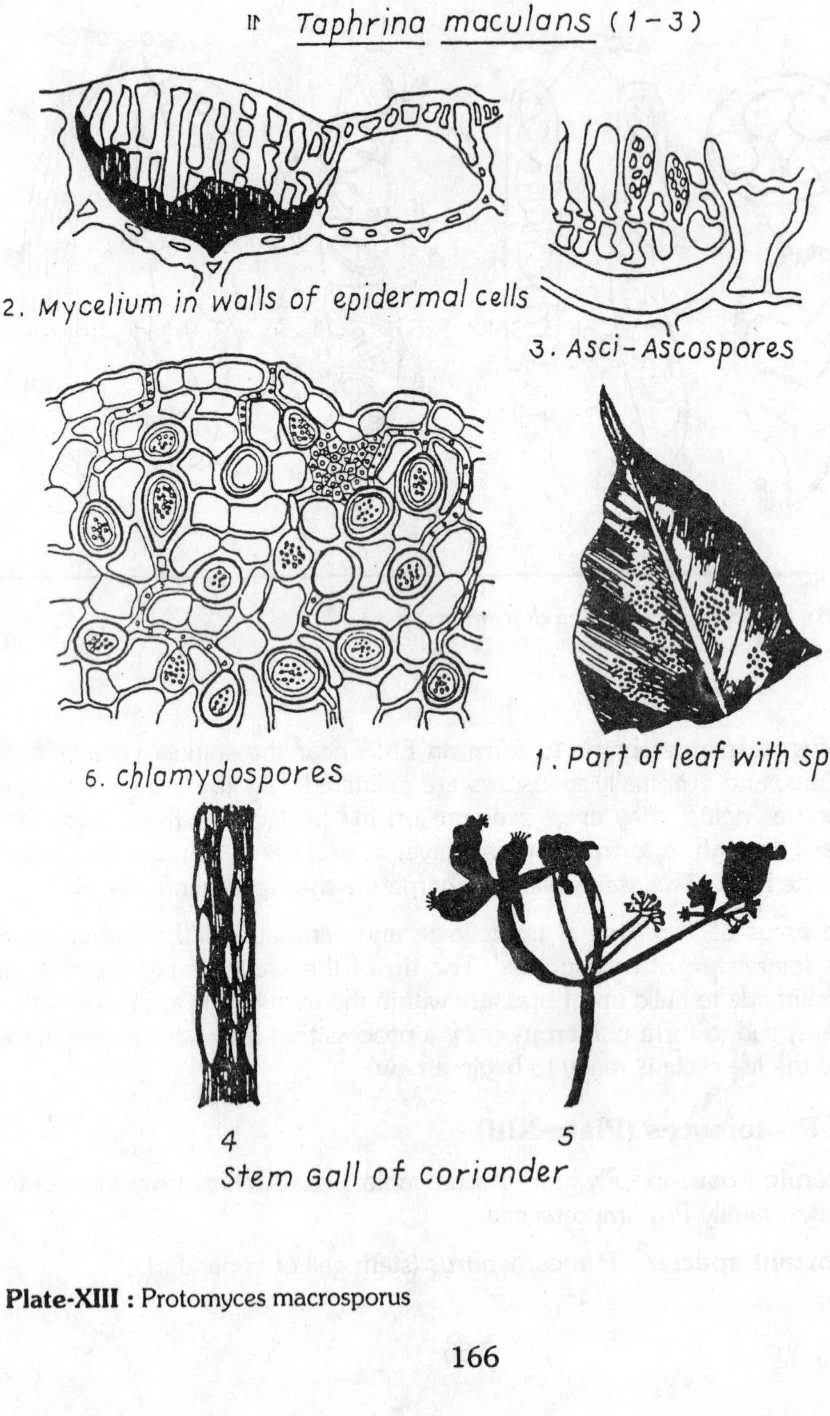
Taphrina maculans (1-3)
2. Mycelium in walls of epidermal cells
3. Asci-Ascospores
6. chlamydospores
1. Part of leaf with spots
4
5
Stem Gall of coriander

Plate-XIII : Protomyces macrosporus

Structures and life-cycle - The mycelium is found only in the galled tissues of the host. The hyphae are intercellular, closely septate, and broad. Branchring is irregular. Scattered cells in the hyphae swell to form ellipsoidal or globose bodies which later develop into resting spores or chlamydospores. As they mature, they become surrounded by thick, 3 layered wall and attain a diameter of 50-60μ. These resting spores germinate by rupturing the exospore. The inner wall is pushed out to form a vesicle which appears in continuation with the mouth of the crack. The protoplasm from the spore passes into the vesicle and gathers towards the periphery. The nucleus divides repeatedly to form 100-200 nuclei. The peripherel cytoplasm now cleaves into several uninucleate masses. These again divide each forming four ellipsoidal spores. On maturity these spores separate and collect in the centre of the vesicle. The latter bursts and the spores are set free. These spores further multiply by budding and infect the host.

Powdery mildew fungi

The fungi belonging to ***Erysiphales*** are obligate biotrophs that cause a major group of plant diseases knownas ***powdery mildews***. These diseases are so named because parts of infected plants, most commonly the leaves, appear to be covered by a white, powdery material. This white appearance is due to conidiophores and conidia they produce. While these conidia are powdery in the sense that they are blown around easily, in mass they are distinctly wet, and therefore, sticky. While the conidia dominate the life cycle of Erysiphales, these fungi also reproduce sexually by forming ascospores. The ascospores are produced in asci borne in ascocarps which are completely closed like cleistothecium but produce their asci in a basal layer, as is the case in many perithecia. The asci have been described as bitunicate but with a poorly development endotunica at the apex.

Over 40000 plant species representing over 40 orders of flowering plants are parasitized by powdery mildew fungi, of these hosts, about 90% are dicots. In their parasitism some members fo Erysiphales are almost omnivorous, as shown by species such as *Erysiphe polygoni* which is reported to have been recorded on 352 host species. On the other hand, a goodly number are known to attack only specific hosts. The various formae specialis of *Blumeria graminis* (previously referred as *Erysiphe graminis*) are highly host specific. Examples include B. graminis f.sp. hordei (on barley), f sp avenae (on oats); f. sp secalis (on rye), and f.sp. tritici (on wheat).

Somatic Structures

With a few exceptions, the somatic mycelium of powdery mildew fungi are produced extensively on host surfaces. Individual hyphal comprising the mycelium may be branched or unbranched and tend to grow closely oppressed to the host surface. Hyphae are thin walled and septate with each cell usually containing a single nucleus. 20 to 50 percent hyphal cells produce simple, unlobed or multilobed appresoria. Microscopic studies have revealed several types of haustoria. The shape varies form globose to pyriform to lobed. The haustorium is separated from the cytoplasm of its host cell by extra haustorial matrix and an extra haustorial membrane.

Asexual Reproduction

A few days after the host infection, the somatic hyphae produce great numbers of hyaline, erect conidiophores. The conidiophores of most species are unbranched. The conidiophore consists of one to three cells. The basal cell is referred as the foot cell while the terminal cells are conidiogenous cells. In the case of *Blumeria graminis* (Fig. 16), the basal cell is swollen, and also functions as conidiogenous cell. Conidia develop singly or in chains. When formed in chain, conidia remain firmly attached end to end with mature conidia at the distal end of the chain and the immature conidia at the proximal end (basipetal development). Conidia are one celled and hyaline (Figure-14, 17) Each is thin walled, uninucleate, vacuolated, oval or cylindrical with rounded sides and ends. Based on characterstics of conidiophores and conidia, the asexual stages of most powdery mildew fungi key out to thé anamorph genus *oidium*. Three additional

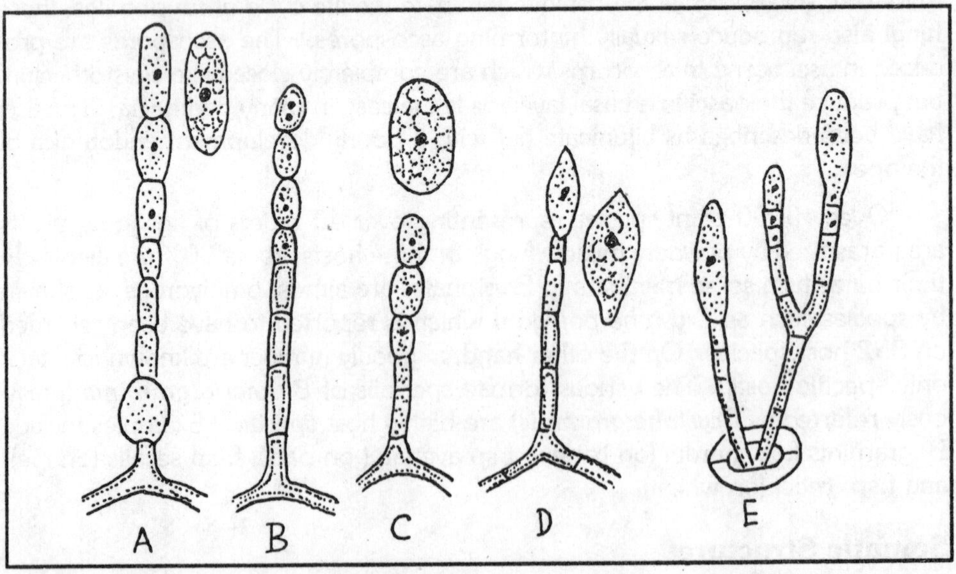
A
B
C
D
E

Figure-14: Conidiophore and conidia of *Bluneria graminis* (A) *E. cichoracearum* (B) *E. polygoni* (C) *Phyllactinia* (D) and *Leveillula* (E).

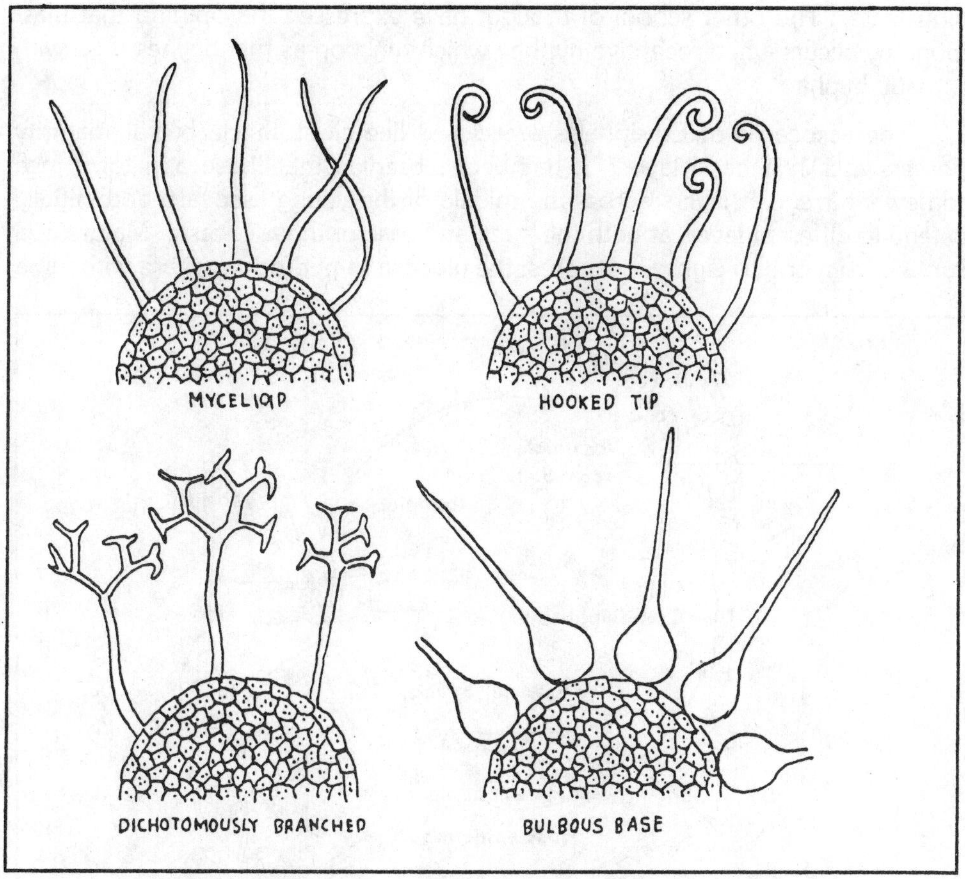
MYCELIOID
HOOKED TIP
DICHOTOMOUSLY BRANCHED
BULBOUS BASE

Figure-15: Appendages of ascocarps of Erysiphaceae

anamorphs namely *ovulariopsis*, *streptopodium*, and *oidiopsis* are also reported.

Sexual Reproduction (Fig. 16)

When conidium production slows down, and eventually ceases, young ascocarps begin to make their appearance on the mycelium. These are at first white, then orange reddish brown, and finally black when mature

Both homothallic and heterothallic species of Erysiphales have been reported. However, details of ascosarp development in these fungi are not resolved. It appears that sexual union takes place, at least in heterothallic species, but the controversy centres around the origin of the sexual nuclei. One school of thought regarded the two hyphal branches that coil around one another as functional gametangia with the antheridial nucleus passing into the ascogonium rendering it binucleate. The other school of thought have expressed the opinion that plasmogamy occurs when receptive hyphae which function as trichogynes, fuse with somatic hyphae.

The ascocarps of Erysiphales are closed like cleistothecia but at maturity possess asci in a basal layer. It has been observed that these asci form in a somewhat irregular layers across the middle of the young ascocarp and initially extend to different levels at both the apex and base of the ascocarp. Mature asci contain from one to eight ascospores; the globose to pyriform asci are without a

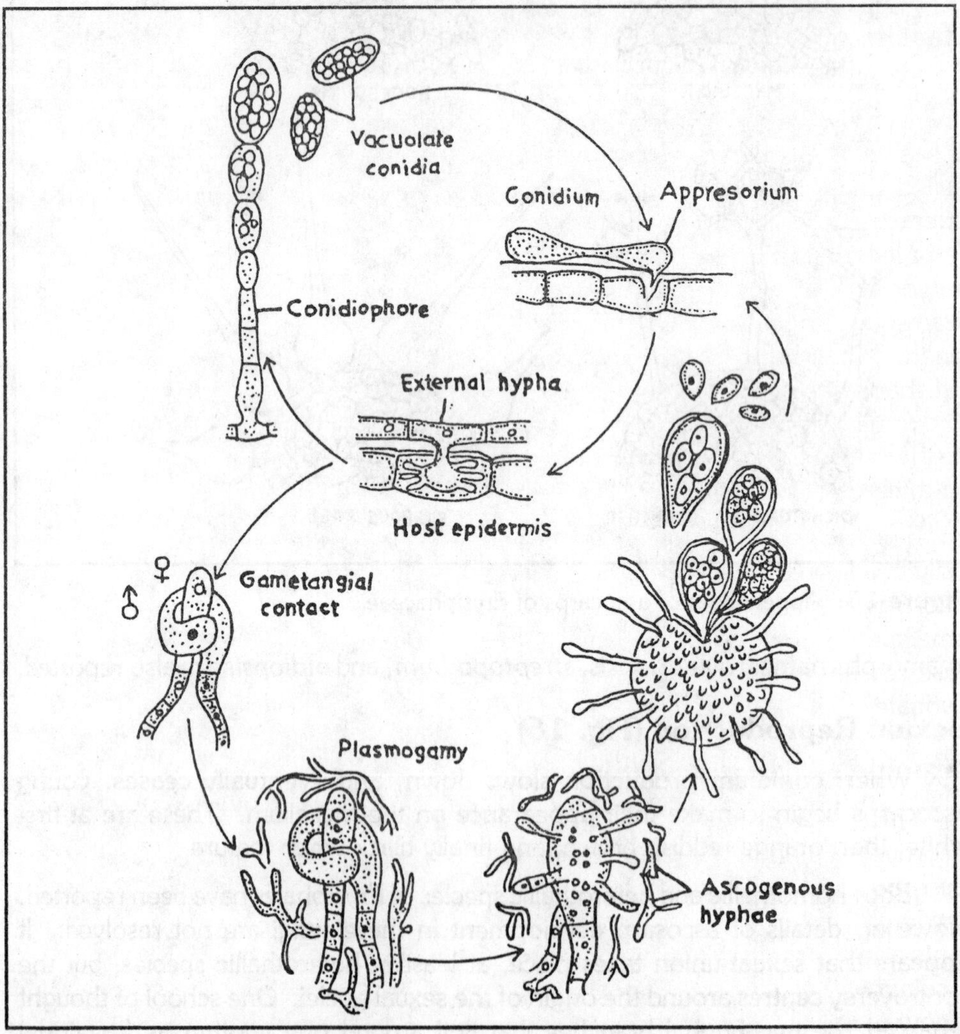
Vacuolate
conidia
Conidium
Appresorium
Conidiophore
External hypha
Host epidermis
♀
♂
a
Gametangial
contact
Plasmogamy
Ascogenous
hyphae

Figure-16: Life cycle of *Blumeria*

definite apical structure. The mature ascocarps of most species of Erysiphales are provided with characterstic appendage (figure-15) that vary considerably in length and character and that, together with the number of asci produced, form the basis of generic separation. Ascocarp appendages fall into four general types: 1. myceloid, which resemble somatic hyphae in being flaccid and indefinite, as in *Erysiphe*, *Blumeria*, *Sphaerotheca*, and *Laveillula*; 2. rigid spear like, with a bulbous base and pointed tip, as in *Phyllactinia*; 3. rigid with curled tips, as in *Uncinula*, *Uncinuliella* , and *Pleochaeta*, and 4. rigid with dichotomously branched tips as in *Microsphaera* and *Podoshaera*.

Classification

Many mycologists placed Erysiphales among the cleistothecial fungi, and others considered them to be Pyrenomycetes or loculoascomycetes. DNA analysis has helped to settle this confusion by showing that *Erysiphales* are not closely related to any of these groups. Saenz etal (1994) found that a member of the order was placed among the early diverging discomycetes and loculoascomycetes. Gargas and Taylor (1994) have found support for a relationship with discomycetes. Yarwood (1973) has recognized seven gevera. Hawksworth etal (1983) recognized some 28 gevera while Braun (1987) and Zheng (1985) have recognized 19 genera. Characters of taxonomic valve include appresorium morphology, size and shape of conidia, the presence or absence of fibrosing bodies in conidia, the production of conidia singly or in chains, the conidiophore characterstics including the number of cells comprising the conidiophore, the nature of foot cell, and the thickness of conidiophore wall.

If ascocarps are available, some of the common powdery mildew fungi can easily be identified to the generic level. In the absence of ascocarps, the identifications is rather difficult. In most cases ascocarps are not always available as some species rarely produce them while others do so very late in the season. For these reasons identification schemes based on conidial characterstics have been advocated.

Parasitic Powdery mildew fungi

The family *Erysiphaceae* (order - *Erysiphales*, Filamentous Ascomycetes, Ascomycota) are true powdery mildews. Among the genera included in the family *Erysiphe* and *Blumeria* are economically important. In India *E. polygoni* causes powdery mildew of peas, several species of *Lathyrus*, several other cultivated plants and their wild relatives. *Erysiphe cichoracearum* causes powdery mildew of cucurbits. Powdery milde of cereals is another dreaded disease. Powdery mildew of wheat (*Blumeria graminis* f. sp. *tritici*), powdery mildew of barley (*B. graminis* f. sp. *hordei*) are ecominically important. Some species of *Sphaerotheca*

(*S. fuliginea* - powderymildew of cucurbits, Podosphaera (*P. leucotricha* - powdery mildew of apple), Uncinula (*U. necator* - powdery mildew of grapevines), *and Phyllactinia (P. dalbergiae* - powdery mildew of *D. sissoo)*.

The Ergot fungi

Family- clavicipitaceae (order - Hypocreales, Pyrenomycetes (ascomycetes with perithecia),filamentous ascomycetes, Ascomycota) members particularly genus - Claviceps causes disease called ergot. The best known species are C. purpurea (ergot of rye) and C. microcephala (ergot of bajra). The infected ears develop several hard pigmented sclerotia which are highly toxic to animals and human beings and cause ERGOTISM.

General characterstics

The members produce bright or darkly pigmented fleshy stomata, often in the shades of orange and yellow or a subiculum, long narrowly cylindrical asci with perforated thickened domelike caps and paraphyses formed on the lateral walls of the ascocarp but not present among basically tufted asci. The ascospores are thread like and of the ascus length. The ascospores when being released, break-up and each functions as an individual spore. They germinate by germtube. The anamorphs are phialidic hyphomycetes that may be formed on the stromata before perithecia are formed, a few are sporodochial or synemmatous. They include *Hirsutella*, *Acremonium*, *Aschersonia*, and *Gibellula*. Members of clavicipitaceae attack plants and arthropods. Genus - *Claviceps* is parasitic on grasses; *Cordyceps* is pathogenic an arthropods and hypogeous basidiocarps; *Torrubiella* is pathogenic on spider; *Epichloe* is endophytic in grasses; and Balansia is endophytic in grasses and sedges.

Claviceps purpurea (Figure-17)

The ascospores released explosively and disseminated by wind, germinate on rye flowers and infect the ovaries. The mycelium that develops soon fills the ovary and within a week numerous conidiophores develop from the peripheral hyphae. These then form hyaline conidia in enormous numbers. A viscous sugary exudate oozes out of the fungus stuffed ovaries as drops called "honey-dew" Insects feed on honey dews and move to healthy flowers and spread the conidial inoculum, and thus ensuring secondary infections. At the end of the crop season, the mycelial mat present in the ovaries develop into hard, pink purplish sclerotia which project out of the glumes (figure-13).

The sclerotia serve as overseasoning organs. In the next crop season, they germinate at the flowering time of the next host crop. During germination, several stromatic masses aris from within the sclerotial body rupturing the rind.

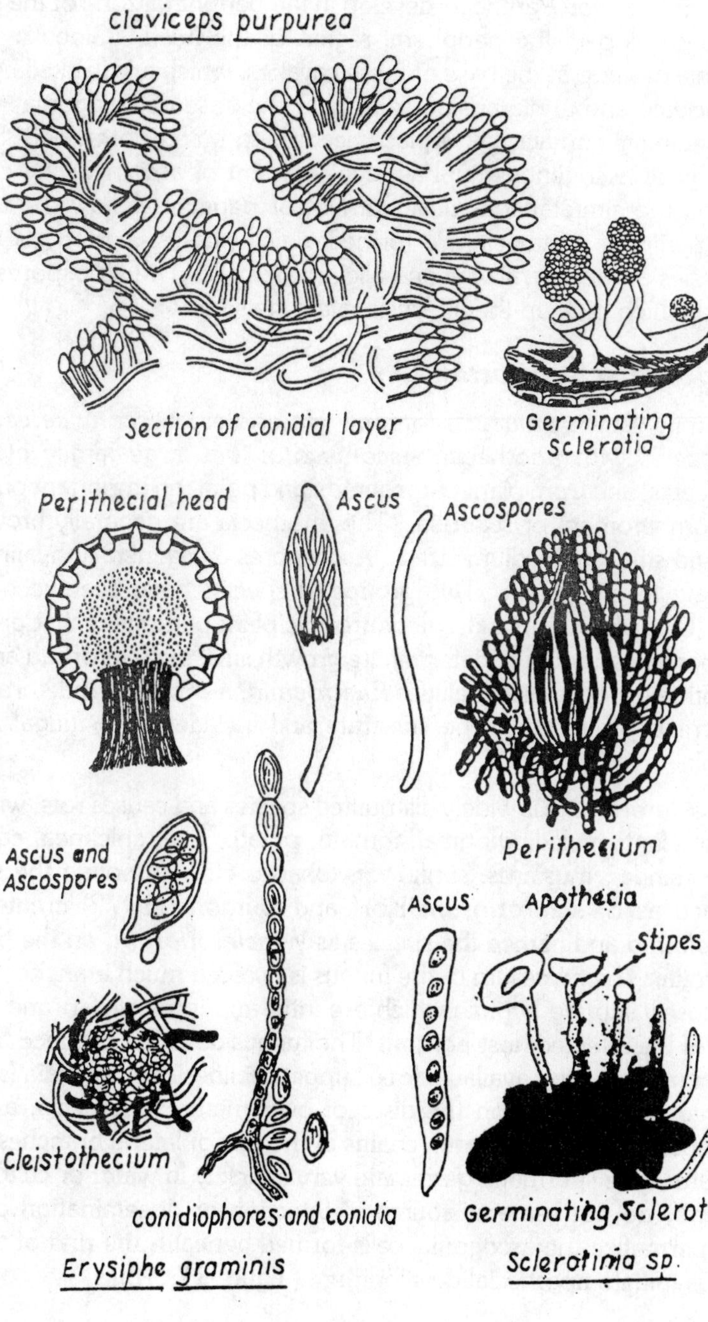
Claviceps purpurea
Section of conidial layer
Germinating Sclerotia
Perithecial head
Ascus
Ascospores
Perithecium
Ascus and Ascospores
Cleistothecium
conidiophores and conidia
Erysiphe graminis
Ascus
Apothecia
Stipes
Germinating Sclerotium
Sclerotinia sp.

Figure-17

These elongate and become globular at the apex. The stromata finally becomes mushroom shaped. Perithecia develop in the peripheral zone of the head. Minute cavaties develop in the peripheral region of the head. Globular multinucleate ascogonia develop at the base of these cavaties, which are flanked by club-shaped multinucleate antheridia arising from near the base of ascogonia. Plasmogamy by gametangial contact, takes place, leading to pairing of nuclei. Subsequently, formation of ascogenous hyphae, development of asci involving crozier formation, and the simultaneous development of peridium follow the usual pattern. Finally pyriform perithecia are formed which possess periphysis in the ostole, paraphyses at the lateral wall and asci at the base. The ascospores are violently released which pick up the life-cycle every season.

Sclerotinia sclerotiorum

Family *Sclerotiniaceae* (order - *Helotiales*, *Discomycetes*, filamentous ascomycetes with apothecia, *Ascomycota*) is a large family of inoperculate dicomycetes, and from plant pathogenic stand point, an important one. Apothecia arise from stromata or sclerotia. The apothecia are generally brown, borne on stalks and small to medium sized. Ascospores are generally hyaline, one called and ovate to elongated. Two groups of species have been recognized in the family. One of these includes necrotrophic plant pathogens that produce melanized tuber like sclerotia of determinate growth such as *Sclerotinia* and *Monilinia*. In the other group, which includes *Rustroemia*, a stroma of indeterminate growth and form is produced in the substrate and includes both fungal and substrate material.

S. sclerotiorum is widely distributed species and causes rots, wilts and blights of many plants including brinjal, tomato, potato, pea, chickpea, cabbage, capsicum, coriander, mustards, sunflower, tobacco etc. Although the species is still described as *S. sclerotiorum* Korf and Dumons (1972) created the genus *Whetzelinina* and named the species as *W. sclerotiorum*, on the basis of nature of sclerotia. The mycelium of the fungus is hyaline, much branched, consisling of large closely septate hyphae which are inter and intracellular and invade all the tissues of the affected host portion. The fungus does not produce true conidia or macrocondia. As the available food supply declines microconidia (spermatia) are produced on sclerotia, on the discs of over mature opothecia, and in culture. These spermatia are formed in chains at the tips of lateral branches of vegetative mycelium. The spermatia germinate very sparsely in water or culture media and apparently do not serve as source of infection or dissemination of the fungus. They spermatize the ascogonial cells formed benealth the rind of the sclerotium and thus initiate apothecial development (figure-17, 18).

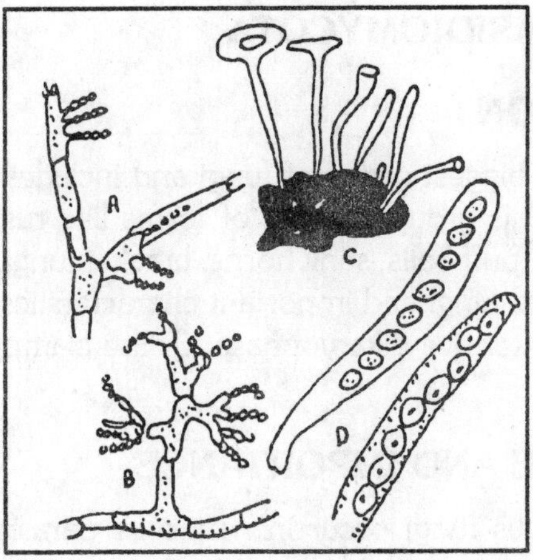
A
B
C
D

Figure-18: *Sclerotinia sclerotiorum*: A-B-endo- and exo-genously produced chains of microconidia (spermatia); C-germinating sclerotium with apothecial heads D-ascus and ascospores.

When the food supply is exhausted and the vegetative growth has ceased the hyphae with thick granular protoplasm and short cells collect in small dense masses which gradually become the sclerotia. At first these are pink but later turn dark-brown to black and become smooth. They vary in size and shape with environmental conditions and with the strain. Sclerotia formed on host surface are usually loaf-shaped or globose while those formed in the pith of the stem are elongated according to space available for growth. They range from 2 to 10 x 2 to 5 mm in the large sclerotial strains while in small sclerotial strains they are usually nearly spherical with a diameter of 0.5 to 1 mm. On germination these sclerotia give rise to several columnar structures (stipes) which develop the funnel-shaped cup (apothecium) at the tip. The mature apothecia are 6-9 mm across, generally borne 6-10 mm above the soil surface and become darker in colour with age. The ascospores are discharged in abundance from these cups. The ascospores are always 8 in each ascus which has an apical pore through which spore discharge occurs with violence. The asci are cylindrical and measure 108 - 153 x 4.5 - 10 microns, the average being 122.9 x 5.9 microns. The ascospores are hyaline, 1-called and ovate. Their size falls within the range of 7-16 x 3.6 - 10 microns. In absence of asexual spores the fungus is disseminated mostly by means of ascospores which are the most common structures of infection. Sclerotia can survive in soil or on plant debris for long.

XIV. PHYLUM : BASIDIOMYCOTA

I. INTRODUCTION

This is the second biggest group of fungi and includes about 1100 genera, and 16000 species. This group consists of forms like rusts, smuts, jelly fungi, mushrooms, toadstools, puff-balls, stink horns, bracket fungi, and birds nest fungi. These are highly evolved fungi and important characterstics are 1. the basidium-basidiospores, 2. the extensive dikaryophase, 3. the clamp connections, and 4. the dolipore septum.

II. OCCURRENCE AND IMPORTANCE

The basidiomycetous fungi occur in nature as parasites and saprophytes. Parasites are rusts and smut which cause serious diseases in plants. Many basidiomycetes are sought eagerly by mushroom lovers the world over. The cultivation of mushrooms for food has developed into an industry.

III. Somatic structures

The mycelium is composed of well developed, septate and branched hyphae which are either white, yellow orange, deep brown, or charcoal-black. In some forms rhizomorphs develop. The mycelium of most heterothallic basidiomycetes passes through three distinct stages of development-the primary, the secondary, and the tertiary- before the fungus completes its life-cycle. THE PRIMARY MYCELIUM, also called homokaryon, usually develops upon the germination of a basidiospore. Mycelial compartments may be multinucleate at first, with the nucleus or nuclei of the basidiospore dividing many times as the germtube emerges from the spore and begins to grow. This phase of primary mycelium is of short duration as the septa are soon formed and divide the mycelium into uninucleate cells. In some other basidiomycetes, septum formation starts upon completion of first division of the spore nucleus or nuclei so that the primary mycelium is septate and composed of uninucleate cells or hyphal compartments from the begining. However, species are known that have primary mycelium that remains multinucleate.

As most basidiomycetes are heterothallic, the formation of SECONDARY MYCELIUM or heterokaryons involves an interaction between two compatible homokaryotic mycelia. This is achieved either through spermatization or more commonly through the fusion of two uninucleate cells of the compatible homokaryotic mycelia. Thus, a heterokaryotic, often binucleate cells are established from this cell. The secondary mycelium is formed in two ways. In the first, the binucleate cell produces a branch into which the nuclear pair migrates; the two nuclei divide conjugately and the sister nuclei separate as the branch is divided into two cells by the formation of a septum. This process continues till the formation of extensive mycelial network in which each cell is dikaryotic. In the second method of dikaryotization, there is a division of nuclei in the binucleate cell followed by a migration of daughter nuclei into the primary mycelium of the opposite mating type. In other words, an 'a' nucleus moves into the 'b' mycelium while 'b' nucleus moves into the 'a' mycelium. The foreign nucleus in each mycelium then divides rapidly and its progeny migrate from cell to cell until both parent mycelia have been completely dikaryotized.

In majority of Basidiomycetes the secondary mycelium is characterized by the presence of hyphal connections known as CLAMP CONNECTIONS (Fig-19) by which the nuclei arising from conjugate division of a dikaryotic cell are separated into the daughter cell. When a binucleate cell is ready to divide, a short hook like projection is formed between two nuclei 'a' and 'b' (fig-19). The out growth is called a CAMP CELL. The two nuclei divide simultaneously. One division becomes oriented obliquely so that one daughter nucleus, 'b' forms in the clamp connection and the other, 'b', forms in the dividing cell. The division of the second nucleus is oriented along the long axis of the dividing cell so that one daughter nucleus, a, forms near one end of the cell and the other a, approaches one of the daughter 'a' nuclei of the other division. A septum forms to close the clamp at the point of its origin and another septum forms vertically under the bridge to divide the parent cell into two daughter cells with 'a' and 'b' nuclei in one daughter cell and 'a' and 'b' nuclei in the other (fig -19)

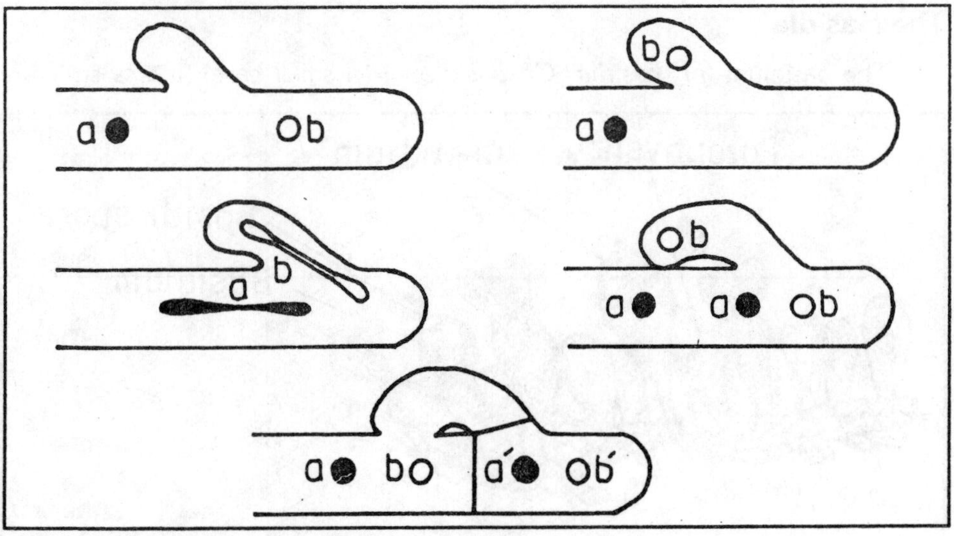
a
Ob
bo
a
b
a
ob
a
a
ob
a
bo
a′
ob′

Figure-19: Development of clamp connection

The TERTIARY MYCELIUM is represented by the organized, specialized tissues that comprise the basidiocarp of the more complex species. Such sporophores originate when the secondary mycelium forms complex tissues.

Studies on the structure of septa have revealed that in Basidiomycetes, the septa are generally laid in the middle surrounded by a double membrane (parenthesome) on each side forming a barrel shaped structure with a minor pore in the centre open at both ends. This type of septum is known as DOLIPORE SEPTUM, through which cytoplasmic continuity is maintained between adjacent cells but migration of nuclei is prevented.

The Basidiocarp

Most-basidiomycetes produce their basidia in fruiting bodies of various types called basidiocarp (*Gr. basidion*=small base, basidium+*Karpos*=fruit). Such fruiting bodies are not formed in rusts and smuts. They may be microscopic in size to several feet in diameter and up to many grams in weight. They could be crust like, gelatinous, cartilaginous, papery, fleshy, spongy, corky, woody, or indeed of almost any texture. Examples include mushrooms, bracket fungi, coral fungi, puffballs, stink horns, and bird's nest fungi. Basidiocarp may be open exposing their basidia or they may be closed where basidiospores are liberated after basidiocarp's disintegration. Basidia are produced in hymanium (layer comparable to ascus wall). Alongwith basidia, are intersperseed sterile hyphal structures called paraphyses. Some sterile hyphae (paraphyses) assume larger and swollen shapes and then are called cystidium (fig-20).

The Basidia

The basidium (*pl. basidia*, *Gr. basidion* = a small base); it is a structure

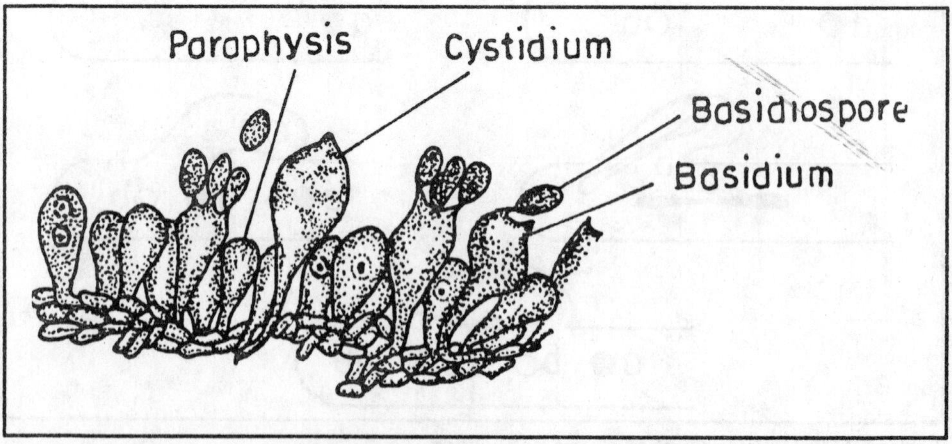
Paraphysis
Cystidium
Basidiospore
Basidium

Figure-20: Hymenium of a basidiomycete

bearing on its surface a definite number of basidiospores (*Gr. basidion*=small base + *spora*) i.e. a meiospore borne on the outer side of a basidium, following karyogamy and meiosis.

Development of Basidium and Basidiospores

Form the terminal of binacleate hypha, a simple club shaped structure develops which further develops to become basidium. After its separation by septal formation, it enlarges and becomes broder (fig-21). The two nuclei present in swollen structure (basidium) undergo karyogamy and the zygote formed undergoes meiosis giving rise to four haploids. During the nuclear division, four small outgrowth structures called STERIGAMATA are formed. Thereafter, vacuoles develop at the base of basidium as contants alongwith nuclei are pushed to periphery where autogrowths have developed. One nucleus then moves into each of the four basidiospores. The basidiospores eventually are discharged. It should be emphasized here that not all basidia bear four spores. Some may produce only two while others produce more that four.

The basidium is divided into three parts: probasidium, metabasidium, and the sterigmata. Probasidium (pl probasidia, Gr. *pro*=before+*basidium*) is the portion of basidium in which karyogamy takes place. Metabasidium (pl. metabasidia, Gr. *meta*=between + *basidium*) it is that portion of basidium where meiosis takes place. Sterigmata (Gr. sterigma=support) is a small hyphal branch or structure, which supports a basidiospore.

Traditionally, two basic types of basidia, termed HOLOBASIDIA (Gr. *holos*=entire+*basidium*) and PHRAGMOBASIDIA (Gr. *phragma*= a fence/

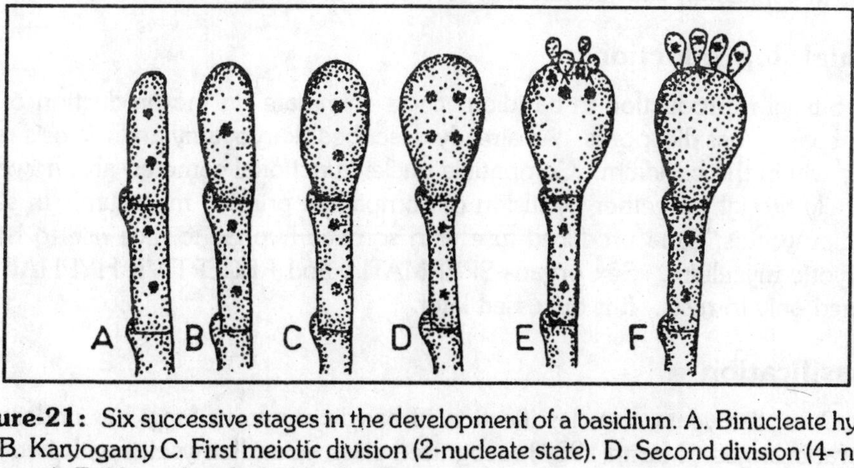
A
B
C
D
E
F

Figure-21: Six successive stages in the development of a basidium. A. Binucleate hyphal tip. B. Karyogamy C. First meiotic division (2-nucleate state). D. Second division (4- nucleate stage). E. Young basidiospores developing on sterigmata and nuclei preparing to migrate into the spores. F. Mature basidium with four uninucleate basidiospores.

septum). Holobasidia are single called; phragmobasidia typically are divided into four cells. Another type of basidium has been described in the rust and smut fungi. The basidium in these fungi begins to develop when teliospores germinate to form a germtube on which the basidiospores are produced. Another type of basidium has also been described. It is HETEROBASIDIUM (Gr. *hetero*=different + *basidium*). It is a term used to refer any type of basidium other than a single celled, club shaped basidium.

The Basidiospores

Basidiospores usually are unicellular, haploid, globose or oval or elongated, or even angular. They may be colourless or pigmented (green, yellow, orange, pink, violet, brown or black). As already discussed, basidiospores usually receive a single nucleus from the basidium, however in some species two nuclei may move in basidiospore. It may also become binucleate due to mitotic division of its nucleus. Basidiospores germinate directly or indirectly. In DIRECT GERMINATION, primary mycelia are formed, while in INDIRECT GERMINATION, secondary spores of buds are formed.

Asexual Reproduction

Asexual reproduction takes place by means of budding, by fragmentation of the mycelium, and by production of conidia, arthrospores, or oidia. Production of conidia is common in smut fungi, where conidia are budded off both the basidiospores and mycelium. The rust fungi produce urediniospores that are conidial in origin and function. Other basidiomycetes produce conidia, sometimes in conjuction with basidiocarp formation.

Sexual Reproduction

Sexual reproduction in basidiomycetes culminate in the production of basidia bearing basidiospores. As already described, karyogamy and meiosis takes place within the basidium. Compatible nuclei (functional gametes) are, however, typically brought together by fusion of compatible primary mycelium. In some basidiomycetes, oidia produced fuse with somatic hyphae to give rise to heterokaryotic mycelium. Sex organs-SPERMATIA and RECEPTIVE HYPHAE are formed only in rusts. It is discussed later.

Classification

Over the years, the classification of basidiomycetous fungi has undergone several changes. Martin(1951) recognized class - Basidiomycetes with two subclasses-Heterobasidiomycetes (3 orders) and Homobasidiomycetes (8 orders).

Smith (1955) described class Basidiomycetae with two sub-classes Heterobasidiomycetidae (6 orders) and Homobasidiomycetidae(7 orders). Alexopoulus (1962) created class Basidiomycetes with two subclasses - Heterobasidio mycetidae (3 orders), and Homobasidiomycetidae with two series - Hymenomycetes (2 orders), and Gasteromycetes (five orders). In 1966, Ainsworth placed basidiomycetous fungi in sub-division-Basidiomycotina with three classes Teliomycetes, Hymenomycetes, and Gasteromycetes. Alexopoulos (1979) placed basidiomycetes in Division Amastigomycota, sub-Division - Basidiomycotina with three sub-classes -Holobasidiomycetidae, Phragmobasidiomycetidae, and Teliomycetidae. Alexopoulos etal (1996) discussed the evolution of basidiomycetes. They have proposed a hypothesis given below, based on informations generated by researchers in areas like electron microscopy, rDNA etc.

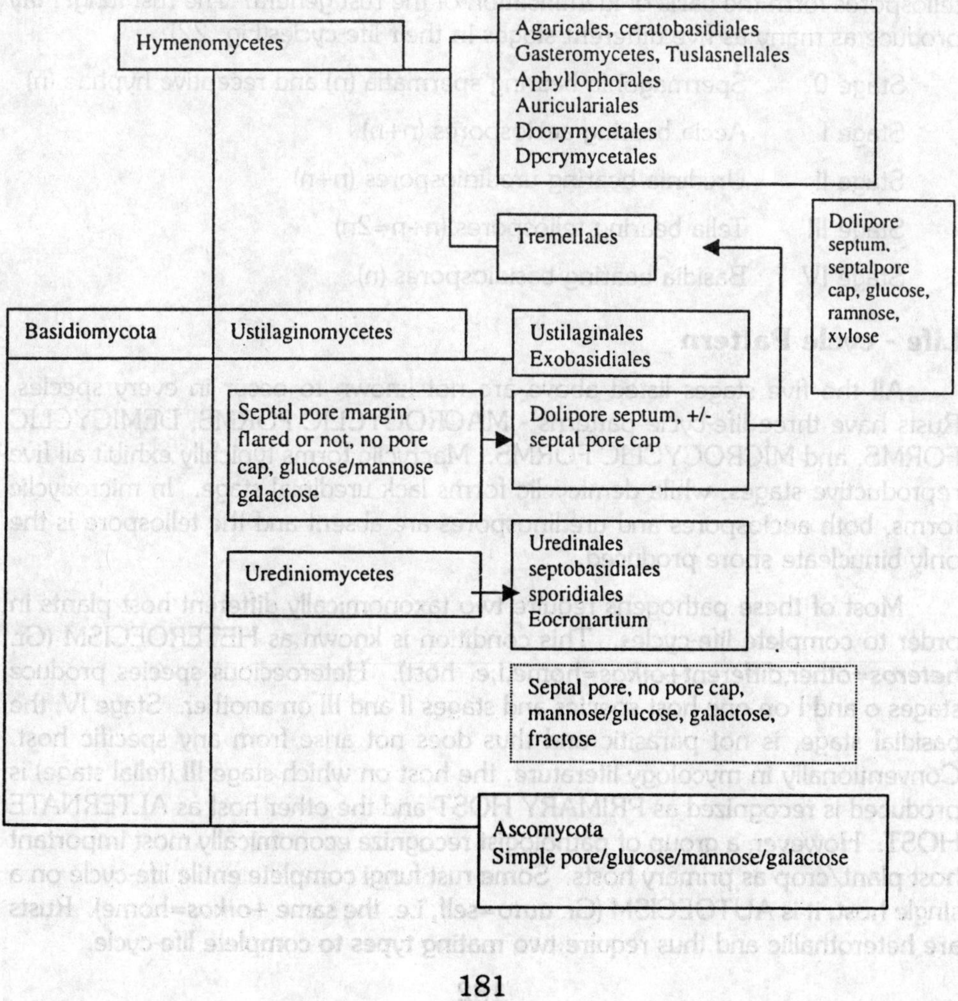
Hymenomycetes
Agaricales, ceratobasidiales
Gasteromycetes, Tuslasnellales
Aphyllophorales
Auriculariales
Docrymycetales
Dpcrymycetales
Tremellales
Dolipore septum, septalpore cap, glucose, ramnose, xylose
Basidiomycota
Ustilaginomycetes
Ustilaginales
Exobasidiales
Septal pore margin flared or not, no pore cap, glucose/mannose galactose
Dolipore septum, +/- septal pore cap
Urediniomycetes
Uredinales
septobasidiales
sporidiales
Eocronartium
Septal pore, no pore cap, mannose/glucose, galactose, fractose
Ascomycota
Simple pore/glucose/mannose/galactose

The Rust Fungi

Phylum - Basidiomycota, class-urediniomycetes, order uredinales.

The rusts are by far the most important amongst all the Basidiomycetes. It is estimated that there are about 5000 species belonging to 140-150 genera. These fungi are parasitic on many important crops, trees, ornamentals etc., and cause "Rust Disease". The basidiocarp is absent, therefore, there is no 'tertiary mycelium'. The hypha usually secondary mycelium, are intercellular in host cells and derive nutrients through 'haustoria'. Clamp connections are rare or absent. The structure in which karyogamy takes place is a specialized structure called Teliospore and therefore, technically part of the basidium. These teliospores originate from the apical cells of dikaryotic hyphae. The structural details of teliospores form the basis of identification of the rust genera. The rust fungi may produce as many as five different stages in their life-cycles(Fig. 22).

Stage 0	Spermagonia bearing spermatia (n) and receptive hyphae (n)
Stage I	Aecia bearing aeciospores (n+n)
Stage II	Uredinia bearing urediniospores (n+n)
Stage III	Telia bearing teliospores (n+n=2n)
Stage IV	Basidia bearing basidiospores (n)

Life - cycle Pattern

All the five stages listed above are not known to occur in every species. Rusts have three life-cycle patterns - MACROCYCLIC FORMS, DEMICYCLIC FORMS, and MICROCYCLIC FORMS. Macryclic forms typically exhibit all five reproductive stages, while demicyclic forms lack uredinial stage. In microcyclic forms, both aeciospores and uredinospores are absent and the teliospore is the only binucleate spore produced.

Most of these pathogens require two taxonomically different host plants in order to complete life-cycles. This condition is known as HETEROECISM (Gr. *heteros*=other,different+*oikos*=home,i.e. host). Heteroecious species produce stages o and I on one host species and stages II and III on another. Stage IV, the basidial stage, is not parasitic and thus does not arise from any specific host. Conventionally in mycology literature, the host on which stage III (telial stage) is produced is recognized as PRIMARY HOST and the other host as ALTERNATE HOST. However, a group of pathologist recognize economically most important host plant/crop as primary hosts. Some rust fungi complete entile life-cycle on a single host, it is AUTOECISM (Gr. *auto*=self, i.e. the same +*oikos*=home). Rusts are heterothallic and thus require two mating types to complete life-cycle.

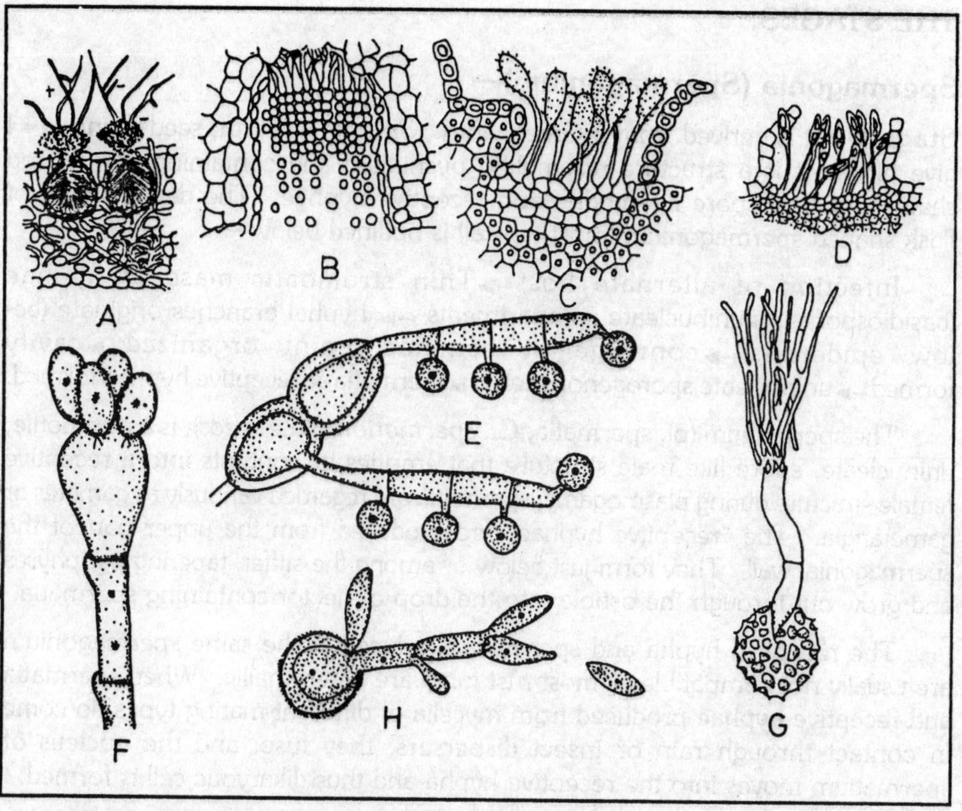
A
B
C
D
E
F
G
H

Figure-22: Different stages in life-cycles of rust fungi

A=	Stage 0	Spermagonia bearing spermatia (n) and receptive hyphae (n)
B=	Stage I	Aecia bearing aeciospores (n+n)
C=	Stage II	Uredinia bearing urediniospores (n+n)
D=	Stage III	Telia bearing teliospores (n+n=2n)
E=		Germinating teliospores
F=	Stage IV	Basidia bearing basidiospores (n)
G=		Germinating basidiospores bearing primary sporidia
H=		Germinating primary sporidia bearing secondary sporidia

Microcyclic forms are, of course, all autoecious. The autoecious species are hought to have developed from heteroecious species. According to "Tranzschel's .aw", telia of such species typically stimulate the habit of the aecia of the preumptive parental species. They are grouped in the same way as aecia. If the ecial mycelium is systemic, then the telial mycelium also will be systemic. In ome cases teliospores are morphologically similar to aeciospores. However, pon germination they produce only basidia and basidiospores. These are reerred as ENDOCYCLIC RUST FUNGI.

THE STAGES.

Spermagonia (Spermagonium)—

Stage O : it is derived from a Greek word. *Sperma*= sperm, seed+*gennao*= I give birth. It is a structure resembling pycnidium and containing minue, rod shaped, or oval spore like bodies and receptive hyphae. The development of flask shaped spermagonium (most typical) is outlined below —

Infection of alternate host→Thin stromoatic mass of hyphae (basidiospores)→uninucleate compartments → hyphal branches originate (below epidermis)→converge at common point organized→cavity formed→uninucleate sporogenous cells→spermatia & receptive hyphae formed.

The spermatium (pl. spermatia, Gr. *spermation*=little seeds), is a nonmotile, uninucleate, spore like male structure that empties its contents into a receptive female structure during plasmogamy; spermatia are regarded variously as gametes or gametangia. The "receptive hyphae" are produced from the upper part of the spermagonial wall. They form just below or among the stiffer, tapering periphyses and grow out through the ostiole into the drop of nector containing spermatia.

The receptive hypha and spermatia produced in the same spermagonium are usually not compatible as most rust fungi are heterothallic. When spermatia and receptive hyphae produced from mycelia of different mating types do come in contact through rain or insect dispersers, they fuse, and the nucleus of spermatium moves into the receptive hypha and thus dikaryotic cell is formed.

Aecia and Aeciospores (*Gr. aikia*=injury-blisters (stage I)

Aecia primordia are formed from primary mycelium prior to dikaryotization. The cells are initially uninucleate and become dikaryotic following spermatization. An aecium consists of a group of typically dikaryotic cells within the parasitized host that gives rise to chains of dikaryotic "Aeciospores" Aeciospores of heteroecious species are, of course, produced on alternate host. They are, however, capable of infecting only primary host.

Uredinia and Urediniospores (Stage II)

Urediniospores constitute the repeating stage of the rusts since several crops of the spores may be produced in one growing season. They are formed in structure called UREDINIA (sing. *Uredinium*; L. *Urere* = to burn) because of their reddish colour. The uredinial cells are formed subepidermally from dikaryotic mycelium originating from the germination of aeciospore. As the spores form, they press against host epidermis, pushing it out and finally rupturing it. Urediniospores are dikaryotic and have a thick wall covered with minute spines.

Telia and Teliospores (stage III)

Telia (sing. Telium) is derived from Greek word *telos* meaning the end. Telia are groups of binucleate cells that give rise to special thick walled cells called. TELIOSPORES. In many rust fungi the old Uredinia actually are converted to telia. Teliospores may be unicellular or composed of two or more cells formed from the tips of binucleate cells of the telium. Each cell of the spore is first dikaryotic, but enventually Karyogamy takes place, rendering each cell diploid. Teliospore morphology varies greatly and is an important taxonomic character. Teliospore may be sessile or stalked; they may be completely free from one another or they may be embedded in a gelatinous matrix or united laterally by fused walls, forming small grups, layers, or columns. They vary in size, shape, colour and in the characterstics of their walls, including the number and location of germpore regions.

Basidia and Basidiospores (Stage IV)

Rust teliopspore is a probasidium. Under congenial conditions each cell of teliospore germinates forming a hyphal structure called PROMYCELIUM. The diploid nucleus migrates into the promycelium and undergoes meiosis. The four resulting haploid nuclei become distributed at equal distances from each other in the promycelium, and septa develop between them to divide the promycelium into four uninucleate cells. Each of these cells then produces a sterigmata at the tip of which a typically pear-or kidney shaped basidiospore develops. The four nuclei now migrate into the developing basidiospores. Eventually, basidiospores are forcibly discharged from their sterigmata.

Upon germination, a basidiospore either gives rise to a germtube-so called direct germination or forms an out growth that functions as sterigmata, forming another spore at its tips. The new spore called either a sporidium (Gr. *spora*=seed, spore+*L. idium* = dimin, suffix) or a secondary spore, is virtually identical in appearance to parent basidiospores. These basidiospore through infection structures once inside the host cells, intercellular homokaryotic mycelia develop and finally spermagonia distinctive of the species develop.

Classification

At the present time there is little agreement regarding the division of *Uredinales* into families. over the years various workers have proposed from as few as 2 to as many as 14 familes. The information that is available, supports the division of the order *Uredinales* into the families *Pucciniaceae*, and *Melampsoraceae*.

Family - Pucciniaceae (Plate-XIV)

The members of this family produce stalked teliospores which may be free from one another or may be surrounded by a common gelatinous matrix. The spores may be one, two or many celled. The wall of the spores are reddish brown, thick, smooth or variously sculptured. The genera of this family are distinguished on the basis of the teliospores, aecial characterstics and the type of life-cycle. Some of the important genera are *Uromyces*, *Puccinia*, *Gymnosporangium* and *Phragmidium*.

Genus : *Puccinia*

It is named after an Italian scientist T Puccini. About 262 species are reported to occur in India. Many species are *autoecious* (*P.helianthi, P. obtegens*). Still several species are *heteroecious*. *P. graminis* is a macrocyclic heteroecious rust completing life-cycle on wheat and barberry. Some common rusts found in India are —

Host	Disease	Pathogen
Wheat	Black/stem rust	*P. graminis tritici*
	Brown rust	*P. recondita*
	Yellow rust	*P. striiformis*
Oats	Brown rust	*P. coronata*
	Stem rust	*P; graminis avenae*
Bajra	Leaf rust	*P. penniiseti*
Sorghum	Leaf rust	*P. purpurea*
Bean	Bean rust	*uromyees appendiculatus*
Gram	Gram rust	*u. cicer arietini*
Pea	Pea rust	*u. pisi*

Typical life-cycle of a Rust fungus

Now you are fully familiar with the many structures produced by rust fungi. It is time to put them together to form a life-cycle. *P. graminis*, a versatile parasite of many cereals and grasses is being used as an example. *P.graminis*, an obligate parasite, is heterothallic, macrocyclic, and heteroecious. It completes life cycle between its grass host (either *Berberis* (barberry) or *Mohonia* species). Stage II and III are produced on the grass host (the primary host) and stage O and I on *Berberis* or *Mahonia* (the alternate host). The two celled teliospores of *P. graminis* after overseasoning germinate and produces a promycelium in which

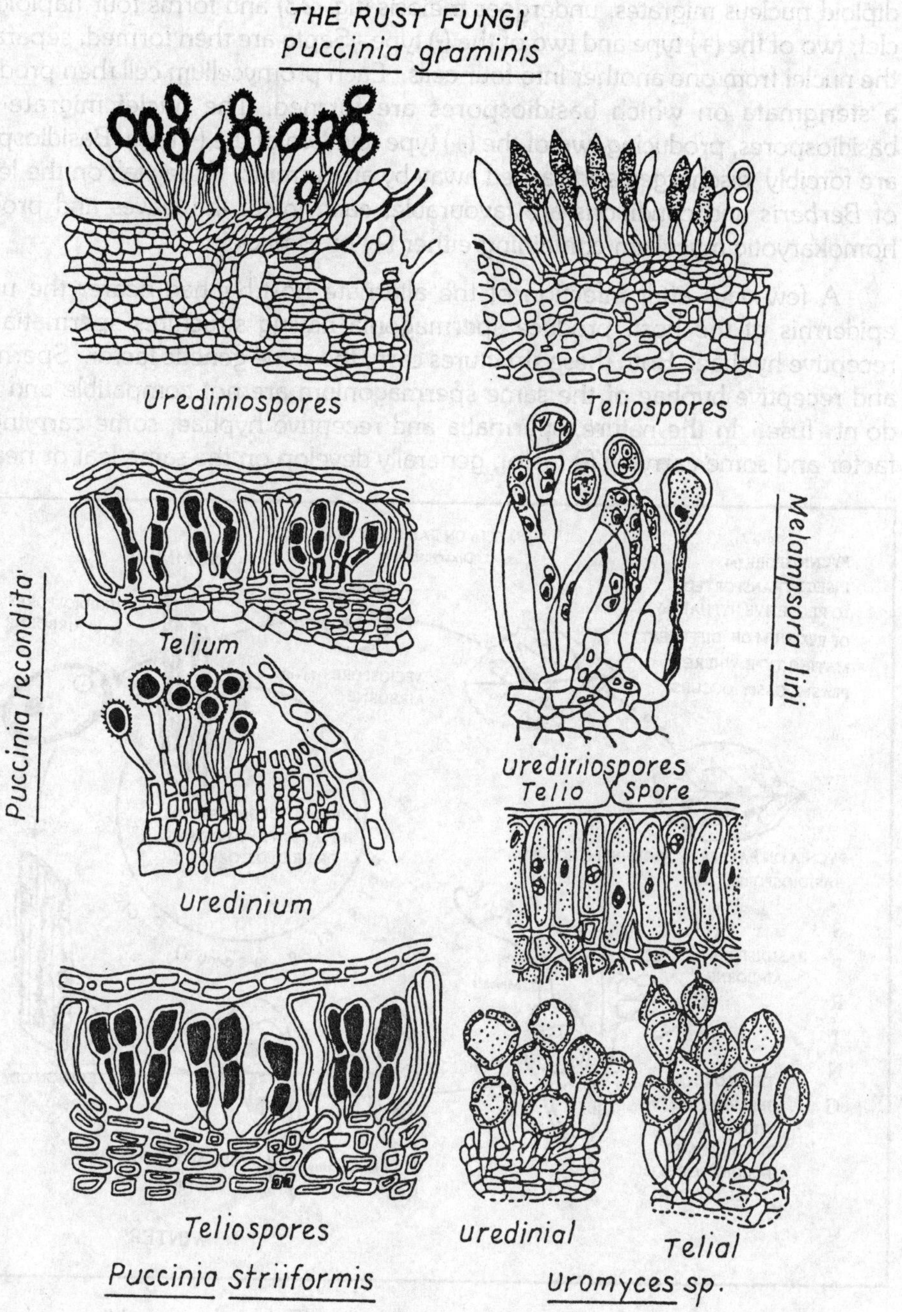
THE RUST FUNGI
Puccinia graminis
Urediniospores
Teliospores
Telium
Puccinia recondita
uredinium
Melampsora lini
urediniospores
Telio
Spore
Teliospores
Puccinia striiformis
uredinial
Telial
Uromyces sp.

Plate-XIV

diploid nucleus migrates, undergoes meiosis (fig -23) and forms four haploid nuclei, two of the (+) type and two of the (-) type. Septa are then formed, separating the nuclei from one another into four cells. Each promycelium cell then produces a sterigmata on which basidiospores are formed. The nuclei migrate into basidiospores, producing two of the (+) type and two of the (-) type. Basidiospores are forcibly discharged and carried away by air currents. If they fall on the leaves of *Berberis* and conditions are favourable, each infect the leaves and produce homokaryotic mycelium containing either (+) or (-) nuçlei.

A few days after infection of the alternate host,hyphae nearer the upper epidermis of the host, produce spermagonia having structures spermatia and receptive hyphae. Both these structures carry the same genetic factor. Spermatia and receptive hyphae of the same spermagonium are not compatible and then do'nt fuse. In the nature, spermatia and receptive hyphae, some carrying (+) factor and some carrying (-) factor, generally develop on the same leaf or near by

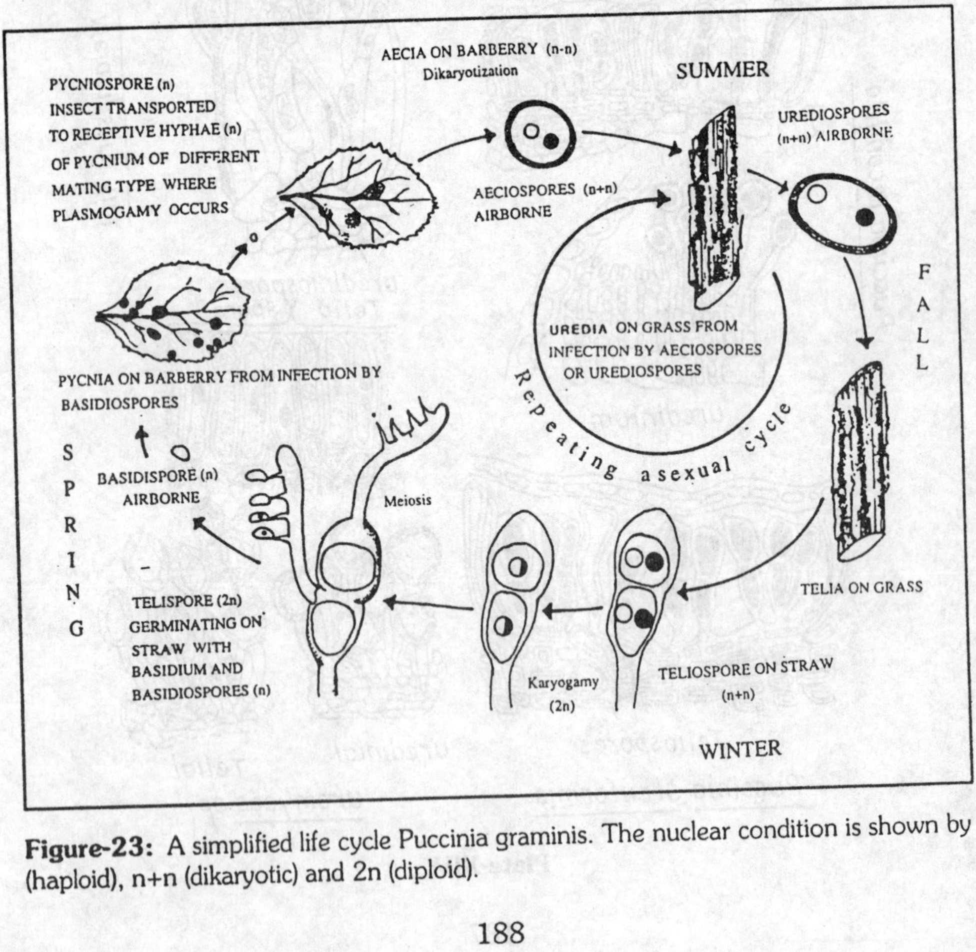
AECIA ON BARBERRY (n-n)
Dikaryotization
SUMMER
PYCNIOSPORE (n)
INSECT TRANSPORTED
TO RECEPTIVE HYPHAE (n)
OF PYCNIUM OF DIFFERENT
MATING TYPE WHERE
PLASMOGAMY OCCURS
UREDIOSPORES
(n+n) AIRBORNE
AECIOSPORES (n+n)
AIRBORNE
F
A
L
L
UREDIA ON GRASS FROM
INFECTION BY AECIOSPORES
OR UREDIOSPORES
PYCNIA ON BARBERRY FROM INFECTION BY
BASIDIOSPORES
Repeating asexual cycle
S
P
R
I
N
G
BASIDISPORE (n)
AIRBORNE
Meiosis
TELISPORE (2n)
GERMINATING ON
STRAW WITH
BASIDIUM AND
BASIDIOSPORES (n)
TELIA ON GRASS
Karyogamy
(2n)
TELIOSPORE ON STRAW
(n+n)
WINTER

Figure-23: A simplified life cycle Puccinia graminis. The nuclear condition is shown by (haploid), n+n (dikaryotic) and 2n (diploid).

leaves. Spermatization now takes place through the agency of insects. The insects attracted by the fragrance of the spermagonial mass, visit it and suck sweet nectar exuding the ostiole. During this process spermatia present in nectar adhere to mouth parts and during subsequent visit to another spermatogium facilitates contact between spermatia and receptive hyphae. The contact and then fusion between (+) factor spermatia and (-) factor receptive hyphae, is spermatization. At contact point between (+) spematia and (-) receptive hyphae, spermatial content passes into receptive hyphae.

Meanwhile, the homokaryotic mycelium continues to colonize leaf tissues. The hyphae that also start developing after fusion between (+) spermatia and (-) receptive hyphae, also move down to aecial primordial cells. The entire structure becomes dikaryotic apparently due to passing through the septal perforations of the mycelium, and thus reaching aecial primordial cells. Besides this, the (+) and (-) factors carrying hyphae also fuse directly to bring dikaryotic conditions.

Dikaryotization leads to formation of aecial stage (aecia, aeciospores). Aeciospore contains both (+) nucleus and (-) nucleus. Thus, first binucleate spore is produced. Epidermis is pressed to rupture by huge density of aeciospores and then disseminated by wind. When they reach their grass hosts, infection results and then dikaryotic mycelium intercellularly formed. Masses of cells aggregate, develop and culminating in the formation of *uredinia* (dikaryotic) and urediniospores (also dikaryotic). Further development results in development of urediniopustules and after epidermis rupture, rust-red pustules are seen. In masses, urediniospores look apparetly rust-red, hence the name rust. The urediniospores behave as asexual spores as they are produced, disseminated, cause new infections till host and environmets favourable. Thus, in rust diseases this stage and this inoculum are of epidemic significance. As the crop nears maturation uredinia begin to form TELIA and TELIOSPORES. Thus, uredinopustules become teliopustules or sori. Thus, from haploid monokaryotic stage—Dikaryotic stage —Diplophase and then — again haplophase.

Genus - *Uromyces (Plate-XIV) (Figure-24)*

The teliospores are brown and simple (1-celled) on a simple pedicel with thickening more at the apex than on the sides. Uredinial, aecial and spermagonial characters similar to *Puccinia*. The species may be heteroecious or autoecious. Mostly parasites on legumes. Common species are —

- *Uromyces Pisi* — Rust of pea ; stages O and I on *Euphorbia cyparissias*, and stages II and III on peas.
- *Uromyces dianthi* — carnation rust; stages O and I on *Tithymalus* sp. ; II and III on carnation.

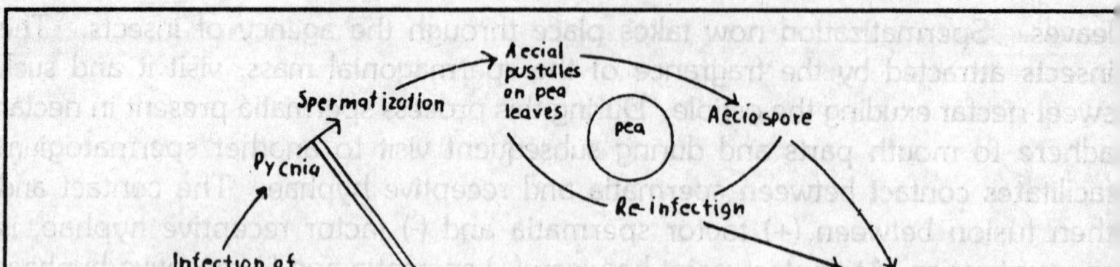
Aecial pustules on pea leaves
Spermatizotion
Pea
Aeciospore
Pycnia
Re-infectign
Infection of pea
Uredia
Bosidiospores
Pea
Infectionof pea
HAPLOPHASE
DIKARYOPHASE
Urediopere
Meiosis
Pea leave and stems
Germination
DIPLOPHASE
Telia
Mature teliospore
young teliospore (binudeotc)
Karyogamy

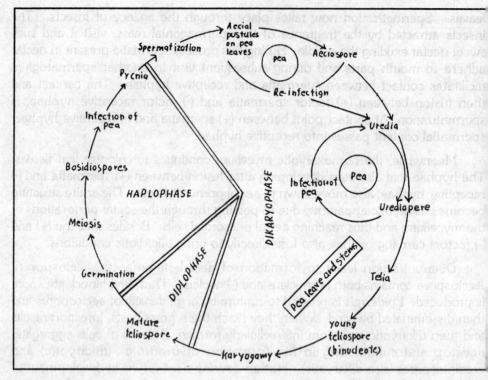

Figure-24: Life cycle of Uromyces fabae

- *Uromyces fabae* — Rust of pea; Autoecious
- *Uromyces phaseoli typica* — Rust of beans; autoecious; all stages on beans.
- *Uromyces cicerisarietini* — Rust of chickpea, stages O and I not seen; II and III on gram

The life-cycle of *uromyces fabae* is shown in the figure -24.

Family — Melampsoraceae

This family is considered to contain all those uredinales in which the teliospores are sessile. The teliospores may be present in one or more layers, may or may not be strongly united laterally. In most cases the telia is subepidermal and teliospores form crusts or columns. *Melampsora* is the most important genus of this family in India.

Genus — *Melampsora*

The teliospores are sessile, single-celled, and formed in a palisade layer. The telia are erumpent or crustose and diffuse. The aecia lack peridium. The uredinia are crumpent and have capitate paraphyses interspersed among urediniospores. *M.lini* causing rust of flax is the best known. The fungus which infects several species of *Linum* is an autoecious rust. The pycnial and aecial stages have not been recorded in nature. However, Prasada (1940) observed these on experimental plants.

Life-cycle (*M.linl*) (Fig-25)

The urediniospores growing locally on collateral hosts, or blown down from hills, initiate infection every year, when flax crop is sown. Bright reddish—yellow uredial spots appear in December—January in circles on leaves and as elongate patches on stem. The urediniospores serve to spread the pathogen. The infected leaves die and fall off prematuraly. Black, crust like telia appear on the stem. The teliospores are released only after the disintegration of the host tissues. They germinate when exposed to freezing temperature and produce four basidiospores on a septate epibasidium, but play no role in the recurrence of disease as they can not infect the host. The teliospores fail to survive the intervening hot summer. *Linum mysorens* is reported to serve as collateral host.

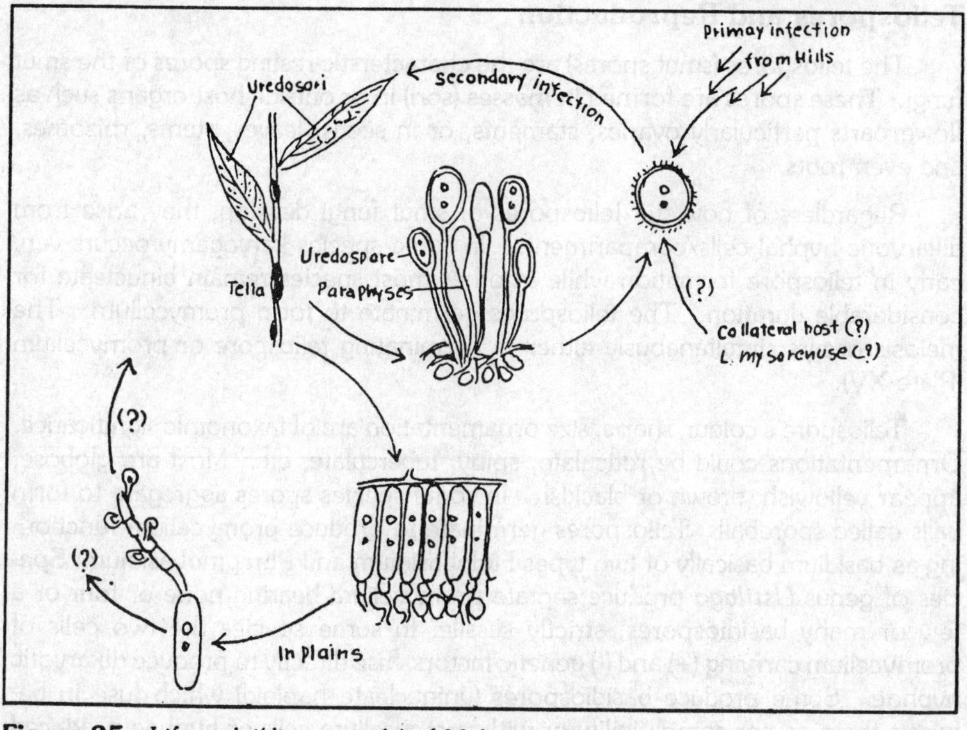
Primay infection
from Hills
Secondary infection
Uredosori
Uredospore
Telia
Paraphyses
(?)
Collateral host (?)
L. mysorenuse (?)
(?)
(?)
In plains

Figure-25: Life cycle (disease cycle) of *Melampsora lini.*

Phylum — BASIDIOMYCOTA

Class — ustilaginomycetes

order — ustilaginales (The smut fungi)

General Characterstics

The ustilaginales are the smuts. They produce their basidiospores on promycelium which grows out of a germinating teliospore; incontrast to rusts, the smuts 1. produce their teliospores from intercalary dikaryotic hyphal cells, 2. bear their basidiospores not on sterigmata but sessile, 3. do not discharge their basidiospores violently, only passive dispersal and 4. produce no specific sex organ.

Occurrence

About 1200 species of smut fungi belonging to more than 50 genera, attacking about 4000 species of angiosperm belonging to over 75 plant families, is the statistic showing wide distribution and potential parasitism.

Teliospores and Reproduction

The teliospores (smut spores) are the characterstic resting spores of the smut fungi. These spores are formed in masses (sori) in or outside host organs such as flowerparts particularly ovaries, staments, or in seeds, leaves, stems, rhizomes, and even roots.

Regardless of how the teliospores of smut fungi develop, they arise from dikaryotic hyphal cells/compartments. In some species karyogamy occurs very early in teliospore formation while in other most species remain binucleate for considerable duration. The teliospores germinate to form promycelium. The meiosis occurs simultaneously either in germinating teliospore or promycelium (Plate-XV).

Teliospore's colour, shape, size ornamentation are of taxonomic significance. Ornamentations could be reticulate, spiny, tuberculate, etc. Most are globose, appear yellowish, brown or blackish. In some species spores aggregate to form balls called sporeballs Teliospores germinate to produce promycelium functioning as basidium basically of two types Holobasidium and Phragmobasidium. Species of genus *Ustilago* produce septate promycelium bearing none or four or a few or many basidiospores, strictly sessile. In some species the two cells of promycelium carrying (+) and (-) genetic factors fuse directly to produce dikaryotic hyphae. Some produce basidiospores (uninucleate haploid) which fuse in between them as per compatibility or with promycelium cells or hyphae produced from promycelium cells. All these lead to creation of dikaryotic state through plasmogamy as listed above and then teliospores, Karygamy, meiosis and finally basidiospores.

In smut fungi like *Neovossia*, *Tilletia*, the promycelium is aseptate hollow swollen structure i.e. holobasidium. The basidiospore are formed at the tip of basidium.

Classification

Historically, the order *Ustilaginales* has been divided into the families — *Ustilaginaceae*, and *Tilletiaceae*. In addition to ultrastructural features, these are separated on the basis of the mode of germination of teliospores. In *Ustilaginaceae* species the promycelium is prostrate and transversely septate. Basidiospores develop both laterally and teminally. In *Tilletiaceae* forms, the promycelium is either unicellular or aseptate and aerial and only terminal basidiospores are produced.

Family - Ustilaginaceae

As noted above, members of this family produce promycelia that are transversely septate. Basidiospores are formed both laterally and terminally directly on promycelium. The promycelium arises directly from a diploid teliospore soon after their formation or only after a period of dormancy. At the time of germination, the spore wall cracks open and the promycelium issues forth as a short germtube. The diploid nucleus soon migrates into the promycelium and undergoes meiosis, and the four resulting nuclei (haploid) get distributed uniformaly in promycelium. In some species meiosis occurs in the teliospore and the nuclei migrate into promycelium. Septa are now laid down separating each nucleus and thus septate promycelium now consists of uninucleate cells. As each nucleus divides mitotically, one daughter nucleus migrates into a bud that develops at the tip of terminal cell or at the side of each of the other promycelial cells. The other nucleus remains in the parental cell. These uninucleate buds are basidiospores or primary sporidia. Some of the common genera belonging to this family are *Ustilago*, *Sphacelotheca* and *Tolyposporium*.

Genus : *Ustilago (Plate-XV)*

The genus includes nearly 300 species, many of them parasitic on crop plants. Some important species and disease caused are —

U. maydis	smut of maize (corn smut)
U. avenae	loose smut of oats.
U. segatum	loose smut of wheat

U. hordei	covered smut of barley
U scitaminae	whip smut of sugarcane

Somatic structure

The vegetative body of *Ustilago* passes through two phases Primary and Secondary. Primary mycelium is uninucleate (monokaryotic) which develops from basidiospores and carry (+) or (-) genetic factor. Secondary mycelium is binucleate formed by diploidization of the primary mycelium.

Reproduction

Ustilago reproduces both asexually and sexually. In asexual reproduction, conidia which are generally hyaline and pyriform, and either uni or binucleate depending upon the genetic factor(s) of the parent mycelium. Budding is quite a common method.

In sexual reproduction recognized sex organs i.e. gametangia are not formed. Heterothallism is common. The creation of dikaryotic condition takes place through-

1. the copulation between two sprout cells by the breaking of the promycelium,
2. the copulation between two monokryotic hyphae carrying opposite but campatible genetic factors,
3. the promycelium gives rise to monokaryotic hypha and two hyphae of opposite strains copulate and develop a dikaryon (e.g. *U. hordei*),
4. in *U.maydis*, the sporidium infects the young host tissues and produces monokaryotic hyphae. Two such hyphae copulate and produce dikaryon,
5. basidial copulation takes place between two cells of the same promycelium both belonging to the different sexual strains,
6. sporidial copulation occurs between the sporidia belonging to opposite phases, and
7. sometimes the sporidium may copulate with the cells of the promycelium.

Smut spores

Smut spores develop only on the dikaryotic mycelium. The spore mother cells enlarge, their walls gelatinize and protoplasm becomes granular. Below the gelatinous covering, a wall develops which persists in the mature spores and separates the spores from each other. The young spore is hyaline having dikaryon. With further development and maturation, the nuclei fuse creating diploid condition. The spore shows a thin inner wall (ENDOSPORE), and dark coloured thick outer wall (EXOSPORE).

Germination

Smut spores germinate by forming a germ-tube called promycelium. The diploid nucleus migrates in the promycelium. Eventually four haploid nuclei are formed. Septa are formed in promycelium, dividing the structure into 4 cells, with a nucleus in each. These nuclei divide by mitosis and subsequently a sporidium is formed in each cell and one of the daughter nuclei migrates in the sporidium and the other divide again and another sporidium germinates to form a primary mycelium or multiply by budding producing several crops of secondary sporidia.

Life - cycle pattern of ustilaginales and life-cycle of *U. maydis* are shown in figure-26 and 27 respectively.

Genus - *Sphacelotheca (Plate-XV)*

This genus produces sori in the inflorescence, predominantly in the ovaries. The sori are covered with a psendomembrane composed of fungal cells. The membrane later flakes away exposing the dusty spore mass and central columella composed of host tissues. Spores are free, single, and develope in centripetal

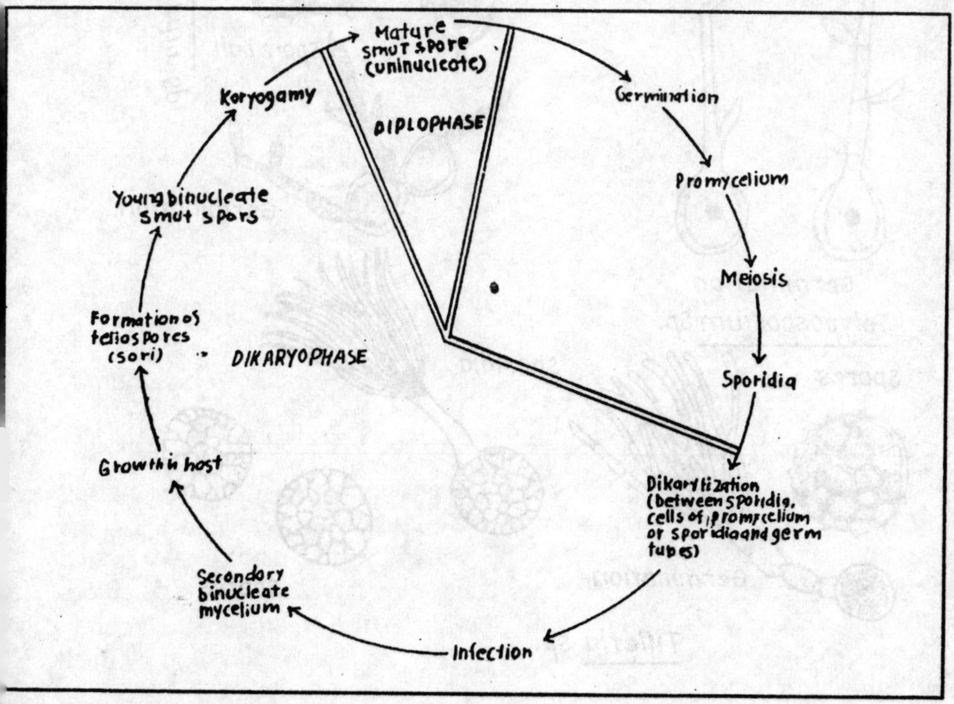

Mature smut spore (uninucleate)
Germination
Promycelium
Meiosis
Sporidia
Dikaryotization (between sporidia, cells of promycelium or sporidia and germ tubes)
Infection
Secondary binucleate mycelium
Growth in host
Formation of teliospores (sori)
Young binucleate smut spors
Karyogamy
DIPLOPHASE
DIKARYOPHASE

Figure-26: Life cycle pattern of Ustilaginales

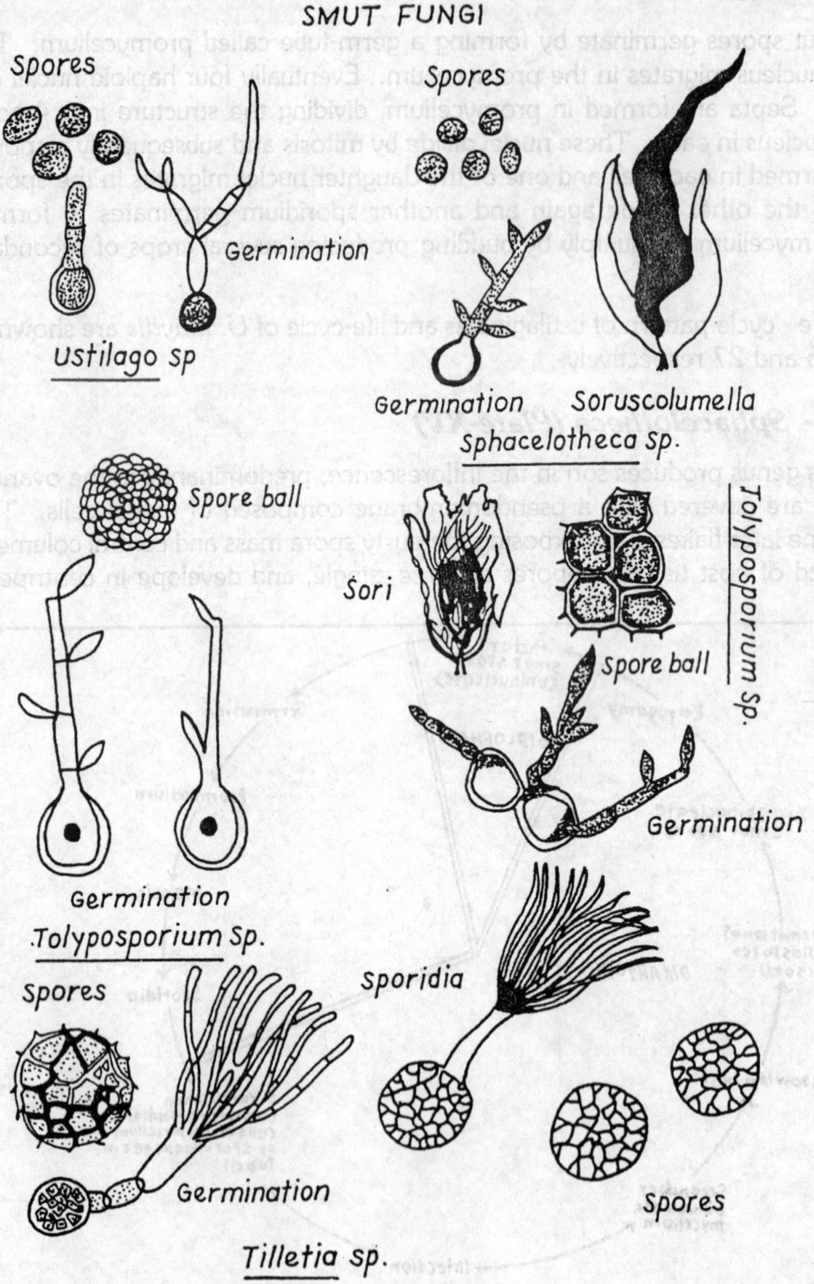
SMUT FUNGI
Spores
Germination
Ustilago sp
Spores
Germination
Soruscolumella
Sphacelotheca Sp.
Spore ball
Sori
Tolyposporium Sp.
Spore ball
Germination
Germination
Tolyposporium Sp.
Spores
Sporidia
Germination
Spores
Tilletia sp.

Plate-XV

manner. They germinate as the spores of *Ustilago*. A comparison of the characters of four smuts found in India is given in Table.

Characters	***Grain smut***	***Loose smut***	***Long smut***	***Head Smut***
Organism	*S. sorghi*	*S. cruenta*	*T. ehrenbergii*	*S. reiliana*
Ear infection	All or most grains smutted	All or most grains smutted	Only about 2 per cent of grains are infected	Entire head smutted into a single sorus
Sori	small, 5-15 X 3-5 mm	small 3-18X2-4 mm	long 40 mm x 6-8 mm	3-4 inch x 1-2 inch
Columella	short columella present	long columella present	columella absent, but 8-10 vascular strands present	columella absent but a network of vascular tissues present
Spores	in singles round to oval, olive-brown, smooth walled, 5-9 μ in diameter	in singles spherical or elliptical, dark-brown, spore walls pitted, 5-10 μ in diameter	always in balls, globose or angular, brownish green, warty spore wall 12-16 μ in diameter	loosely bound into balls, spherical or angular, dull brown, minutely papillate 10-16 μ in diameter.
Viability of spores	Over 10 years	about 4 years	about 2 years	up to 2 years
In culture	yeast-like growth with sporidia	in colonies with sporidia and resting spores 40 x 50 μ in diameter	incolonies with masses of sporidia	in colonies germ tubes and sporidia
Spread	externally seed-borne	externally seed-borne	air – borne	soil-borne and seed-borne

Genus - *Tolyposporium*(Plate-XV)

The main characteristic of this genus is the presence of spores in permanent balls. The sori develop in the inflorescence, usually in the ovaries, appearing black, somewhat granular spore masses. The spore ball are dark, composed of numerous spores, permanently united into rather irregular shaped balls of medium size. The spores are bound by ridged folds or thickenings of their outer walls. They germinate *in situ* by means of septate promycelium producing terminal and lateral sporidia.

Tolyposporium ehrenbergii causes long smut of sorghum. Sori are formed in individual and often only in few ovaries of the inflorescence. These sori are

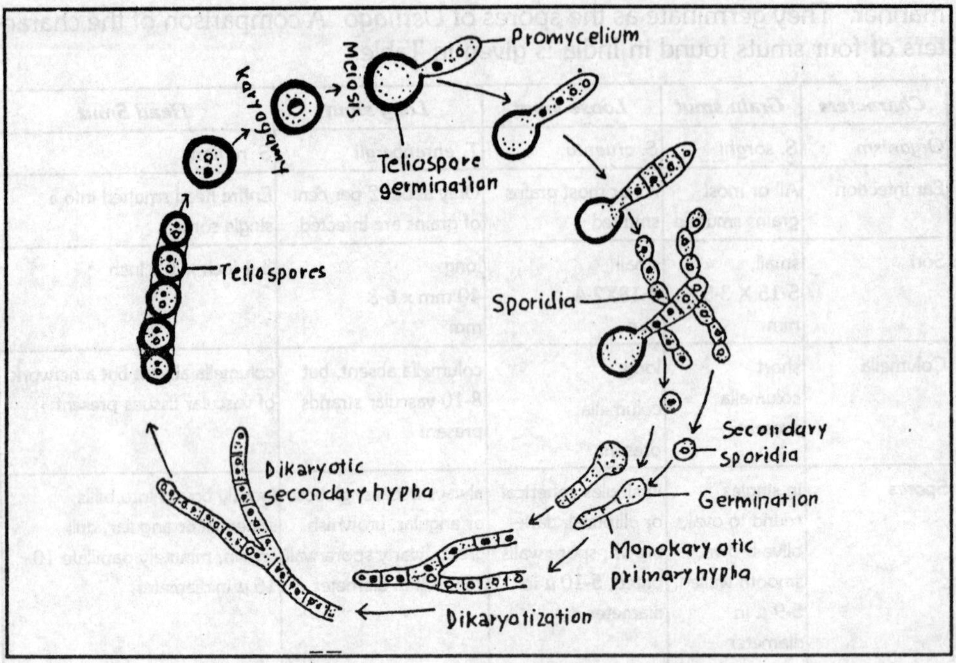
Promycelium
Meiosis
Karyogamy
Teliospore germination
Teliospores
Sporidia
Secondary sporidia
Germination
Dikaryotic secondary hypha
Monokaryotic primary hypha
Dikaryotization

Figure-27: Life cycle of Ustilago maydis, which has both saprophytic and parasitic mode of life

surrounded by healthy grains and are vary prominent being up to 4 cm long, cylindrical, slightly curved with tapered ends and yellowish. The thick pseudomembrane, composed of fungal cells, ruptures at the apex to release the brownish-green spore balls and expose a bundle of 8-10 dark-brown filaments. The spore balls are irregular, oblong or subglobose opaque, dark-brown, persistent, composed of 50 or more spores, and measure 35 to 240 microns. In India the size of spore balls is reported to be 30 to 120 (avarage 67.5) microns with less than 50 spores each. Individual spores are angular, globose, papillate on the free side and dark-brown in colour. They measure 9 to 15 (average 12.0) microns. The epispore is covered with dark-brown warts on the free surface and smooth inside the ball. These spores have no dormancy and germinate *in situ* by a septate promycelium bearing terminal and lateral sporidia which are spindle-shaped, 8-25 microns long and often arise directly from the spore.

Tolyposporium penicillariae causes smut of pearl millet (bajra). Pear-shaped or oval sori develop in few scattered ovaries of the inflorescence. They are larger than the seed and 3-4X2-3 mm in size. The sori are covered by a tough membrane composed of host tissue. Inside, there is a coarse, black powder formed mostly of roundish spore balls. The spore balls are of unequal size, 40 to 140 microns in diameter and consist of several spores pressed together more or less permanently. Individual spores are oval to ellipsoid or spherical, 6.5 to 12.5 microns in diameter and light-brown in colour. The epispore is smooth. The spores germinate *in situ* by a 4 celled promycelium. Each cell of the promycelium separates, falls off and buds out sporidia.

Family - *Telletiaceae (Plate-XV)*

The members of this family differ from those of *Ustilaginaceae* in the method of teliospore germination. The diploid nucleus in teliospore urdergoes meiosis just before germination. Following zero to two synchronous mitotic divisions, the haploid nuclei migrate into the promycelium. The promycelium remains aseptate. Thereafter, a number of basidiospores, typically eight, are formed at the tip of promycelium. Each nucleus in the promycelium moves into the basidiospore initial. However, once a nucleus has entered a spore initial, it divides mitotically and one of the danghter nuclei returns back to promycelium. These extra nuclei either degenerate or enter anucleate initials.

Genus — *Tilletia(Plate-XV)*

Tilletia caries and *T. foetida* cause the common bunt of wheat, whereas *T. contraversa* causes dwarf bunt, and *T. indica* causes karnal bunt of wheat. The first two species are similar in their life-histories and disease development. The biology of *T. contraversa* is different and some what similar to that of *T. indica*. The pathogens produce teliospores with different sets of wall markings. The life cycle of *T. caries* and *T. foetida* have been given in Fig. 28.

The mycelium is hyaline. During sporulation most cells are transformed into spherical, brownish teliospores, while the rest of the mycelial cells remain sterile and hyaline. On germination of a teliospore a basidium is produced, at the end of which 8 to 16 basidiospores develop in *T. caries* and *T. foetida*, where as 14 to 30 basidiospores develop in *T. contraversa*, and 32 to 48 in *T. indica*. The basidiospores usually called 'primary sporidia', fuse in pairs through production of lateral branches between compatible mating types and appear as H. shaped structure. The nucleus of each primary sporidium divides, and through exchange of one of the nuclei, the two fused primary sporidia become dikaryotic. When the primary sporidia geminate, they produce dikaryotic secondary sporidia. These produce dikaryotic mycelium, which can penetrate the plants and cause infection. After systemic development through the plant, the mycelium again forms teliospores in the Kernels.

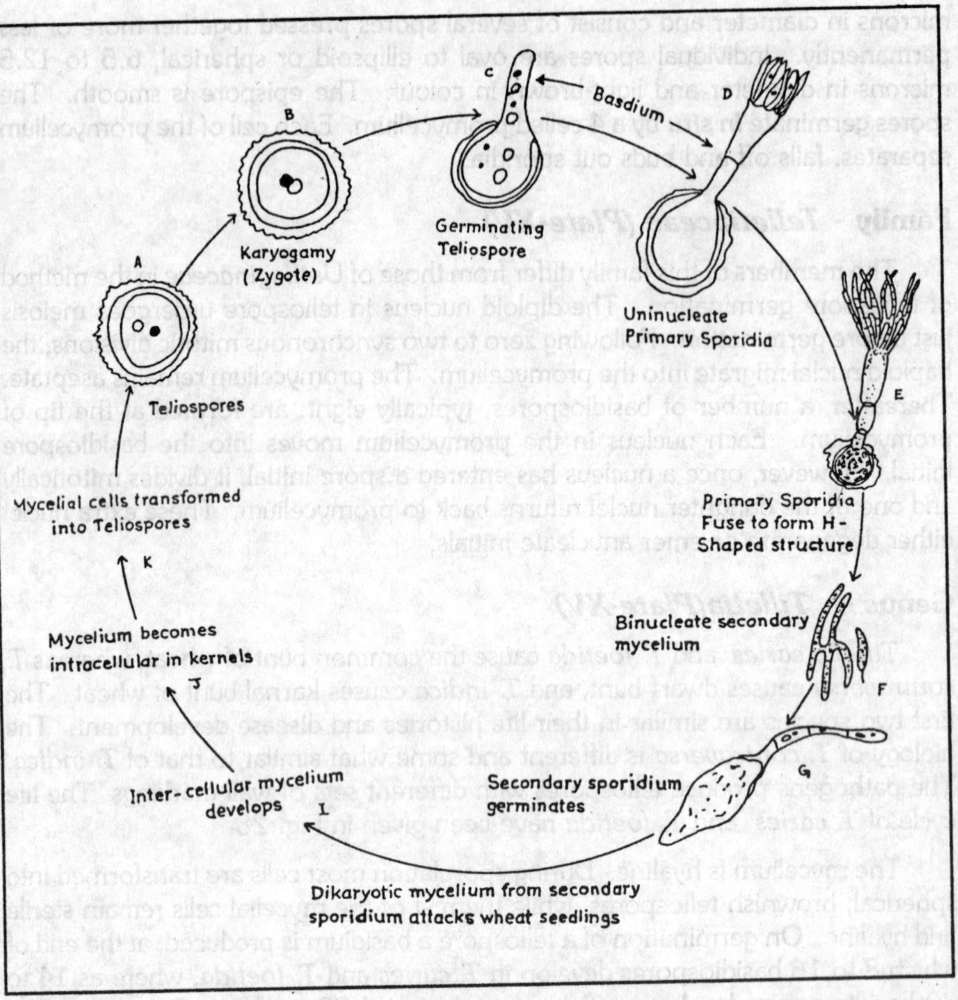
C
Basdium
D
B
Germinating
Teliospore
Karyogamy
(Zygote)
A
Uninucleate
Primary Sporidia
Teliospores
E
Mycelial cells transformed
into Teliospores
Primary Sporidia
Fuse to form H -
Shaped structure
K
Binucleate secondary
mycelium
Mycelium becomes
intracellular in kernels
J
F
G
Inter-cellular mycelium
develops
I
Secondary sporidium
germinates
Dikaryotic mycelium from secondary
sporidium attacks wheat seedlings

Fig. 28 : Life cycle of *Tilletia caries* and *T. foetida*

Life cycle of *T. indica* (karnal bunt of wheat)

T. indica is a heterothallic fungus. Teliospores are diploid (2N), thick walled, globose to subglobose. After a period of dormancy lasting 2 to 6 months, they germinate at the soil surface under favourable conditions. During germination process, the diploid nucleus (2N) undergoes meiosis followed by several mitotic divisions to produce haploid (IN) nuclei. When a teliospore germinates, a promycelium (basidium) emerges and produces as many as 180 primary sporidia (basidiospores) (Gill etal, 1993, Singh 1994). Each primary sporidium contains a single haploid nucleus. The primary sporidia germinate to produced mycelia.

Mycelia in turn produce large numbers of secondry sporidia.At the time of flowering , primary and secondary sporidia are splashed or blown on to the surface of glumes enclosing developing kernals. According to Dhaliwal etal,(1988) spordia from the soil surface are splashed or blown on to the surface of lower leaves, germinate, colonize the leaf surface, and produce secondary sporidia. These secondary sporidia are splashed or blown on to upper leaves, and in this progressive manner, pathogen finally reaches the head. On the surface of the glumes of developing head, sporidia germinate and peneterate through stomates.Mycelia grow downword to the base of the glumes, and up into the developing kernals(Goates,1988).

Apparently, the fungs is restricted to the pericarp, where it grows exclusively intercellulary. As the kernals mature, large numbers of teliospores can be produced and at the time of harvest teliospores are redeposited onto the soil surface to perpetuate the dissemination and survival of the pathogen.

Nuclear cycle of *T. indica*

The life-cycle of *T. indica* is partly dictated by the heterothalic nature of the fungus. When primary and secondary sporidia (each having a single haploid nucleus) germinate, they produce monokaryotic (IN) hyphae. At some unkwon point in the life-cycle of the pathogen, monokaryotic hyphae of different, compatible mating types fuse to produce dikaryotic (IN+IN) hyphae. This dikaryotyzation most likely occours on the plant surface just prior to peneteration, or within the plant tissue after penetration by the hyphae. At some later time, fusion (Karyogamy) of compatible nuclei occours within the dikaryotic hyphae, and teliospares develop (Funtes- Davila and Duran, 1986).

Genus- Neovossia

It is a break-up genus from *Tilletia*, which bears numerous filiform basidiospores (= primary sporidia). H- shaped structures are not formed .In addition, the teliospares bear a long pedicel- like appendage called atricules. The teliospores are formed singley from the end cells.

Deuteromycetes

Those fungi that do not reproduce sexually have been referred to as deuteromycetes or imperfect fungi. The conidial stages of most of these fungi are very simila- to conidial stages of seveal Ascomycetes. It is presumed that, with relatively .ew exceptions the imperfect fungi represent conidial stages of Ascomycetes whose ascigerous stages either are rarely formed in nature, and have not been found, or have been dropped from the life-cycle in the evolution of these organisms. In fact, in some cases sexual stages have been found in nature or have been produced in culture many years after they were first described as imperfect fungi. In such cases, the organisms can be classified in the ascomycete genera in which the characters of the ascigerous stage place them. In a few cases, the perfect stages whcih have been discovered have proved to be basidiomycetes.

Somatic Structure : With the exception of asporogenous yeasts, the thallus of the fungi imperfectii consists of well developed, septate, branched hyphae. The cells are usually multinucleate. The septa are perforated, permitting the streaming of cytoplasm and nucleus migration from one cell into the next.

Sporulation : In general, species which produce conidia on conidiophores arising directly from the somatic hyphae sporulate more rapidly than those in which conidia are produced in spore fruits such as pycnidia, sporodochia, or synnemata. Fungi which produce conidia on more or less loose, cottony hyphae are often called Hyphomycetes. Such conidia bearing hyphae may be simple or variously branched. They may be little different from the somatic hyphae, or they may be characteristically marked and provided with sterigmata or specialized branches on which they bear conidia. Types of conidiophores and conidial production have been studied in depth. The major types of conidial production in these fungi include :

1. Conidia budded off the conidiophores singly or in chains (blastospores),
2. Conidia produced singly from successively formed new growing points; some times forming false chains,
3. Thick-walled conidia produced singly from the apex of the conidiophores or its branches, sometimes forming false chains (chlamydospores),
4. conidia produced from the apex of a phialide (phialospores),
5. cònidia produced by growth of the apical region of the conidiospores, forming chains which merge imperceptibly with the conidiophores (meristematic arthrospores),
6. conidia produced from pores of the conidiophore (porospores),
7. conidia produced by fragmentation of the conidiophore (arthrospores)
8. conidia borne singly at apex and laterally on conidiophores which elongate at the base, and
9. conidia borne singly at apex in such a way as to form false chains with the conidia attached end to side.

Genus - *Colletotrichum* (Plate XVI) (Deuteromycetes, coelomycetes, Melanconiales, Melanconiaceae, Amero-, Hyalo-, Phialospores). This genus is a facullative parasite causing anthracnose diseases on leaves, youngtwigs and fruits

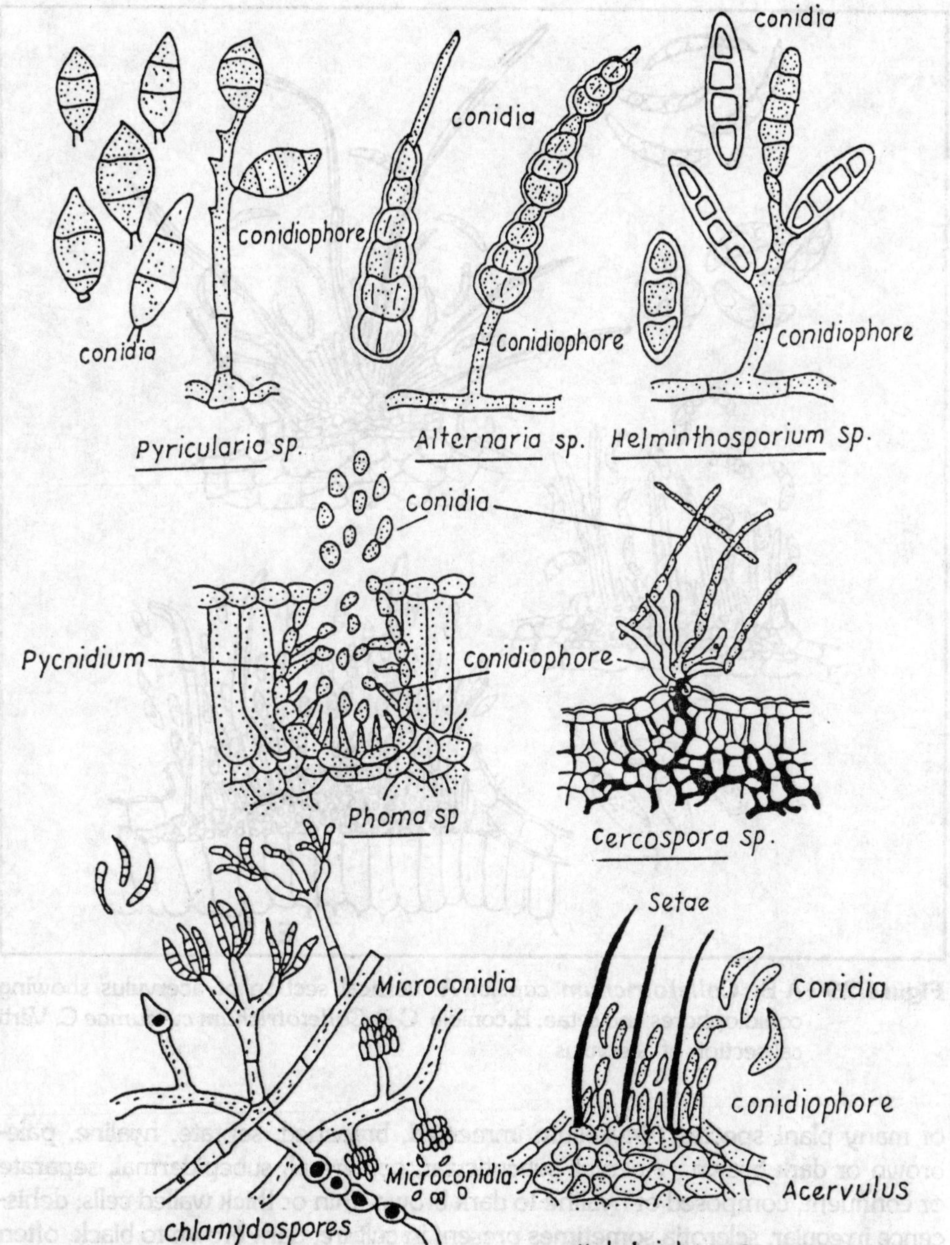
conidia
conidiophore
conidia
Pyricularia sp.
conidia
Conidiophore
Alternaria sp.
conidia
conidiophore
Helminthosporium sp.
conidia
Pycnidium
conidiophore
Phoma sp
Cercospora sp.
Setae
Microconidia
Conidia
conidiophore
Microconidia.
Acervulus
Chlamydospores
Fusarium sp.
Colletotrichum sp.

Plate-XVI : Fungi imperfectii

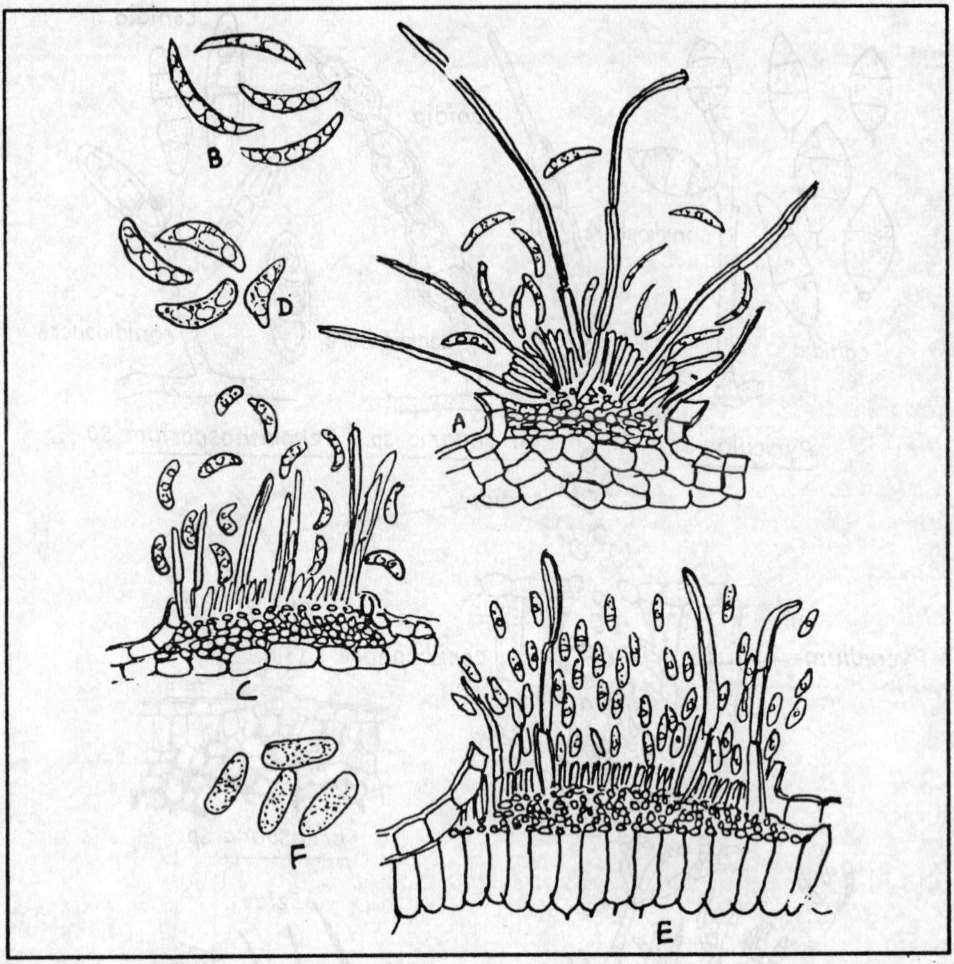
B
D
A
C
F
E

Figure 29 : A-B. *Colletotrichum capsici* A. vertical section of acervulus showing conidiophores and setae; B.conidia C-D.*Colletotrichum curcumae* C. Vertical section of acervulus

of many plant species. Mycelium immersed, branched, septate, hyaline, pale-brown or dark brown. Acervuli subcuticular, epidermal, subepidermal, separate or confluent, composed of hyaline to dark brown, thin or thick walled cells; dehiscence irregular, sclerotia sometimes present in culture, dark-brown to black, often confluent, occassionally setose, setae in acervuli or sclerotia brown smooth septate, acutely pointed at the apex. Conidiophores develop from upper cells of the stroma in a dense, even stand, simple or branched only at the base, aseptate or septate, short, hyaline to brown, conidiogenous cells phialidic, enteroblastic, hyaline, aseptate, straight to falcate, smooth, thinwalled, sometimes apex drawn into a cellular appendage. Appresoria brown, entire or with crenate to irregular margins, simple or repeatedly germinating to produce complex columns or several closely connected appresoria.

Important species *Colletotrichum capsici* Dieback and ripe fruit rot of chillies, other hosts chickpea, cotton, brinjal, tomato, turmeric

Colletotrichum coccodes Hosts belong to solanaceae, leguminosae, and cucurbitaceae. The species causes black dot of potato, and tomato and anthracnose of fruits of tomato and chilli.

Colletotriclum graminicola Attacks sorghum spp,, Maize, *Triticum*, *Secale* and other cultivated and wild Gramineae causing red stalk rot of internodal stem tissues, anthracnose, leaf spots and seedling blight.

Colletotrichum truncatum Attacks groundnuts, pigeonpea, soybean, pea, cowpea, *Phaseolus* spp. causing anthracnose and leaf spots.

Colletotrichum lini Anthracnose of linseed

Colletotrichum falcatum (Glomerella tucumanensis) Red rot of sugarcane

Colletotrichum gloeosporioidies (Glomerella cingulatia) anthracnose of mango, citrus and apple

Colletotrichum lindemuthianum (Glomerella lindemuthianum) Anthracnose of beans.

Life cycle of *C.Lindemuthianum [Figure 30]*

Imperfect stage (conidial Anamorph)

The fungus produces septate and branched mycelium which changes in colour from hyaline to nearly black upon maturity. Some hyphae aggregate to form stromata, which produce conidiospores and setae, Unicellular hyaline conidia are produced on the conidiophores. The conidia usually contain a clear, vacuole-like body near the center. Conidial shape may be oblong, cylindrical, kidney-like, or sickle-like with rounded or slightly pointed ends. Condiophores are hyaline, erect, short, and unbranched. Setae which are pointed and stiff with septate brown hairs may appear among the conidiophores. A conidium may germinate in six to 10 hours and produce one to four germtubes, which form appresoria at their tips during pathogenesis (fig 30). Penetration of host cell is achieved by a small thread which develops from appresorium. Upon entering the cell, the hyphal thread enlarges and penetrates the cell layers. This process continues for

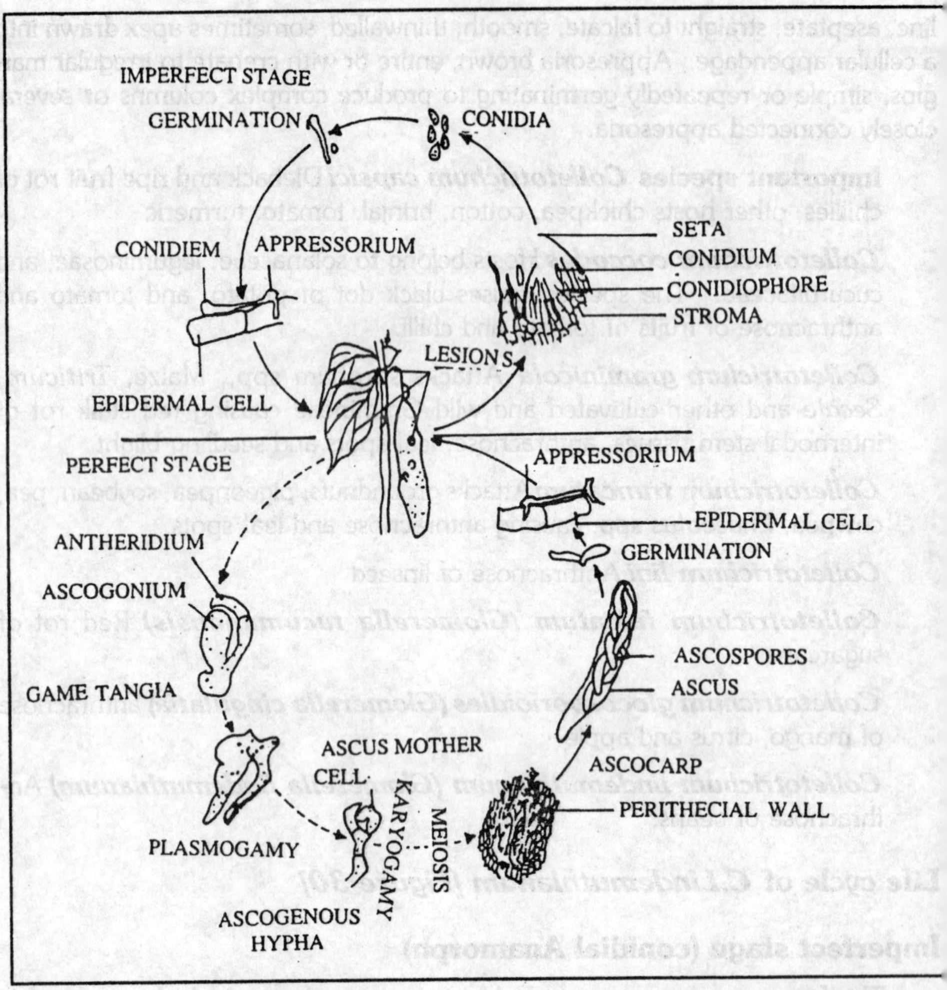
IMPERFECT STAGE
GERMINATION
CONIDIA
CONIDIEM
APPRESSORIUM
SETA
CONIDIUM
CONIDIOPHORE
STROMA
LESIONS
EPIDERMAL CELL
PERFECT STAGE
APPRESSORIUM
EPIDERMAL CELL
GERMINATION
ANTHERIDIUM
ASCOGONIUM
ASCOSPORES
ASCUS
GAME TANGIA
ASCUS MOTHER
CELL
KARYOGAMY
MEIOSIS
ASCOCARP
PERITHECIAL WALL
PLASMOGAMY
ASCOGENOUS
HYPHA

Fig. 30 : Life Cycle of *Colletotrichum sp.*

several days without killing the cells. Strands of mycelium increase beneath the epidermis and produce cavaties that become acervuli containing the spore masses A single acervulus may contain 3 to 50 or even more conidiophores, depenting on the size of the lesions. Each acervulus is composed of a stromatic layer of interwoven hyphae from the surface of the acervulus. Conidiophores arise, and conidia are produced. The conidia are dispersed by splashing raindrops to nearby plants, and the cycle can repeat several times in a growing season.

Perfect stage (Teleomorph)

The perfect stage is rarely formed in nature. The fungus was originally named *Glomerella lindemuthianum*, but now it has been renamed *G. cingulata*. The fungus produces perithecia which contain hyaline, filiform paraphyses and asci. Each ascus contains eight ascospores which may be allantoid or ellipsoid in shape. Ascospores are ejected from the tip of the ascus. The ascospores germinate and infect the plant in a manner similar to that of the conidia if sufficient moisture is available.

Genera — *Helminthosporium*, *Drechslera* and *Bipolaris*[Plate .VI]

These three genera belong to series Porosporae of Hyphomycetes according to classification proposed by Hughes (1953) as modified by Tubaki (1963) and Barron (1968) and to family *Helminthosporiaceae* of Hyphomycetes as proposed by Subbraminian (1962). In the traditional classification the genus *Helminthosporium* is placed under *Moniliales*, family *Dematiaceae* (dark conidia), phragmosporae. Perfect states where known belong to ascomycetes.

In *Helminthosporium* conidiophores are determinate, ceasing growth with production of the apical conidia, relatively short and simple or, spariangly branched. The branching is not dichotomous. In *Drechslera* and *Bipolaris* the conidiophores are simple or sparingly branched, indeterminate continuing growth sympodially. Conidia lack apical prolongation, and are cylindrical or fusiform, sometimes curved but then concolourous. In *Drechslera* conidia are cylindric and germinate form all the cells, while in *Bipolaris* conidia are fusiform germinating from end cells only. Most of the important plant pathogenic genera, especially those attacking Graminaceous hosts, are now considered *Drechslera*.

Helminthosporium

Colonies effuse, dark and hairy. Mycelium immersed , stromata usually present, conidiophores often in fascicles, erect, brown to dark brown. Conidia develop laterally, often in verticils, through pores beneath the septa of the conidiophores while the tip of the conidiophores continues to grow but growth ceases with the formation of terminal conidia. Conidia subhyaline to brown, usually obclavate, pseudoseptate and frequently with a dark-brown to black protruding scar at the base.

Drechslera

Conidiophores brown, simple or less sparingly branched, indeterminate, producing conidia singly at the apices through small pores while continuing growth sympodially from a point just below and to one side of the apex from which a second pore is then produced. A succession of spores produced in similar fashion acroginously form the sympodially extending conidiophore. Conidia cylindrical, multiseptate and dark, germinate from any or all cells.

Bipolaris

The genus is characterized by germination of conidia from the end cells. Only conidiophores brown, producing conidia through an apical pore and forming a new apex by growth of the subterminal region. Conidia fusoid, straight or curved, germinate by one germtube from each end cell; exosporium smooth, rigid and brown, endosporium hyaline, amorphous, separating cells of the mature phragmospore. Indeterminate conidiophores extend sympodially producing a succession of dark transverselly septate porospores. Chiefly parasitic on *Graminae*. Perfect state where known occur in ascomycetous fungus *Cochloibolus*. The type species of this genus was *Bipolaris maydis* which was transferred from *H maydis* However, later publications of CMI placed this species as *D maydis* with perfect state as *cochliobolus heterostrophous*. *Helminthosporium* species transferred to *Drechslera* are given in table—

***Helminthosporium* sp.**	**Drechslera sp**	**Ascigerous state**
H. carbonum	*D. zeicola*	*Cochliobolus carbonum*
H. gramineum	*D. graminea*	*Pyrenophora graminea*
H. heveae	*D. heveae*	-
H. incurvatum	*D. incurvata*	-
H. maydis	*D. maydis*	*C. heterostrophous*
H. nodulosum	*D. nodulosus*	*C. nodulosus*
H. oryzae	*D. oryzae*	*C. miyabeanus*
H. sacchari	*D. sacchari*	-
H. sativum	*Bipolaris sativus*	*C. sativum*
H. stenospilum	*D.stenospila*	-
H. teres	*D.teres*	*P. teres*
H. turcicum	*D.turcica*	*Trichometasphaeria turcica*

Life - cycle

The organism survives in the diseased plant debris, seeds, propagative materials, etc. in the form of mycelia and spores. Under congenial conditions (environment and nutrition), the fungus gets activated and breeds asexually leading to production of conidial inoculum. They are disseminated by wind or rain. In those species that produce perithecial stage, the ascospores are released and function as inoculum. After reaching appropriate host organs, they germinate by forming germtube and penetrate host cells and colonize them. Within a few days, conidiophores and conidia are produced. The conidia produced now function as secondary inoculum and cause secondary infections. This cycle runs several times. In those species where sexual reproduction is known to occur, antheridia and ascogonia are produced. After plasmogamy, ascogenous hyphae and finally ascus mother cells are produced . After karyogamy and meiosis, asci and ascospores are produced in perithecia.

Genus - Alternaria[Plate XVI]

Hyphomycetes (Moniliales - Dematiacae)

Spores - Dictyo, Phaeo, Poro, Muriform

The genus generally grows as a saprophyte on the decaying plant parts and also in the soil. Some species are parasitic on higher plants. The leaf spots produced by *Alternaria* species are usually characterized by concentric rings hence the name "target spot". The mycelium is composed of slender, septate, profusely branched hyphae. The mycelium is hyaline at first but becomes brown at maturity. In parasitic species the hyphae are intercellular at first and become intracellular later on and the cells are usually multinucleate.

Reproduction[Figure 31]

Alternaria reproduces mainly asexually. The perfect stage is known for *A. tenuis*, the type species, it is *Pleospora* of ascomycetes. *Lewia* is another teleomorph of *Alternaria* species causing leaf spots and blights. The asexual reproduction takes place by means of conidia which are produced on specialised branches of hyphae and the conidiophores. The conidiophores are dark, septate, sometimes inconspicuous and simple or branched. They come out through stomata of the host or from dead or damaged host tissues. The production of conidia in *Alternaria* species is extra ordinarily vigorous. The conidia may be borne singly or in short chains of 3-5 conidia or in long chains. The conidia are large, dark coloured, generally obclavate and are bottle shaped with a broad cup like base and a narrow beak. Each conidium is multi septate with a number of transverse (5-10) and a few longitudinal septa. The conidia when mature, get detached from the conidiophore and are disseminated by wind. Upon getting a suitable substrate and favourable conditions, each conidium germinates giving one or more germ tubes. The germ tube enters usually through the stomata of the host leaf.

Important Plant Parasitic Species

Species	Diseases caused
Alternaria brassicae	Leaf spot of crucifers
A. brassicicola	Leaf and pod spots of crucifers
A. Carthami	Leaf spot of safflower
A. citri	Black rot of oranges
A. longipes	Brown spot of tobacco
A. porri	Purple blotch of onion
A. solani	Early blight of potato
A. triticina	Leaf blight of wheat

Genus - *Cercospora [Plate XVI]*

(Hephomycetes, Moniliales, Dematiaceae, phragmo, Phaeo-, Sympodulo-)

The genus *cercospora* includes about 700 species which are parasites causing leaf spot or leaf blotch. Distinct necrotic spots are produced on leaves, stems, fruits, etc. Mycelium consists of fine to coarse, hyaline to coloured and septate hyphae, external as well as internal in the host. Young hyphae hyaline with slight tinge of yellow, older hyphae colour tends to be shade of olive or brown. Hyphae aggregate below the stomatal cavity to form stromata. Conidiophores emerge through stomata or directly through epidermis, single or in fascicles, arising from the stomata or directly from vegetative hyphae, pale to dark pigmented, septate or continuous, simple or branched, some times tortuous, smooth or geniculate. Spore scars visible or prominent. Conidia borne singly and terminally but become lateral by sympodial development of the conidiophores, acircular to obclavate or cylindrical, rarely clavate, hyaline to darkly pigmented, thin walled, smooth without appendages, usually multiseptate, straight or strongly curved sometimes sinuous, base sharply obcomic to truncate and the tip acute to obtuse. The diagnostic features of the genus are long hyaline or pigmented phragmospores (conidia) borne in acropetal succession from a usually simple sympodially extending pigmented conidiophore.

Important species

Cercospora beticola	Leaf spot of beet
Cercospora apii	Early blight of celery
Cercospora cruenta	Leaf spot of *Phaseolus mungo*
Cercospora arachidicola	Early leaf spot of groundnut
Perfect state	Mycosphaerella arachidis

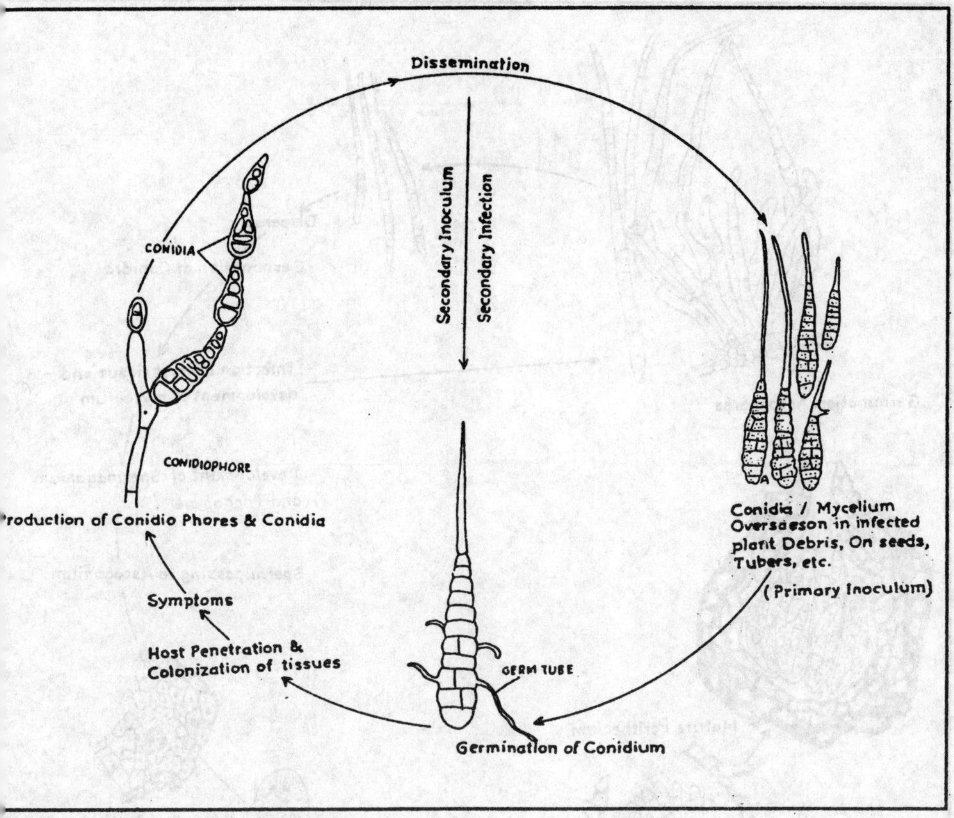
Dissemination
Secondary Inoculum
Secondary Infection
CONIDIA
CONIDIOPHORE
Production of Conidio Phores & Conidia
Symptoms
Host Penetration &
Colonization of tissues
GERM TUBE
Germination of Conidium
Conidia / Mycelium
Oversaeson in infected
plant Debris, On seeds,
Tubers, etc.
(Primary Inoculum)

ig. 31 : Life cycle of *Alternaria spp*

ife Cycle [Figure 32]

In most species of *Cercospora* only asexual reproduction is reported which kes place by means of comidia. These conidia after dispersal, reach the host nd germinate (Fig. 32). After host infection and colonization conidiophores and onidia are produced. This way several crops of conidia are produced and infecon continues till the conditions are favourable. Genus *Mycosphaerella* is invaribly reported as perfect stage of some species of *Cercospora*. The occurrence of erfect stage in nature is rather rare. In many species of *Mycosphaerella*, plasogamy is probably accomplished by means of spermatization. The spermatia re borne in spermagonia which resemble pycnidia. They are formed in the asal spermogonial cells and are pushed out. They are rod shaped and minute. all probability they function as male cells. Spermatia fuse with tips of trichogynes. he ascocarp cavity is dissolved within the stroma by the growing tuft of asci. arr (1958), studying *M. tassiana* and *M. typhi* found no typical ascogenous

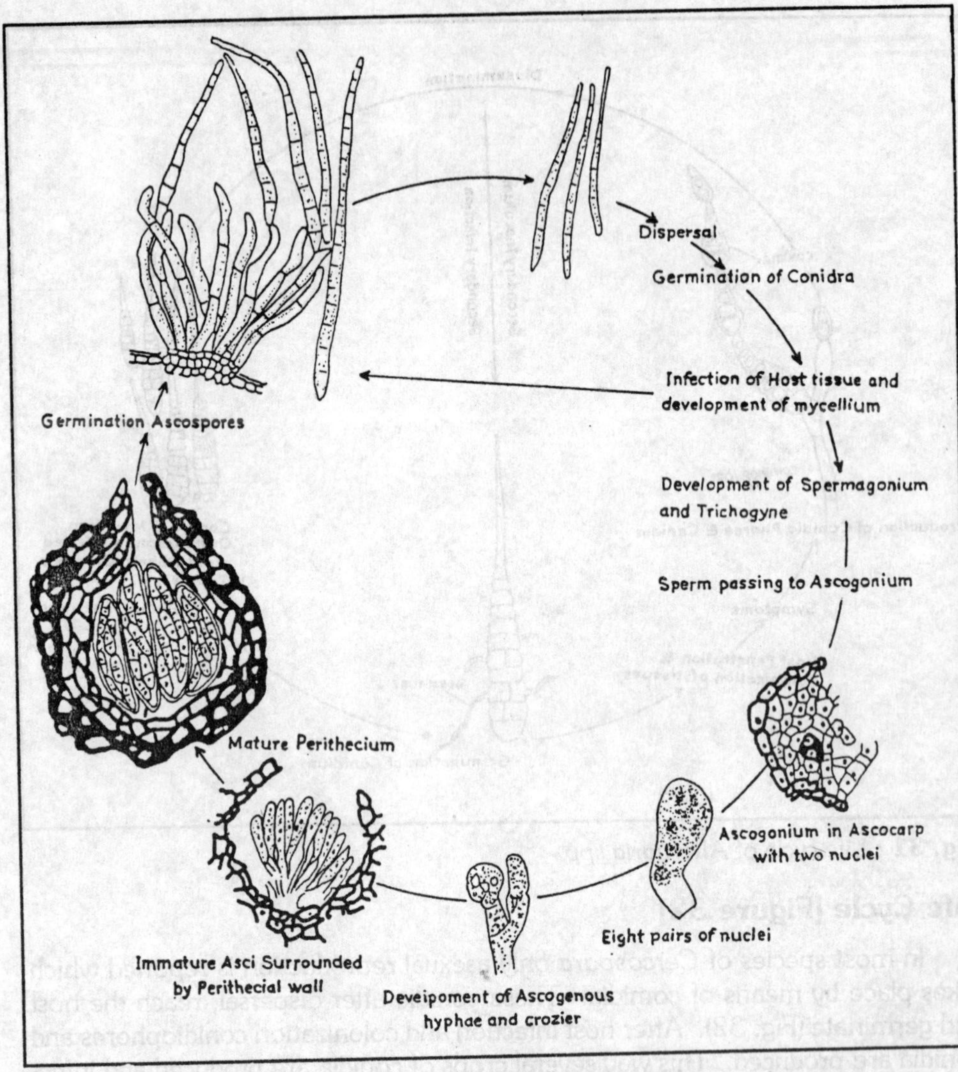
Dispersal
Germination of Conidra
Infection of Host tissue and development of mycellium
Development of Spermagonium and Trichogyne
Sperm passing to Ascogonium
Ascogonium in Ascocarp with two nuclei
Eight pairs of nuclei
Deveipoment of Ascogenous hyphae and crozier
Immature Asci Surrounded by Perithecial wall
Mature Perithecium
Germination Ascospores

Fig. 32 : Life cycle of Cercospora spp. having Mycosphaerell as perfect stage

hyphae. A number of multinucleate cells develop at the base of locule. Each develops a protrusion into which two nuclei pass. This cell may become crozier like or may develop in to an ascus by enlarging. Karyogamy and meiosis take place before spores are formed.

Genus - *Pyricularia [Plate XVI]*

Pyricularia forms hyaline or pale grey bi or tri celled, pyriform conidia (with a rounded base and basal papilla) singly at the tip of the conidiophore and its successive growing points. The conidiophores are simple, erect, septate and hyaline or grey.

P. oryzae causes the most destructive "blast" disease of rice plants. There is much controversy about the name of the pathogen, as two generic names have been used *Pyricularia* and *Dactylaria*. The specific name of the fungus on rice was first described as *Pyricularia oryzae*. Comparing the various characteristics of *P. oryzae* and *P. grisea*, it has been concluded that there are not enough differences between these two to maintain them as different species. The morphology of the perithecial stage (Perfect stage) produced by mating of different rice isolates is similar to *Magnaporthe grisea* (ascomycetes) and therefore, *M. grisea* is retained for the teleomorph stage of rice blast fungus.

Rhizoctonia and *Selerotium*

The fungi *Rhizoctonia and Sclerotium* are soil inhabitants and cause diseases on may hosts by affecting the roots, stems, tubers, corns, and other plant parts. These two fungi were known as sterile fungi because they were thought to produce only selerotia and to be incapable of producing either asexual or sexual spores. The two were distinguished from one another by their mycelial characteristics and by the fact that *Rhizoctonia* sclerotia have a uniform texture throughout where as *sclerotium* sclerotia are internally differentiated into three areas. It is known now that atleast some species within these two genera produce basidiospores as their sexual spores, and, therefore, they are Basidiomycetes. On the other hand, some fungi previously thought to belong to *sclerotium*, for example, *S. bataticola,* are now known to produce conidia (*Macrophomina*) and some (e.g. S. *oryzae*) produce ascospores (*Magnaporthe*). Others, however, such as *S. cepivorum*, causing white rot of onion, still have no known spore stage. In any case, the spores of *Rhizoctonia* and *Sclerotium* either are produced under specific conditions in the laboratory or are extremely rare in nature and therefore of little value in diagnosing the fungus. For these reasons, these fungi continue to be considered as mycelia sterila, and since, for all practical purposes, they behave as such they continue to be referred by the names *Rhizoctonia* and *Sclerotium*.

Rhizoctonia

The pathogen, *Rhizoctonia* sp., consists of a large, diverse, and complex group of fungi. All *Rhizoctonia* fungi exist primarily as sterile mycelium and some times, as small sclerotia that show no internal tissue differentiation. Mycelial cells of the most important *Rhizoctonia, R. solani,* contain several nuclei (multinucleate *Rhizoctonia*), where as mycelial cells of several other *Rhizoctonia* species contain two nuclei (binucleate *Rhizoctonia*). The mycelium, which is colourless when young but turns yellowish or light brown with age, consists of long cells and produces branches that grow at approximately right angles to the main hypha, are slightly constricted at the junction, and have a cross wall near the junction. The branching characteristics are usually the only ones available for identification of the fungus as *Rhizoctonia*. Under certain conditions the fungus produces sclerotia like tufts of short, broad cells that function as chlamydospores, or eventually the tufts develop into rather small, loosely formed brown to black sclerotia. As mentioned earlier, *Rhizoctonia* species infrequently produce a basidiomycetous perfect stage. The perfect stage of the multinucleate *R. solani* is *Thanatephorus cucumeris* where as that of binucleate *Rhizoctonia* is *Ceratobasidium*. A few multinucleate *Rhizoctonia spp. (R. zeae* and *R. oryzae)* have *Waitea* as their perfect basidiomycetous stage.

It is now very much clear that *R. solani* and other species are "collective" species, consisting of several more or less unrelated strains. The *Rhizoctonia* strains are distinguished from one another because *Anastomosis* (fusion of touching hyphae) occurs only between isolates of the same *anastomosis* group. After anastomosis, an occassional heterokaryon hypha may be produced. In the vast majority of anastomoses, however, five to six cells on either side of the fusion cells become vacuolated and die, appearing as a clear zone at the junction of two colonies. This "killing reaction" between isolates of the same anastomosis group is the expression of somatic or vegetative incompatibility. Such somatic incompatibility limits out breeding to a few compatible pairings. The existence of the anastomosis groups in *R. solani* represents genetic isolation of the population in each group.

Although the various anastomosis groups are not entirely host specific, they show certain fairly well defined tendencies: isolates of anastomosis group 1 (AG1) cause seed and hypocotyl rot and aerial (Sheath) and web blights of many plant species; isolates of AG2 cause canker of root crops, wire stem on crucifers and brown patch on turf grasses; isolates of AG3 affect mostly potato, causing stem cankers and stolon lesions and producing black sclerotia on tubers; isolates of AG4 infect a wide variety of plant species, causing seed and hypocotyl rot on almost all angiosperm and stem lesions near the soil line on most legumes, cotton, and sugar beet. Six more antastomosis groups are known within *R. solani* and many more in other *Rhizoctonia*.

Sclerotium

The fungus, *Sclerotium sp.*, produces abundant white, fluffy, branched mycelium that forms numerous sclerotia but is usually sterile, that is, does not produce spores. *S. rolfsii*, occassionally produces basidiospores at the margin of lesions under humid conditions. Its perfect stage is *Aethalium rolfsii*. As mentioned earlier, the species of *S. bataticola*, occassionally produces conidia in pycnidia and is now known as the imperfect fungus *Macrophomina phaseolicola*. A third *Sclerotium* species, *S. cepivorum*, which causes white rot disease of onion and garlic, in addition to sclerotia also produces occassional conidia on sporodochia, these conidia, however, seem to be sterile.

CHAPTER VIII

BACTERIA

I. INTRODUCTION

Bacterial plant pathology is a branch of plant pathology dealing with plant diseases caused by prokaryotes such as eubacteria, actinomycetes, phytoplasmas (MLOS), and spiroplasma. The loss assessment of bacterial plant diseases is an immature field. The available data on yield loss are very limited. Table 1 shows gross loss in 1976 due to prokaryotes in U.S. (Kennedy and Alcorn, 1980). In another assessment, global loss in 1976 was estimated to be $ 19.6 billion. It is obvious that great financial loss are a result of reduced crop yields, and additional expenses for control are annually incurred.

Table 1 : Loss estimates for plant pathogenic prokaryotes in 1976 (Kennedy and Alcorn, 1976)

Prokaryote	Disease	Loss (million of dollors)
Ralstonia solanacearum	Bacterial soft rot of tobacco and tomato	9.1
P.syringē pv. glycinea	Blight of soybean	65.0
P. syringe pv. syringae	Leaf blight of wheat	18.0
X. campestris pv. solanacearum	Blight of cotton	15.0
A. tumifaciens	Crown gall	23.0
E. amylovora	Fire blight of pear	1.7
E. amylovora sub sp. carotovora and sub sp. atroseptica	Soft rot and/or black leg of potato	11.0

Clavibacter spp	Wilt of aifalfa	17.0
	Wilt of corn	3.0
	Ratoon stunt of sugarcane	10.0
Xylella fastidiosa	Phony peach	20.0
	Pierce of grape	3.0
Spiroplasma citri	Stubborn of citrus	1.0
Phytoplasmas (MLOs)	Pear decline	1.6
	lethal yellowing of Coconut	3.0

II. HISTORICAL EVENTS

The history of phytobacteriology can also be traced to the late 19th century. **Table 1** lists some of the noticeable milestones. T.J. Burrill (1880) proved by an inoculation test that fire blight of pear was caused by a bacterium which he named *Micrococcus amylovorus* in 1882. In 1883, J.H. Wakker reported a yellowing disease of *hyacinths* by *Bacterium hyacinthi*. E.F. smith (1897 - 1901) established the concept of bacterial plant pathogens by producing convincing evidences. It can be said, therefore, that phytobacteriology was established in 1901 as a specific discipline of plant pathology.

III. MORPHOLOGY, STRUCTURE AND COMPOSITION

The structure of the bacterial cell envelop is important for pathogenicity because host recognition is determined by the interactions between the host cell wall and extracellular polysaccharides or lipopolysaccharides of bacteria.

(A) Morphology :

Eubacteria have three basic phases. They are SPHERICAL, ROD and SPIRAL, also referred as COCCUS (pl.cocci), BACILLUS(pl. bacilli) and SPIRILLUM (Pl. spirilla), respectively. Most plant pathogenic bacteria are rod shaped and divide by binary fission .Most plant pathogenic bacteria fall within the range of 1.0-5.0x 0.5-1.0 μ m. *Streptomyces* also called ray fungi and possibly a link between bacteria and fungi, is characterised by the formation of highly branched mycelium, 0.5-2.0 μ m in width and the formation of chains of spores at tips of aerial hyphae.

The gross morphology and structure of a representative gram - negative bacteria is shown in figure1.

(B) Cell envelope - structure

The cell envelope of gram -ve bacteria consists of the cell wall, the cell membrane, and the outer membrane. The outer membrane is absent in the gram +ve bacteria .

Table 1: Historical Events in Bacterial Plant Pathology

Name	Year	Event
T.J. Burrill	1882	Publication of fire blight Pathogen *M. amylovorus*
J. Omori	1896	Soft rot pathogen of wasabi *Bacillus* spp.
E.F. Smith	1901	Validation of bacteria as Plant pathogens
C.O. Jensen	1910	Demonstration of evidence between crown gall of Plants and cancer of animals
S. Hori	1912	Publication of soft rot pathogen
G.H Koons and J.E. Kotila	1925	Isolation of bacteriophages of *B. Carotovorus*
R.M. Klein	1954	Basic process of transformation in crown gall
A.C. Braun	1955	Wild fire toxin and its action
Y. Doi *et.al.*	1967	Discovery of MLOs (Phytoplasmas)
A.K. Chatterjee and M.P. Starr	1972	Conjugative gene transfer in bacteria
P.B. New and A. Kerr	1972	Biocontrol of crowngall (*A. radiobacter* Strain 84)
I. Zaenen *et.al.*	1974	Demonstration of Ti plasmids in *A. tumifaciens*
D.W. Dye *et.al.*	1980	Introduced pathover system in taxonomy

(a) The Cell wall

It is the rigid frame work responsible for cell shape maintenance and it's intigrity by protecting from external stresses. Cell wall of gram negative bacteria ranges from 10 to 15 μ m in thickness and from 5to10 % in dry weight of the total cell envelope. These values in grams +ve ranges from 20-80 μ m and 40-90% .The basic component is a macromolecule called PEPTIDOGLYCAN. Chemically it is a hetro polymer having straight chains of alternating units of N-acetylglucosamine and N-acetylmuramic acid and short peptide for cross linking. The cell wall of gram positive bacteria contains polysaccharides, teichoic acid and sometimes proteins.

(b) Outer membrane

The cell envelope of gram -ve bacteria has the outer membrane outside of the cell wall. It has a complex chemical structure consisting of phospholipids , lipopolysaccharides and proteins.This structure (1) provides channels for passive

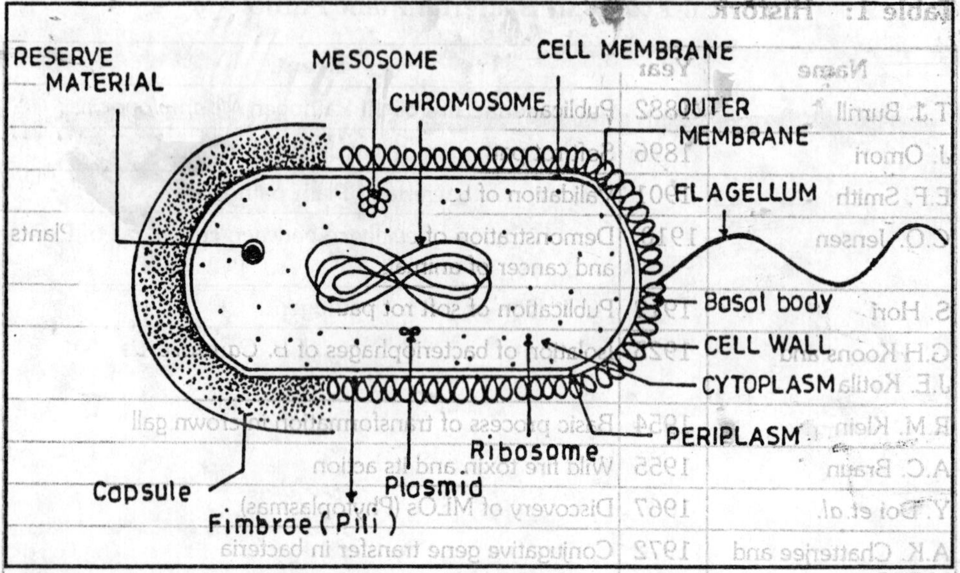
RESERVE MATERIAL
MESOSOME
CELL MEMBRANE
CHROMOSOME
OUTER MEMBRANE
FLAGELLUM
Basol body
CELL WALL
CYTOPLASM
PERIPLASM
Ribosome
Plasmid
Capsule
Fimbroe (Pili)

Fig. 1 : Schematic structure of Gram Negetive Bacteria

diffusion of nutrients or hydrophilic solutes (2) establishes permeability barrier to toxic substances particularly antibiotics (3) provides receptor sites (4) fascilitates pairing during conjugation, and (5) endows hydrophilicity

(c) Cell Membrane

It is a unit membrane located inside the cell wall. It consists of a lipid bilayer of 50 - 75 % protein and 20 to 35 % lipid. The cell membrane contains various enzymes for energy yielding metabolism. In some bacteria, such as *Clavibacter, Streptomyces* and *Bacillus* a part of cell membrane envaginates into the cytoplasm to form complex membrane infoldings called MESOSOMES. Mesosomes are associated with respiratory activity, nuclear division, septum formation, spore formation, and secretion of hydrolytic enzymes.

(d) Periplasm

It is a matrix of polypeptide and saccharides. It contains various enzymes such as cellulases and pectinases.

(e) Gram-Reaction

Gram staining was discovered empirically in 1884 by Christian Gram. The steps of gram staining are as follows:

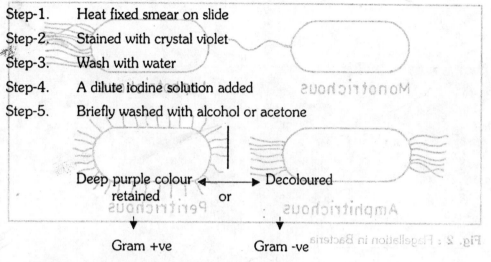

Step-1. Heat fixed smear on slide

Step-2. Stained with crystal violet

Step-3. Wash with water

Step-4. A dilute iodine solution added

Step-5. Briefly washed with alcohol or acetone

Deep purple colour retained ⟷ or ⟷ Decoloured

↓ Gram +ve   ↓ Gram -ve

Counter stain with safranin or carbol fuchsin

There is another simple test for routine study.

Suspend bacterial cells in 3% KOH solution

↓

No change ⟷ vicidity develops

gram +ve   gram -ve

(C) External Structure

(a) Flagellum (pl. Flagella)

Most plant pathogenic bacteria are motile by means of flagella except for *E. stewartii* and many coryneform bacteria. The flagellum has a helical shape. The base of the flagellum forms a slightly wider shaped hook penetrating the cell envelope and connecting with basal body and 30 - 40 nm in diameter. The arrangement as well as number of flagella are of taxonomic importance. Flagellar arrangements (Figure - 2) are grouped as 1. MONOTRICHOUS (a single flagellum at one end of the cell), 2. AMPHITRICHOUS (one flagellum at each end of the cell), 3. LOPHOTRICHOUS (two or more flagella at one or both ends of the cell), and 4. PERITRICHOUS (large number of flagella all over the cell).

(b) Pilus or FIMBRIA

Many filamentous appandages are formed on the surface of majority of gram--ve bacteria. They are called pilli or fimbrae. It is 3 to 14 nm in diameter

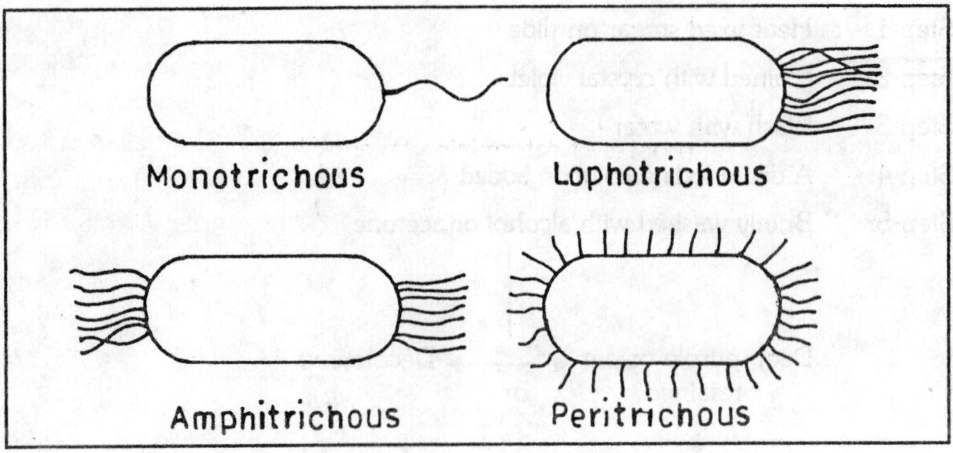
Monotrichous
Lophotrichous
Amphitrichous
Peritrichous

Fig. 2 : Flagellation in Bacteria

and 0.2 to 20 μ m in length. According to function and morphology fimbrae fall in 6 categories for example, SEX PILLI act as specific bridge for gene tranfer in conjugation. Another act as adhesive organelles.

(c) Extracellular Polysaccharides (EPS)

Bacteria often produce EPS out side the cell envelope. The small and dense layer with defined boundry is the CAPSULE. The water soluble and less adherent layer without definite boundry is SLIME LAYER.

(D) Internal Structure

(a) cytoplasm

The cytoplasm contains ribosome granules and enzymes, coenzymes, intermediate metabolic products and inorganic substrates. It is the basin for protein synthesis and active metabolism.

(b) Chromosome

In prokaryotes the terms NUCLEUS, GENOME, and CHROMOSOME can be used interchangeably. The nuclear material within the cell is not separated by a membrane as in eukaryotic cells. There is basically one chromosome per cell but two or more can be seen in the exponential growth phase. The chromosome is highly folded and forms a compact mass and remain attached to cell mambrane at replication site.

(c) Plasmids

Bacteria often harbour small, extracellular, replicons (replication units). Such self replicating entities are PLASMIDS. Bacterial cells generally express new genetic characterstics by acquisition of plasmids. They are generally covalently closed circular DNA. Plasmids can be self transmissible (conjugative) or non self replicating (nonconjugative). Plasmid borne resistance against antibiotics and heavy metal compounds is known in plant pathogenic bacteria. Some plasmids with unknown phenotypic traits are called cryptic/ resident/ or indigenous plasmids.

(d) Endospores

Bacillus and *Cladosporium species* produce "endospores" which are dormant resting cells. They show resistance against stresses like heat, UV irradiation, desiccation and toxic chemicals.

(e) Reserve Materials

Reserve materials include polymers of inorganic phosphate (bolutin), hydroxybutyrate, starch, glycogen etc.

(f) Pigments

water soluble or insoluble pigments, with colours such as blue, green, yellow, red, brown etc. under aerobic conditions, are produced by bacteria.

Examples

1. ***Pseudomonas fluorescens*** — Fluorescent pigments, pyoverdin, siderophores (Fe III chelators)

2. ***Erwinia chrysanthemi*** — Water soluble blue pigment; *E. rhapontici* Iron chelators, *E. herbicola* yellow carotenoid pigments, etc.

IV. TAXONOMY

Prokaryotes are different from eukaryotes in the following properties :

1. The nucleoplasm is never separated from the cytoplasm by a unit membrane
2. No basic protein is present in the nucleoplasm
3. No mitosis during cell division
4. 70s ribosomes are dispersed in the cytoplasm but not present in form of endoplasmic reticulum
5. Cytoplasm streaming is not observed
6. Culture consists of single cells without forming organs, and
7. Mitochondria is absent

(A) Content of Taxonomy

Taxonomy encompasses the study of classification, identification, and nomenclature of bacteria. Classification is the grouping of organisms in some orderly fashion. Identification is a comperative process of determining taxonomic position of unknown bacteria with those of known bacteria. Nomenclature requires following the rules in the "International code of nomenclature of bacteria" in naming any bacterial taxa.

(B) Concept of Bacterial Species

Bacterial species are the taxonomic groups of strains defined on the basis of common phenotypic and genotypic charactersitics.

(a) Taxospecies

It refers to a group of strains that have a number of phenotypic characterstics in common and can be grouped into distinct phenotypic groups for e.g. *E. amylovora* or *R. solanacearum* are the taxospecies.

(b) Nomenspecies

It is the species with binomial names given in accordance with the rules. The names may, therefore, be claimed for another taxa on the basis of some criteria. for example, *X. campestris* is a nomenspecies superseded by *X. campestris* pv. *campestris*

(c) Molecular species

This term is applied to a group of strains with a high degree of nucleic acid (DNA or RNA) homology. The sequence of NA establishes and preserves the identity of a species.

(d) Genospecies

It is a group of strains that can accomplish genetic exchange in some way. Since genetic exchage occurs not only at the species level but also at genus level or even at the family level.

(C) Methods of classification

Several methods of classifications have been tried from time to time. They include "conventional taxonomy, numerical taxonomy, molecular taxonomy, (DNA base composition, DNA -DNA homology, DNA - r RNA homology, RFLP, protein profiles, isozymes, gene transfer), chemotaxonomy, pyrotaxonomy and serological taxonomy".

(a) Higher Ranks of Prokaryotes

The higher ranks listed below are mainly based on the descriptions proposed by Murray in "Bergey's Manual of systematic Bacteriology" (1981). According to his classification, the kingdom prokaryote is divided into four taxa on the basic of the structure and chemical characterstics of cell envelope.

Division 1. Gracilicutes

Prokaryotes that have a gram-type negative cell envelope consisting of an outer membrane, a peptidoglycan layer and a unit membrane with fatty acid glycerol ester type lipid. Endospore is not formed. Usually gram reaction is negative. GRACILICUTES is divided into two classes - Proteobacteria and Oxyphotobacteria on the basis of phylogenetic principles. It consists of four groups or subclass called alpha, beta, gamma and delta. All gram –ve plant pathogenic bacteria are included in proteobacteria and scattered in four subclasses.

Division 2. Firmicutes

Prokaryotes with a gram type positive cell envelope consisting of a thick peptidoglycan and unit membrane but without an outer membrane. Gram reaction is generally, but not always positive. Some produce endospores. Firmicutes is further divided into two classes of FIRMIBACTERIA and THALLOBACTERIA. *Bacillus* and *Clostridium* are included into Firmibacteria. Actinomycetes and related bacteria such as *Streptomyces, Clavibacter, Curtobacterium, Arthrobacterium, Rhodococcus,* and *Nocardia* are included in *Thallobacteria.*

Division 3. Tenericutes

Division 3 is the prokaryotes that lack a cell wall. The cells are enclosed by a unit membrane. They are highly pleomorphic. Tenericutes includes class *Mollicutes* in which plant pathogenic MLO (Phytoplasmas) and *Spiroplasma* belong.

Division 4. Mendosicutes

The prokaryotes which have a cell envelope without conventional peptidoglycan or lack wall material are included. Cell walls are made entirely of heteropolysaccharides and protein macromolecules. Gram reaction is positive or negative. This division has a single class - Archaeobacteria. No plant pathogenic prokaryotes belong to this division.

Relationships between higher taxa and plant pathogenic bacteria are shown in Table 2.

(D) Plant Pathogenic Genera (PLATE-I)

Family : Pseudomonadaceae

Genus : *Pseudomonas*

Pseudomonas are straight or slightly curverd rods, 0.5-1.0 x 1.5 - 5.0 µm, gram -ve, motile with one to several polar flagella and aerobic. Strictly respiratory type of metabolism. The species of *Pseudomonas* is divided into five RNA groups.

Genus: *Xanthomonas*

Straight rods of 0.1 - 0.7 x0.7 - 1.8 µm, motile with one polar flagellum, gram -ve, obligately aerobic, and strictly respiratory type of metabolism.

Xylophilus

Straight or slightly curved rods, 0.4 - 0.8 x 0.6 -3.3 µm, gram -ve, motile with one polar flagellum, oxydase negative, strictly aerobic. Type species is X. *ampelinus* which causes bacterial necrosis and canker of grape vines, and it is the only species included.

Rhizobacter

Straight or slightly curved, 0.9-1.3x2.1 -2.5 mm, gram -ve, motile with polar flagella or lateral flagella or nonmotile, aerobic, white or yellowish white, tough or viscid colonies on agar plates. Type species is R. *daneus* causing bacterial carrot gall. The genus includes only one species.

Rhizomonas

Straight or slightly curved rods, 0.43-0.53 x0.92 -1.34 µ m gram -ve, motile by one lateral, subpolar or polar flagellum or nonmotile, obligatery aerobic, have oxidative metabolism, colonies white/yellowish, smooth/wrinkled. The type species is R. *suberifaciens* cause corky root of lettuce. The genus includes only one species.

Acidovorex

Straight to slightly curved rods, 0.2-0.7 x 1.0-5.0 µm, gram -ve, motile by a single polar flagellum, oxidase positive, no pigment produced, four plant pathogenic species proposed (A. *avenae* subsp. *avenae,* subsp. *cattleyae* subsp. *citrulli* and A. *konjaci*)

Table 2 : Classification of higher ranks of prokaryotes and affiliation of plant pathogenic bacteria

Kingdom	Division	Class	Attribute	Plant Pathogens	
				Family	Genus
	Gracilicutes	Proteobacteria	Nonphotosynthetic	Enterobacteriaceae	*Erwinia*
				Pseudomonadaceae	*Acidovorex*
					Pseudomonas
					Rhizobacter
					Rhizomonas
					Xanthomonas
					Xylophilus
				Rhizobiaceae	*Agrobacterium*
prokaryotae				Unidefined	*Xylella*
			Photosynthetic without O_2 generation		
		Oxyphotobacteria	Photosynthetic with O_2 generation		
	Firmicutes	Firmibacteria	simple gram +ve		*Bacillus*
					Clostridium
Prokaryotae		Thallobacteria	Gram +ve branching bacteria		*Streptomyces*
					Arthrobacter
					Clavibacter
					Curtobacterium
					Rhodococcus
	Tenericutes	Mollecutes	wall less prokaryotes	spiroplasmataceae	*Spiroplasma*
				uncertain affiliation	MLOs (phyto plasmas)
	Mendasicutes	Archaeobacteria	unusual walls, membrane lipids		

Family : Rhizobiaceae

Genus : *Agrobacterium*

Gram negative rods, 0.6-1.0x1.5-3.0 μ m, motile by 1-6 peritrichous flagella, aerobic, type species *A. tumefaciens*, mainly soil inhabitants.

Family : Enterobacteriaceae

Genus : *Erwinia*

Straight rods, 0.5 -1.0 x 1.0 - 3.0 μ m, gram –ve, motile with peritrichous flagella except *E. stewertii*, a *facultatively* anaerobic, oxidase negative, catalase positive and type species is *E. amylovora*.

Irregular, Nonsporing, Gram +ve Rods. *Clavibacter*

Gram +ve, pleomorphic rods, nonmotile, aerobic obligatery, oxidase negative, includes serveral sub sp. - *michiganensis, sepidonicus, nebraskensis*. In addition, *C. tritici C. xyli* sub sp. *xyli* sub sp. *cynodontis* are also included in this genus.

Arthrobacer

Both forms of rod and coccoid occur in life cycle, gram +ve, type species - *A. globiformis*.

Curtobacterium

- small, irregular, gram +ve, motile by lateral flagella or nonmotile, obligately aerobic, type species- *C. citreum*, Four pathovars of *C. flaccumfaciens* (pv. *betae*, pv. *flaccumfaciens*, pv. *oortii* and *poinesettiae*).

NOCARDIFORMS

Rhodococcus

Rods to branched substrate mycelium, gram +ve, partially acid alcohol fast, aerobic, type species - *R. rhodochrous, R. fascians* causes fasciation of sweet potato.

STREPTOMYCETES

Streptomyces

Extensively branched vegetative hyphae, 0.5-2.0 μ m in diameter, forms chains of spores, aerobic, highly oxidative, type species *S. albus, S. scabies, S. acidiscabies, S. ipomoeae* etc included in the genus.

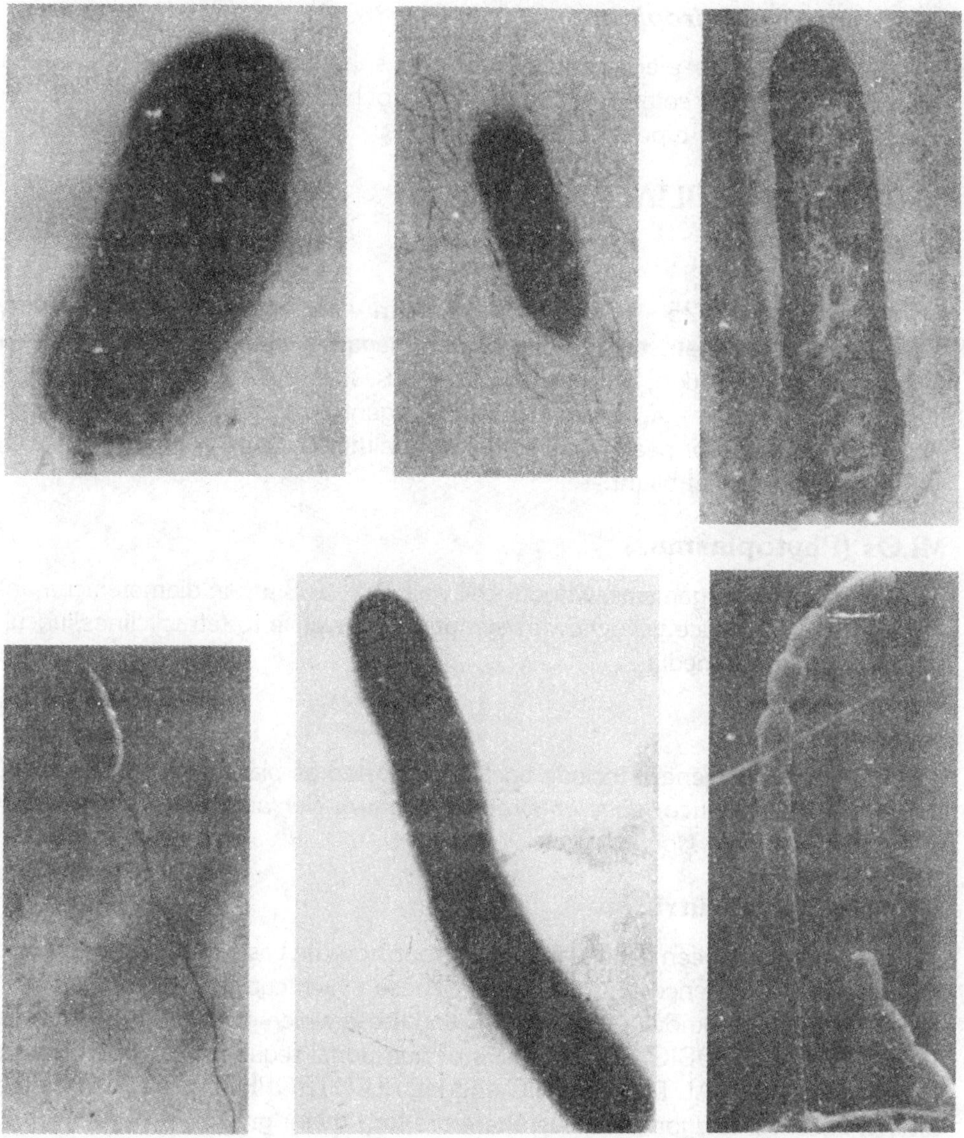

Plate-1. Electron micrographs of important genera of Plant Pathogenic bacteria (A) *Agrobacterium*, (B) *Erwinia*, (C) *Pseudomonas* (D) *Xanthomonas* (E) *Xylella* (F) *Streptomyces*

Family : **Spiroplasmataceae**

Genus: ***Spiroplasma***

Cell wall absent, pleomorphic (helical to spherical or ovoid), motile with flexional, twitching, or rotatory movement and not by flagella, facultatively anaerobic, colonies diffused, type species - *S. citri*.

UNDEFINED AFFILIATION

Xylella

straight rods, 0.25 - 0.35 x 0.9 - 3.5 μm with long filamentous strands, gram -ve, nonmotile strictly aerobic, oxidase negative, colonies smooth opalescent or umbonate rough, nutritionally fastidious, transmitted by grafting or leaf hopper, type species - *X. fastidiosa*, causal agent of pierce's disease of grape vine, phony disease of peach, leaf scorches of almond, plum, oak, mulberry and mapple, and cirus leaf blight.

MLOs (Phytoplasmas)

Polymorphic organisms, without cell wall, 0.2 - 0.8 μm in diameter, present in sieve tubes, produce yellow dwarf symptoms, sensitive to tetracyclines, uncultivated on artificial media.

Other Bacteria

Several other genera include bacteria reported as plant pathogenic. They are *Acetobacter, Gluconobacter, Bradyrhizobium, Serratia, Enterobacter, Bacillus, Clostridium, Nocardia*, etc.

(E) Bacterial Nutrition

Bacteria have been divided into two groups on the basis of their existence in the presence or absence of the oxygen. Those which can live only in the presence of oxygen are known as AEROBIC and those which grow in the absence of oxygen as ANAEROBIC. On the basis of nutritional requirements, the bacteria can be grouped as AUTOTROPHIC and HETEROTROPHIC. Both are further divided into two categories. Thus, there are four major groups—

(a) Photoautotrophs — Those using light as the energy source and Co_2 as the principal carbon source. They include photosynthetic bacteria.

(b) Photoheterotrophs — Those using light as the energy source and an inorganic compound as the principal carbon source. This group includes some of the purple and green bacteria.

(c) Chemoautotrophs — Those using a chemical energy source and Co_2 as the principal carbon source. Energy is derived by oxidation of reduced inorganic compounds. Only certain bacterial types belong to this group. Because of their ability to grow in strictly mineral media in the absence of light, these are also called *chemolithotrophs* (*lithos* = rock).

(d) Chemoheterotrophs — Those utilizing a chemical energy source and an inorganic compund as the principal carbon source. Most bacteria including phytopathogens belong to this group.

However, often the mode of nutrition within and between the groups overlap. The terms OBLIGATE and FACULTATIVE are often used meaningfully to know, identify and designate principally based on presence or absence of nutritional versatility. Plant pathogenic bacteria are basically facultative saprophytes because (1) when living host plants unavailable, they survive as saprophytes on dead organic matter, (2) they can be grown on synthetic media, (3) the degree of saprophytic survival ability/competitive survival ability.

(F) External Environment - Cell Response

Natural forces- temperature, moisture, pressure, light, H-ion concentration do affect the "state of balance in cell's protoplasmic material". This leads to distrubed metabolic processes which results in poor growth, reproduction, reduced population and reduction.

(a) Temperature - Bacteria can survive between 0 to 85°C or may be even more. Bacteria can broadly be grouped into three categories on the basis of growth temperature range - (1) Psychrophilic - Min. 0°C Max. 30°C opt 15°C; (2) Mesophilic - Min. 2-25°C, Max. 30 - 50°C, opt.20-45°C, and (3) thermophilic Min. 20 - 45°C, Max. 60 - 93°C, opt. 55°C.

(b) Moisture - Bacteria are more aquatic than terrestrial and thrive under moist conditions. Water helps in absorption of nutrients and motility.

Other factors such as light intensity, pressure, chemicals, solutions, metabolites affect biological activities both positively and negatively.

V. REPRODUCTION AND GENETICS

A. Binary Fission

Under favourable conditions bacteria multiply rapidly by binary fission which is the most common method of reproduction found in them. The dividing cell elongate considerably and the genome divides into two (Fig 3). In the middle of the cell, two ingrowths appear followed by the cell wall formation. Finally two

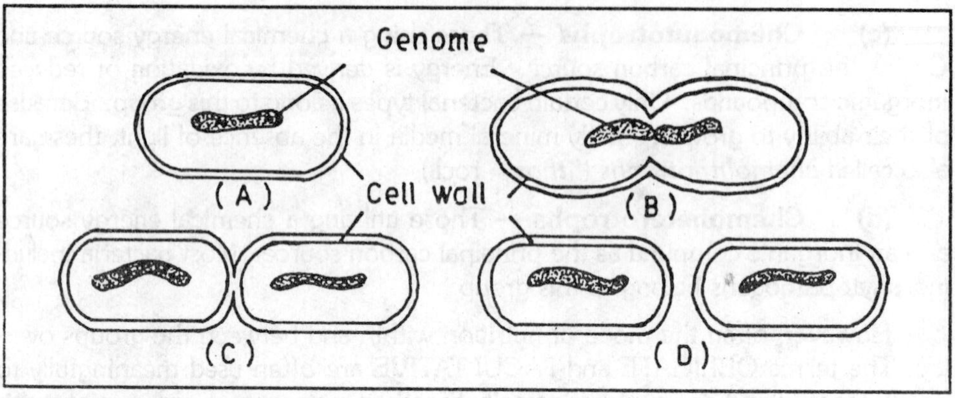
Genome
(A)
Cell wall
(B)
(C)
(D)

Fig. 3 : Binary Fission in Bacteria

daughter cells are formed. The daughter cells again divide to form more daughter cells. Sometimes, cells may remain attached together and division continues forming a chain of cells. It has been estimated that within 6 hours, more than 700,000 new cells are formed.

(a) DNA Replication

The bacterial "chromosome" (circular DNA molecule) divides (replicates) resulting in the production of two circular chromosome (fig 4). The molecule is duplicated at a site that transverses the whole molecule. This site is called REPLICATING FORK. It is that point where the two strands become four in each of the two daughter circular 'chromosomes' one strand is derived from the parent and the other one is new. This type of replication in which one strand is old and the other one is newly synthesized is called the SEMICONSERVATIVE REPLICATION. This mechanism of replication was proposed by Cairns in 1963 and is known as Cairn's model.

(b) Cell division

A peripheral ring of plasmamembrane invaginates and grows centripitally to form a double membrane septum (Fig. 5). Wall material is deposited between the two membranes of the septum. Subsequently, septum formation and the cross wall synthesis occur simultaneously. The cell division is followed by separation of the daughter cells by dissolution of the matrix between the newly synthesized walls. Higgins and Shockman (1971) proved that mesosomes are involved in DNA replication and formation of septum and the wall.

B. Genetics

Bacteria have a great capacity to adapt to diverse environments through

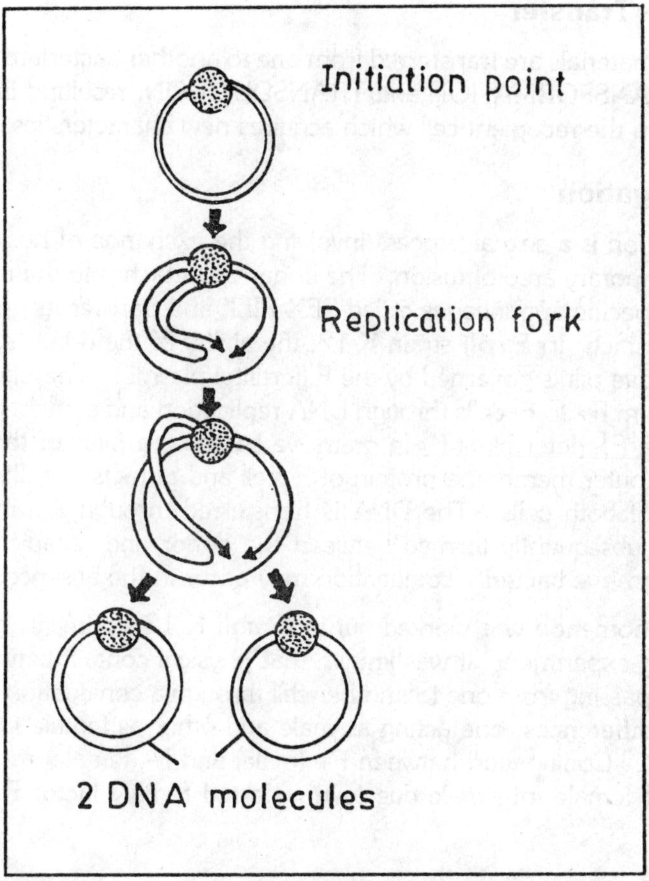
Initiation point
Replication fork
2 DNA molecules

Fig. 4 : Replication (Duplication) of the bacterial chromosome

gene transfer, mutation, or metabolic regulations. In this respect, plasmids are particularly important as they transfer themselves or mediate transfer of chromosomal DNA to other bacteria and thus altering their genetic properties.

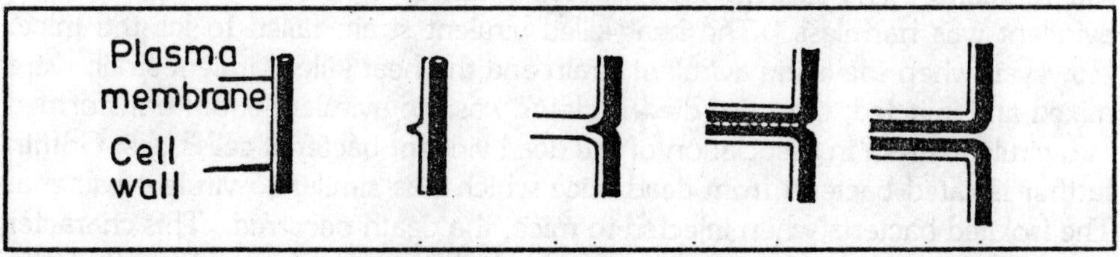
Plasma
membrane
Cell
wall

Fig. 5 : Septum and cell wall formation during cell division

(a) Gene - Transfer

Genetic materials are transferred from one to another bacterium by CONJUGATION, TRANSFORMATION and TRANSDUCTION, resulting in genetic recombination in the recepient cell which acquires new characterstics.

(b) Conjugation

Conjugation is a sexual process involving the exchange of nuclear material through a temporary area of fusion. The donar cell attaches to the recipient cell, usually with specific appandages called SEX PILI, and transfer its gene by direct cell to cell contact. In *E.coli* strain K-12, the ability of the donor strain to produce conjugative pili is governed by the F (fertility) plasmid. The plasmid can be transferred from F+ to F- cells through DNA replication and converts the latter to F+ by its own F+ determinant. In gram -ve bacteria, a pilus of the F+ cell attaches to the outer membrane protein of F- cell and retracts, resulting in wall to wall contact of both cells. The DNA is transferred through a transmembrane pore that is subsequently formed between the donor and recepient cell envelopes. In gram +ve bacteria, conjugation may occur in the absence of pili.

The phenomenon was worked out in *E. coli* K 12 strain. Through "interrupted mating experiment" it was known that physical contact between cells involves, DNA passing from one to another cell through a conjugation tube, strains show sexual differences, one acting as male and other as female i.e. the donor and recepient. Conjugation between F+ (male) and F- (female) resulting in the conversion of female into male due to transferred fertility factor F is shown in Fig. 6.

(c) Transformation

Griffith observed transformation in 1928 and this observation initiated a biological revolution. He was working with *Pneumococcus pneumoniae*, the causal agent of pneumonia. He had both virulent and avirulent strains which differed in cultural characters. The virulent one was capsulated and formed a smooth colony on medium while the avirulent was non-copsulated and formed rough colony. The virulent strain caused death of mice on injection while the avirulent was harmless. The heat killed virulent strain failed to kill the mice. However, when the living avirulent strain and the heat killed virulent strain were mixed and injected, the mice died. How? was the avirulent strain transformed into virulent strain in association of the dead virulent bacterial cells? Yes. Griffith further isolated bacteria from dead mice which was similar to virulent bacteria. The isolated bacteria when injected to mice, the death occurred. This character was transmitted to the progeny as heritable character. He called the

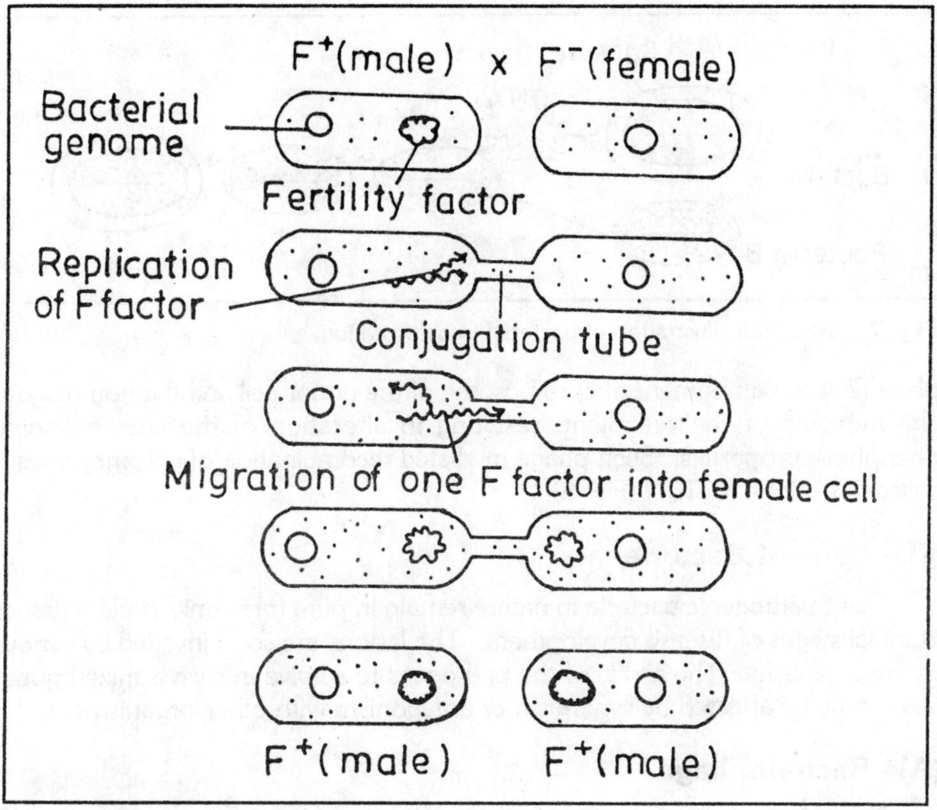
F+(male) x F−(female)
Bacterial genome
Fertility factor
Replication of Ffactor
Conjugation tube
Migration of one F factor into female cell
F+(male)
F+(male)

Fig. 6 : Conjugation in bacteria

phenomenon TRANSFORMATION.

The transformation process consists of three steps : (1) external binding of the DNA fragment onto the outer membrane, (2) penetration of the DNA fragment through the cell envelop, and (3) gene expression in the state of independent replicon or integrated into the chromosome. Schematic illustration of bacterial transformation is shown in (Fig. 7).

(d) Transduction

When prophages are released from host chromosomes and initiates independent replication, they may carry a very small portion of the host chromosome, resulting in the assemble of temperate phase particles with the DNA consisting of phage and host cell DNA. The phage particle is a type of defective phage because it does contain a full complement of the DNA's part that is left on the host chromosome when this phage infects other cells, crossing over takes

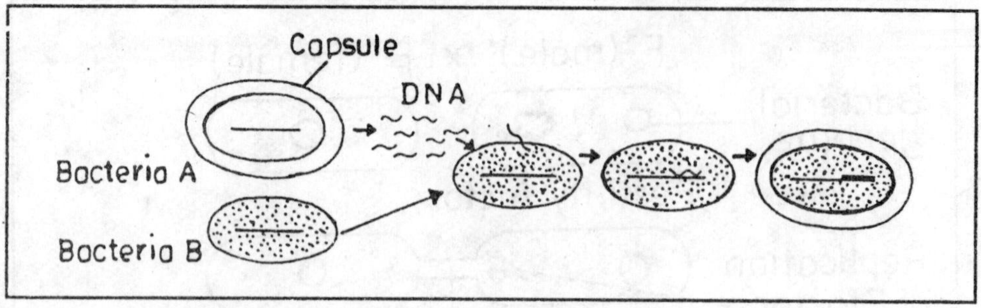
Capsule
DNA
Bacteria A
Bacteria B

Fig. 7 : Schematic illustration of bacterial transformation

place between a fragment of chromosome of the donor cell and the homologous chromosome of the recepient, resulting in alteration of the corresponding phenotypic properties. Such phage mediated recombination of chromosomes is called TRANSDUCTION (Fig. 8)

VI Lysis of Bacteria

Plant pathogenic bacteria in nature remain in pure form only in plant tissues at initial stages of disease development. The lesions are soon invaded by various macroorganisms. The ability of the pathogens to survive in such a mixed population may be affected by synergism or antagonism with other organisms.

(A) Bacteriophage

One of a groups of viruses that infect specific bacteria, usually causing their disintegration or dissolution. Phages are morphologically classified into three

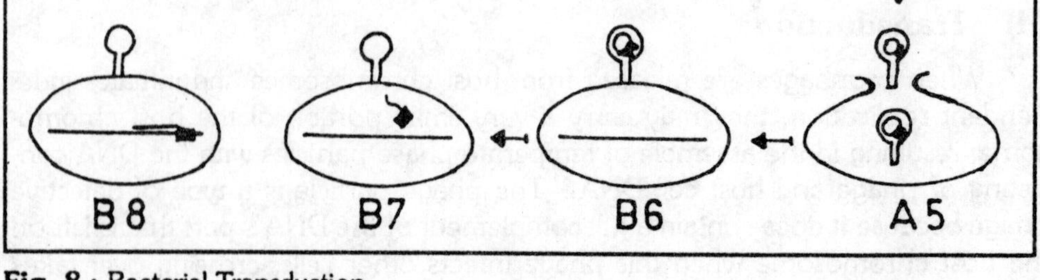
A1
A2
A3
A4
B8
B7
B6
A5

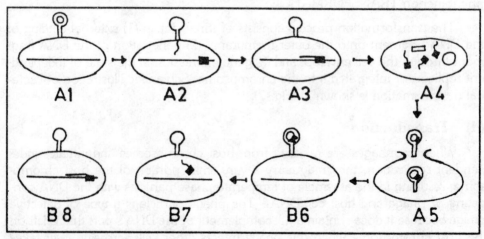

Fig. 8 : Bacterial Transduction

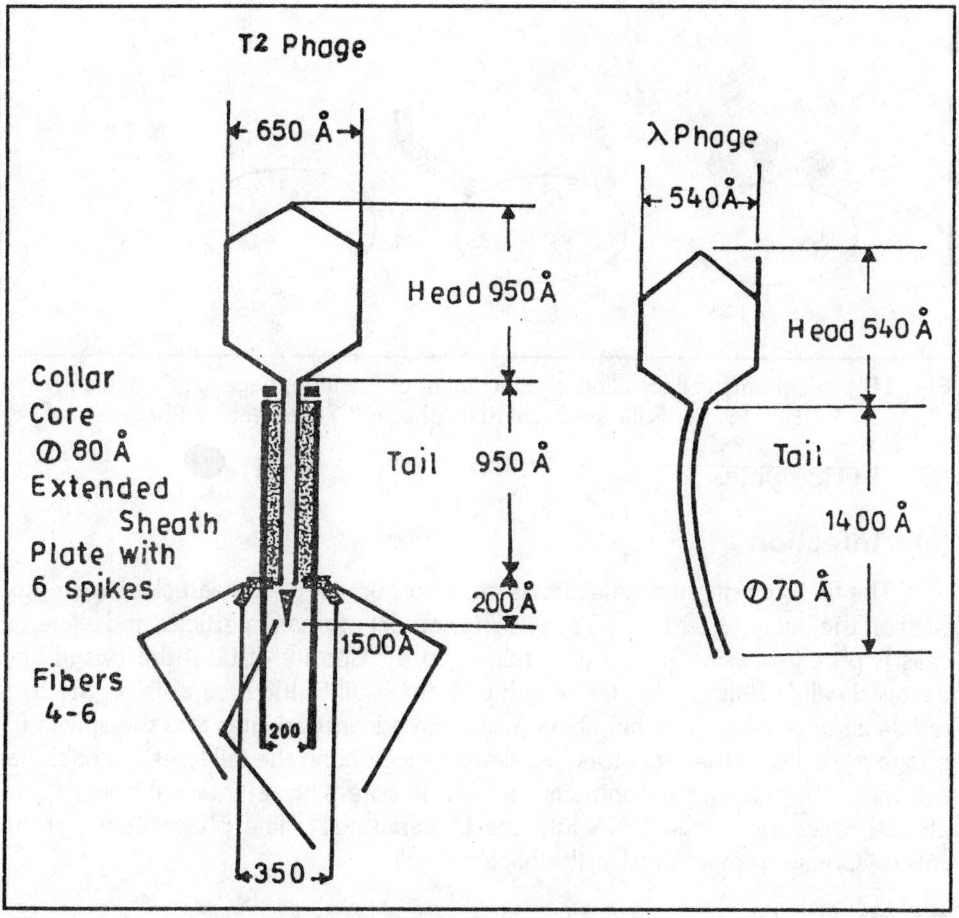
T2 Phage
650 Å
Head 950 Å
Collar
Core
Ø 80 Å
Extended
Sheath
Plate with
6 spikes
Tail 950 Å
200 Å
1500 Å
Fibers
4-6
200
350
λ Phage
540 Å
Head 540 Å
Tail
1400 Å
Ø 70 Å

Fig. 9 : Shapes and Dimensions of T2 and λ phages

categories: tadpole, spherical, and filamentous. The heads of tadpole-shaped phages are actually icosahedral and measure 50 - 100 μm in diameter and the tails 15 - 170 x 10-20 μm. The size of spherical phages is generally 20-26 μ m in diameter and that of filamentous phages is about 800-1400 μm in length and about 4 - 7 μm in width. The typical structrue of a todpole shaped phage is shown in figure 9. Most phages of plant pathogenic bacteria have DNA but RNA phages are also known. The base composition of DNA varies with the phages, although most phage NA consist of the usual bases of thymine, adenine, cytosine and guanine.

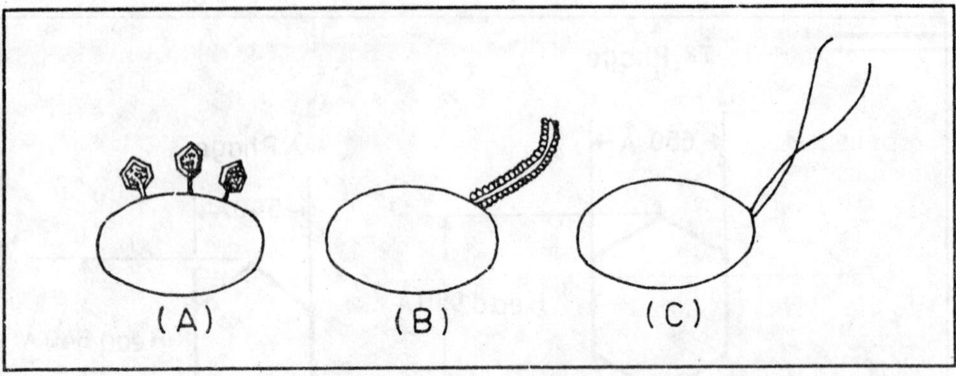
(A)
(B)
(C)

Fig. 10 : Schematic illustration of attachment of Bacteriophage. (A) Tadpole Phage; (B) F Pilus specific spherical RNA phage; (C) Filamentous Phage.

(B) Lytic-cycles

(a) Infection

The first step of phage infection is the adsorption of phage particles to receptor sites on the bacterial surface (Fig 10). The contact of phage particles and bacterial cells is purely physical and is determined by the density of both the phage and bacterial cells. Phage particles adsorb with their tail to the receptors on the bacterial cell envelope. The tail fibers probe the adsorption sites and the spikes fix phage particles to the receptors. As isozyme located in the tail lyses the bacterial cell wall, the tail sheath contracts so that the core with an internal hole is penetrated, injecting phage DNA into the bacterial cell. The phage coat portein called "Ghost" remain outside the bacterial cell.

(b) Reproduction

Phage progeny in the reproduction process do not acquire infectivity until mature phage particles are assembled in the infected host cell. The noninfective form is called *vegetative phage.* In the early stages of phage reproduction, infective phage particles can not be recovered from bacterial cell lysates. This stage is called *eclipse phage.* Replication of phage NA, formation of structural proteins,and biosynthesis of enzymes required for synthesis of these viral components as well as the assembly of phage particles are perfomed on a regular time schedule according to phage genes. Some genes start working immediately after infection while others are transcripted late in the viral growth cycle. The phage head is likely assembled through the process in which viral NA is first pushed into prohead protein coat and then is pulled in as consequence of ATP hydrolysis.

(c) Release

Bacterial cells containing mature phage particle usually burst to release phage progeny. This is LYSIS. Along with mass lysis of bacterial cells, turbidity sharply drops in liquid cultures and PLAQUES are formed on agar plates. The lytic cycle of virulent phages normally is completed within an hour after infection. (Fig 11).

B Temperate phage and lysogeny

Temperate phages differ from virulent phages in that phage DNA injected into the cells is integrated into the host chromosome, and thereafter it replicates synchronously with the host genome. The phage genome in this state is called PROPHAGE and the phages that have the capacity to undertake a prophage state are called TEMPERATE PHAGE. Such relationships occurring between temperate phages and the bacteria are called LYSOGENIC BACTERIA.

Lysogenic bacteria successively transfer their phophage DNA to the daughter cells during cell divisions without releasing any free phages. The prophage is, however, occasionally released from the host chromosome in some host cells and initiates the lytic cycle as in virulent phages. This is called SPONTANEOUS INDUCTION. Thus the cultures of lysogenic bacteria usually contain a small number of phages.

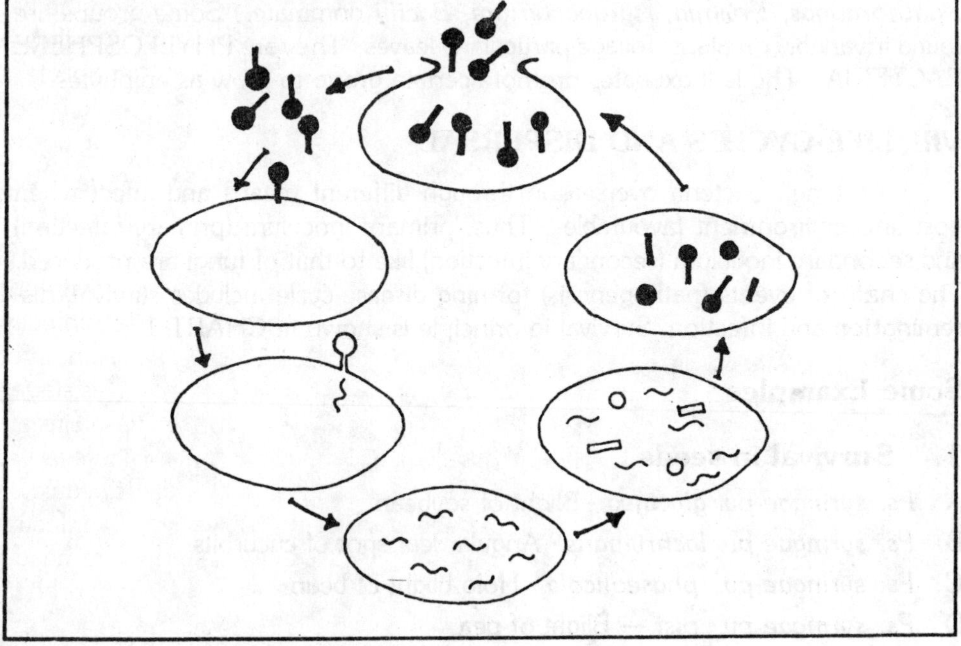

Fig. 11 : Life-cycle of virulent phages

VII. ECOLOGY

It is a science where interactions, within, inbetween and/amongst, the populations, communities i.e. all biotic and abiotic factors, is studied. In the on going process of evolution, each species, genus, populations, communities, flora, fauna, etc. adapt to certain given conditions. There are several systems in which certain groups interact and play a role of energy conversions and its transfer at different trophic levels is ECOSYSTEM.

In relation to soil where majority reside wę have broadly two groups-Soil inhabitants and soil invaders. In the latter category all those bacteria that are active include those that attack plants and live as parasite, then on refuge mixed in soil and/or in soil. This cycle is repeated almost seasonally. SOIL INHABITANTS are those which entirely and exclusively carry their activities within the soil system either actively or passively. Some have made plants root zone as their home and are called RHIZOSPHERE BACTERIA. Rhizosphere is the zone where roots release/ exude/ secrete energy rich organic compounds and this attracts microorganisms. Those that grow fast, breed fast, produce varieties of enzymes and toxins, produce antibiotics, tolerate metabolites, naturaly dominate the root zone to grab the nutrients. Bacteria have these properties and therefore have approximately over 40% of the total rhizosphere populations. *Pseudomonas, Xanthomonas, Erwinia, Agrobacterium, Bacilli* dominate. Some groups are found invariably on plants foliage particularly leaves. They are PHYLLOSPHERE BACTERIA. The leaf exudates promote certain group to grow as epiphytes.

VIII. LIFE-CYCLES AND DISPERSAL

Like fungi, bacteria overseason through different means and infect when host and environment favourable. Thus, primary inoculum (primary infection) and secondary inoculum (secondary infection) like to that of fungi are produced. The chain of events (pathogenesis) forming disease-cycle includes survival, dissemination and infection. Survival in principle is shown in CHART 1.

Some Examples

1. Survival in seeds

A *Ps. syringae pv. glycinea* - Blight of soybean

B *Ps. syringae pv. lachrymans* - Angular leaf spot of cucurbits

C *Ps. syringae pv. phaseolicola* - Holo blight of bean

D *Ps. syringae pv. pisi* — Blight of pea

E *X. campestris pv. campestris* — Black rot of crucifers

BACTERIA

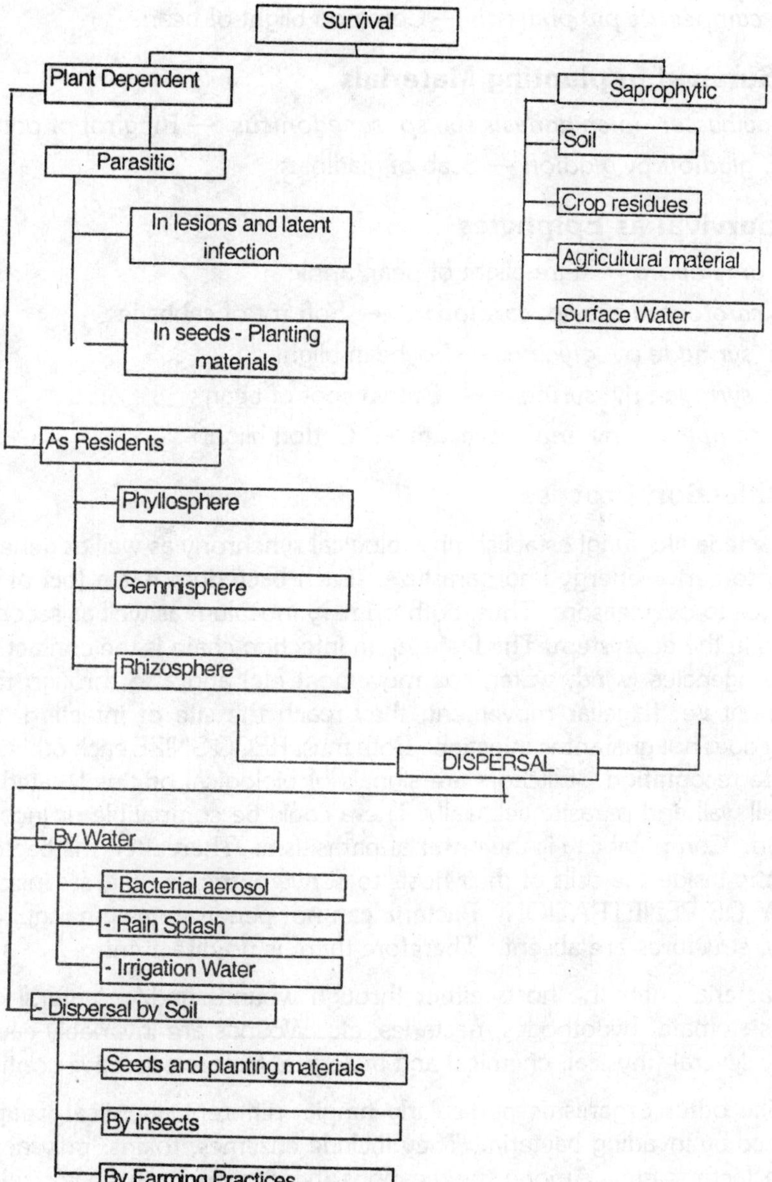

Survival
Plant Dependent
Saprophytic
Soil
Parasitic
Crop residues
In lesions and latent infection
Agricultural material
Surface Water
In seeds - Planting materials
As Residents
Phyllosphere
Gemmisphere
Rhizosphere
DISPERSAL
By Water
- Bacterial aerosol
- Rain Splash
- Irrigation Water
- Dispersal by Soil
Seeds and planting materials
By insects
By Farming Practices

CHART-1

F *X. campestris pv. malvacearum* — Blight of cotton

G *X. campestris pv. phaseoli* — Common blight of bean.

(2) Survival in planting Materials

A *Clavibacter michiganesis* sub *sp. sepedonicus* — Ring rot of potato.

B *Ps. gladioli* pv *gladioli* — Scab of gladiolus

(3) Survival as Epiphytes

A *E. amylovora* — Fire blight of pear, apple

B *E. carotovora* sub sp. *carotovora* — Soft rot of cabbage

C *Ps. syringae* pv. *glycinea* — Soybean blight

D *Ps. syringae* pv. *syringae* — Brown spot of bean

E *X. campestris* pv. *malvacearum* — Cotton blight

IX Infection Process

Bacteria like fungi establish physiological synchrony as well as genetical synchrony to derive energy and parasitize. Each bacterium has a foci of infection and place to overseason. Thus, both primary inoculum as well as secondary are existing in the ecosystem. The first step in infection chain is the contact. Several passive agencies (wind, water, soil movement etc) and also through the active movement i.e. flagellar movement, they reach the site of infection. A mere contact does not guarantee infection. Both must RECOGNIZE each other. Ellicitors facilitate recognition. Ellicitors are signals of biological origins located both on host cell wall and parasite cell wall. There could be compatible or incompatible reaction. Compatability is the onset of parasitism. Thereafter, the bacterial cells find entry inside the cells of their hosts to derive nutrients. Ingress inside host is ENTRY OR PENETRATION. Bacteria can not penetrate as the required force, energy, structures are absent. Therefore there is no direct entry.

Bacteria enter the hosts either through wounds and/or natural openings such as stomata, hydothodes, nectaries, etc. Wounds are invariably caused/created by several physical, chemical and biological factors operative continuously.

Like other organisms particularly fungi, different chemical weapons are produced by invading bacteria. They include enzymes, toxins, polysaccharides, growth factors, etc, Among the weapons the enzymes play major role. They produce pectolytic enzymes for dissolution of pectin, pectic acid and release polymers. These hydrolytic enzymes include methyl esterases, polygalacturunase transeliminases. To digest proteins proteinases, to digest cellulose - cellulases and for hemicellulose – hemicellulases, etc. are produced.

Among the toxins, both host specific and host non-specific are produced. Some examples are TABTOXIN or WIDLFIRE toxin by *P. syringae pv. tabaci*. It is a potent toxin. Besides, enzymes and toxins growth regulators, polysaccharides, ethylene etc. are also produced and play role in pathogensis.

After full host tissue colonization signs and symptoms appear. Bacteria, in general produce following specific symptoms —

1. Crown Gall bacterium - *Agrobacterium spp:* crown gall, twig gall, cane gall, hairy roots, etc.
2. *Clavibacter spp* — Ring rot, canker, wilt, spots, fasciation etc.
3. *Erwinia spp* - Blight, wilts, soft rots etc.
4. *Pseudomonas spp* - leaf spots, galls, wilt, blight, cankers, etc.
5. *Xanthomonas spp* - leaf spots, rot, cankers, blights, etc.
6. *Streptomyces* - Scab, soft rots, root nodules, etc.

X. DIAGNOSIS OF BACTERIAL DISEASES

Quick and correct identification of the disease (S) in question, is the first step in devising and applying effective control. The process of identification is called diagnosis and the science (theoretical and practical) is called diagnostics.

A Field Diagnosis

Distribution of disease in the field, record of the problem in the past, cultural practices adapted, application of agro-chemicals, varieties/genotypes, rotation are the necessary inputs for reaching some conclusion as to the identity of the disease.

B Plant Diagnosis -

1. Symptoms and signs - (a) classify symptoms into wilting, necrotic spot, blight, soft rot, hypertrophy, malformation, etc.

(b) Examine for water soaked affected tissues,

(c) Examine for exudation and ooze.

2. Microscopic -By using scraping, teasing or sections, alongwith proper staining, the presence of bacterial cells will bring you closer to identification of the cause and disease.

3. Isolation - Isolate the bacterium following standard technique, purify the isolate, multiply, prove pathogenicity and simultaneously study cellular characters and gram staining.

C. Diagnosis by Bacteriophage

This technique is useful for examination of seed materials. Macerate seeds in a liquid medium to release bacterial cells. Add known numbers of phage particles and incubate overnight. A significant increase in the numbers of phage particles indicates the presence of host-bacterium.

D. Serological Methods

Among various procedures immuno fluorescence and ELISA have been most widely used for the diagnosis of phytopathogenic bacteria.

E. Gene Diagnosis

Rapid and accurate identification is possible by NA hybridization by dot-blot or colony hybridization techniques.

XI. Management of Bacterial Disease

A. Quarantine

Many bacterial pathogens introduced into abroad have caused serious epidemics in new areas where they were previously unknown. For example fire blight of apple and pear, which was prevalent in the US as early as 1780's, was introduced in U.K. in 1950's and caused first out break in 1958. The disease was subsequently introduced into continental Europe in the 1960's Citrus canker was introduced in 1910's from Asia to Gulf states of North America and caused a severe epidemic in Florida. An eradication programme was launched from 1914 to 1934. It was one of the most intensive programmes in the history of plant pathology. Categories of seed borne bacteria on the basis of potential hazard in the international quarantine are given below :

Category A

Definition -Dangerous pathogen, not known in regions of introduction, considerable epidemic potential.

Precaution - Complete prohibition against introduction.

Examples- *Curtobacterium flaccumfaciens pv. flaccumfaciens (French bean, soybean) Clavibacter yathayi (Cocksfoot)*

X. campestris pv. papeacricola (poppy)

X. campestris pv. sesami (sesame)

Category B

Definition - Dangerous pathogen, not present, moderate epidemic potential

Precaution - Seed testing, Tested seeds must be free

Examples - *Ps. syringae* pv. *glycinea* (soybean)

X. campestris pv. *campestris* (crucifers)

X. campestris pv. *vesicatoria* (tomato)

Category C

Definition - other plant pathogens of importance to the planting value of seed.

Precaution - Adequate testing.

Examples - *Clavibacter michiganensis* subsp. *michiganensis* (tomato)

Ps. syringae pv. *lachrymans* (cucumber)

Ps. syringae pv. *phaseolicola* (french bean)

Ps. syringe pv. *pisi* (pea)

X. campestris pv. *phaseoli* (French bean)

B. Farming Practices - Disease Management

a. Eradication - Eradication methods are employed against the pathogen, to the host plants or alternate/collateral/alternative hosts. Practical procedures include fumigation, heat treatment, flooding, solarization, burning or removal of residues.

b. Field Hygiene - Removal of diseased plants or plant parts from field is important to reduce inoculum density. Defoliated leaves or pruned twigs of fruit trees and straw of cereals are preferably burnt. Infected woody stems such as tomato, brinjal and tobacco should not be ploughed into the soil because they may remain partly undecomposed and thus harbour the inoculum.

c. Disinfection of Seeds and planting Materials - Several bacterial pathogens survive in or on propagative materials including seeds. Necessary physical (temp. heat, radiation), chemical or biological methods should be used.

d. Disease Escape - By proper selection of cultivars and some modifications in cultivation period particularly date of sowing/planting, many bacterial pathogens can be managed. It is now proved that to easily contain bacterial leaf blight of rice, plant early maturing varities and start cultivation early enough torrential rains.

e. Rotation - This is an unique method to free the soil of sickness particularly nutritional and microbiological. Rotation is ineffective against bacteria having wide - host range, or when the orgamism is soil inhabitants (e.g. *R. Solanacearum, Agrobacterium, Erwinia*) or when the bacterium is both seed and soil borne.

f. Irrigation - Controlled irrigation both in respect of amount and timings provide good management. In common scab of potato (*S. scabies*), irrigation for several weeks after tuber initiation is effective in reducing the desease. It is experimentally proved that maintaining the soil-moisture defecit of 0.6 for 3-4 weeks in the tuber initiation period controls the disease.

g. Plant Density and Nutrition - Adequate plant density and balanced fertilization are obviously important for growing healthy plants. High N levels, and dense plant density bring conginial conditions for infection as plants get predisposed. Pathogens proliferate. Dense plant canopy reduces air movement and drying of foliage.

C. Biocontrol - A commercial biocontrol is fully established. Crown gall of woody trees caused by *Agrobacterium tumifaciens* is controlled by strain K 84 of *A. radiobacter*. This strain produces Agrocin 84. The bio-control is a competitive inhibitor and for successful control must be applied before the contact of the pathogen and the roots. Thus, root dipping is the important delivery method. Once root niches are occupied, *A. tumifaciens* gets no space and nutrients, thus eliminated.

D. Chemical Control - Several chemical compounds such as streptomycin, Kasugamycin, oxytetracycline, novobiocin, copper compounds etc. are available as spray materials to control bacteria infecting above ground plant parts. Sodium hypochlorite and oxolinic acid as seed disinfectant. Chloropicrin and thiram as soil disinfectant.

Fastidious Vascular Bacteria

The fastidious vascular bacteria (previously known as RLO) that cause plant diseases can not be grown on simple culture medium in the absence of host cells. Fastidious phloem-limited bacteria were observed first in 1972, in the phloem of clover, and later in citrus plants affected with the greening disease. In 1973, fastidious xylem - limited bacteria were observed in the xylem vessels of grape plants affected with pierce's disease and of alfalfa affected with alfalfa dwarf. Subsequently, similar organisms were observed in the xylem of plants affected with one of more than 20 other diseases. For example, peach affected with the phony peach disease, sugarcane affected with ratoon stunting, and in plum leaf scald, almond leaf scorch, and elm leaf scorch.

They are generally rod shaped cells 0.2 to 0.5 μ m in diameter by 1 to 4 μ m in length. They are bounded by a cell membrane and a cell wall. They have no flagella and are gram (-ve) negative. Several such xylem limited bacteria have been placed in genus *Xyella*. Only the xylem inhabiting bacteria causing sugarcane ratoon stunting and Bermuda grass stunting are gram (+ve) positive, and they are classified as members of *Clavibacter*. None of the phloem -inhabitng fastidious bacteria has been grown in culture so far, but all the xylem - inhabiting fastidious bacteria can be grown in culture. All gram (-ve) negative xylem inhabiting fastidious bacteria are transmitted by xylem - feeding insects, such as sharpshooter leaf hoppers and spittlebugs. So for, no insect - vector is known for the gram positive xylem inhabiting fastidious bacteria.

The symptoms of diseases caused by fastidious xylem inhabiting bacteria often consist of marginal necrosis of leaves, stunting, general decline, and reduced yield. Among the most important diseases caused by fastidious xylem-limited, gram (-ve) negative bacteria are Pierce's disease of grape, citrus variegation chlorosis, phony peach disease, almond leaf scorch, and palm leaf scald. They are all caused by forms of the bacterium *Xyella fastidiosa*. The important ratoon stunting disease of sugarcane is caused by the xylem - limited, gram-positive bacterium *Clavibacter xyli* subsp. *xyli*. Phloem limited bacteria are so far known to cause citrus greening disease and some minor diseases of clover and periwinkle.

L- Forms of Bacteria :

Bacteria aften produce variants that fail to produce cell walls. The progeny of such variants comprise populations of wall less bacteria, called L form or L phase bacteria, that are morphologically indistinguishable from the phytoplasmas observed in plants. They are usually produced under laboratory conditions when penicillin or other substances that inhibit cell wall formation are added to the culture medium. They can apparently also develop in living organisms during treatment with certain antibiotics. L - form bacteria are either unstable and revert back to the original form when the substance inhibiting cell wall synthesis is removed from the medium, or they are stable, that is, they are unable to revert to the original bacterium, The only plant pathogenic bacteria reported to produce L-forms are *A. tumifaciens* (crown gall disease) and *E. carotovora* pv. *atroseptica* (black leg of potato). The L forms of *A. tumifaciens* retained the pathogenicity of the parent bacterium, produced tumours identical to those produced by the bacteria, and could be reisolated and cultured from such tumors.

Phytoplasmas and spiroplasmas

PHYTOPLASMAS

The organisms observed in plants and insect vectors, that is, the mycoplasmas, which do not include the spiroplasmas, resemble the mycoplasmas of the genera *Mycoplasma* or *Acholeplasma* in all morphical aspects. Genetically, phytoplasmas ae more closer to *Acholeplasma* than to *Mycoplasma*. They lack cell walls, are bounded by a "unit membrane", and have cytoplasm, ribosomes, and strands of nuclear material. Their shape is usually spheroidal or ovoid or irregularly tubular to filamentous, and their size are comparable to typical mycoplasmas (Fig-12).

Phytoplasma (and spiroplasmas) are generally present in the sap of phloem sieve tubes and are transmitted by leaf hoppers. Some are transmitted by psyllids and plant hoppers. They also grow in the alimentary canal, hemolymph, salivary glands, and intracellularly in various body organs of their insect vectors. The vectors don't transmit phytoplasmas immediately after feeding. It begins to transmit them after an incubation period of 10 to 45 days, depending on the temperature. The incubation period is required for the multiplication and distribution of the organism within the insect, In spite of attempts, culture of phytoplasmas has not yet been possible. They are sensitive to antibiotics, particularly those of the tetracycline group.

Important diseases caused include aster yellows, apple proliferation, apri-

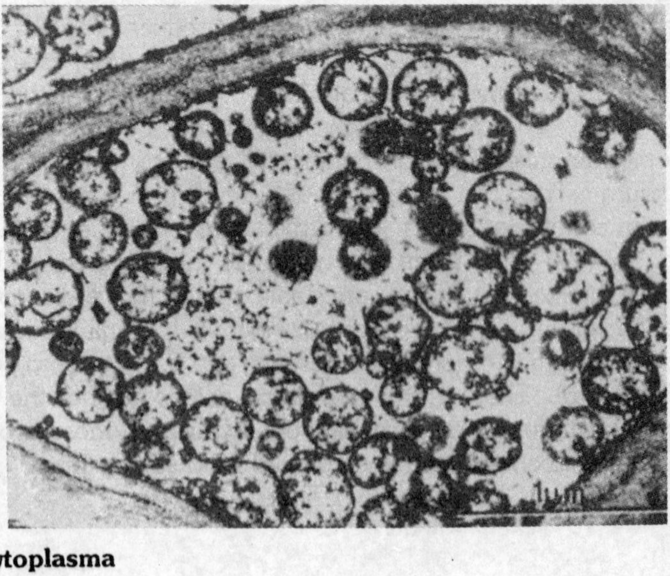
1um

Fig. 12 : Phytoplasma

cot chlorotic leafroll, coconut lethal yeilowing, elm yellows, aster yellows, grapevine yellows, peach X disease, peach yellows, pear decline, etc.

SPIROPLASMAS

Spiroplasmas are helical mollecutes (Fig -13) . They are known to cause the stubborn disease in citrus plants and the brittle root disease in horse radish (*spiroplasma citri*), stunt disease in corn plants, and a disease in periwinkle. *S. citri* has also been found in many other plants such as crucifers, lettuce, and peach. *S. citri* and the corn stunt spiroplasma also infect their leaf hopper vectors. Spiroplasmas are cells that vary in shape from spherical to slightly ovoid, 100 to 240 μ m or larger in diameter, to helical and branched nonhelical filaments spiroplasmas can be cultured on nutrient media. They produce mostly helical froms in liquid media. They multiply by fission. They lack a true cell wall and are bounded by a unit membrane. The helical filaments are motile, moving by a slow undulation of filaments and probably by a rapid rotary or screw motion of the helix. They have no flagella. Colonies on agar have a diameter of about 0.2 μ m, some have a typical fried-egg appearance, but others are granular. They are resistant to penicillin but sensitive to tertacycline.

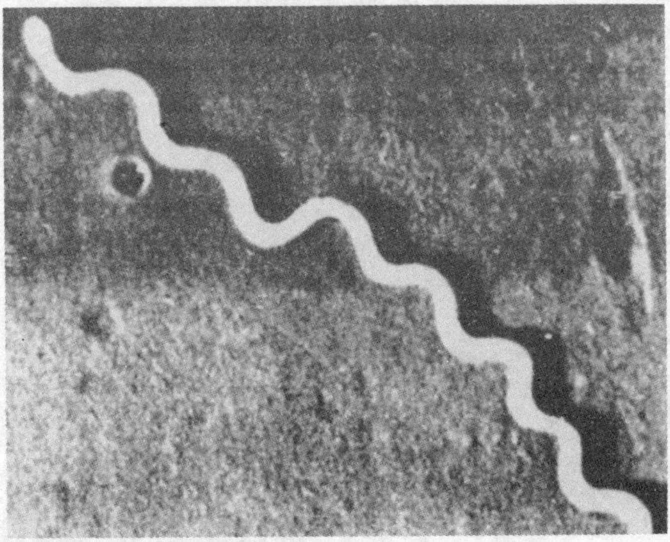

Figure-13: Spiroplasma

CHAPTER IX

VIRUSES

I. Introduction

Although virus diseases of plants came to be recognized as distinct from those caused by other pathogens towards the end of the 19th century, the first record of the symptoms of a plant virus dates back to the early part of the 17th century. Compared to the uniformly coloured blooms of virus-free plants, the patterns of colour-breaking were so attractive that it soon became a craze to possess bulbs producing 'broken' tulip flowers. The demand went to such an extent that a single such bulb used to fetch a fortune. The highly profitable trade, however, collapsed when some Dutch tulip growers found out that the stripping could be induced in a normal tulip plant by tuber grafting.

II. History - Major Events

As a matter of fact, many of the plant diseases now known to be caused by viruses, had been encountered long ago but only a few received serious attention before the end of nineteenth century. The first breakthrough was made in 1886 by Adolf Mayer, while studying the highly contagious, mysterious disease of tobacco which he called *mosaic* disease. He found that healthy plants could be infected by injecting their leaf veins with sap taken from diseased plants. He found also that boiling the sap inactivated the unknown agent. His overall conclusion was that the disease was the manifestation of a becterium. In 1892, Dimitrij Ivanovsky published the results of his study on the mosaic disease of tobacco. He confirmed Mayer's report of the sap-transmissible nature of the disease and further showed the sap to remain infectious even after passage through bacteria-proof filter. However, he claimed the incitant to be a microbe. It was left to Dr. Martinus Beijerinck, to realise the causal agent to be something novel. His results not only confirmed the earlier findings but also showed that the incitant could diffuse into an agar gel. Based on all the findings, he concluded in

1898 that the mysterious pathogen was not a bacterium but a *contagium vivum fluidum*. He thought the contagium to be able to reproduce itself in living plants and referred it as a *virus*. It must have required clarity of thought and conviction to overcome the very strong bias for bacteria prevailing in those days. Beijerinck is, therefore, rightly considered as the Father of Virology.

In the meanwhile, Hashimoto, a Japanese rice grower, suspected a link between the occurrence of leafhoppers and incidence of rice dwarf disease. In 1894, he experimentally demonstrated the causal relationship but could not identify the leafhoppers and publish his findings.

Although the true nature of viruses remained obscure till 1935, there were some important discoveries that included the first aphid vector (1916); symptomless carriers (1918); mutability of viruses and development of virus strains (1925,1926) development of quantitative assay of tobacco mosaic virus (TMV) by local lesions (1929); antigenic property of plant virus to induce antibodies in mammals (1928,1929); and use of indicator plants for separation and identification of viruses (1931).

The long awaited breakthrough concerning the nature of plant virus was made in 1935. Stanley (1935), used the newly introduced methods for isolation and characterization of enzymes and other proteins for purifying TMV. He succeeded in isolating it as a proteinaceous crystalline substance. The substance was highly infectious and remained so after repeated recrystallization. He was later awarded the Nobel Prize for his achievement. However, his studies did not recognize the distinct nucleoprotein nature of the virus which was first shown by Bawden and Pirie (1937), who reported TMV to contain 5 percent RNA and 95 percent protein. In 1939, Kausche, Pfankuch and Ruska took electron microscope picture of TMV, being the first of a plant virus.

Research on the composition and properties of viruses during the last four decades has immensely enhanced knowledge of the structure, function and biological behaviour of viruses. While a large number of viruses have been detected and characterized, quite a number of diseases that used to be considered as viral disorders, are now known to be caused by other agents such as phytoplasmas (Doi *et al.*, 1967), viroid (Diener, 1971) and rickettsia(Goheen *et al.*, 1973). Techniques have been developed to produce virus-free clones of numerous vegetatively propogated crops. The involvement of new groups of vectors such as nematodes and fungi has been discovered and knowledge of virus-vector relationships considerably enhanced.

III. Economic Impact of Plant Viruses

Virus diseases affect most economic crops, causing reduction in yield or quality of the produce. The extent of losses depends on many factors and may vary widely. Viruses seldom kill plants outright but virus epidemics during vulnerable stages of crop growth may cause crop failure. The vegetatively propagated crops such as potato, sugarcane, banana, strawberry, citrus and other fruit trees may suffer severe losses if raised from infected propagation material. The progressive deterioration of vegetative stocks have resulted in elimination of many commercial virieties.

There are numerous examples of severe crop losses due to virus diseases. Nearly 75 percent of the sweet orange trees were wiped out by the citrus tristeza virus in Sao Paulo State, Brazil within a period of twelve years. Cacao swallen shoot virus has been a menace to the cacao plantations in West Africa. The devastating nature to the disease is evident from the fact that in Ghana more than 100 million cacao trees have been cut down since 1945 just to stop the spread of the disease. The nematode borne grapevine fanleaf virus, may reduce yield of grapes by 10 to 50 percent. The sugar beet industry in the Western United States was once on the verge of ruin due to beet curly top virus and the disaster was averted by the introduction of tolerant varieties. Sugar beet crops in Europe and America have suffered serious losses due to sugar beet yellows for many years. Total losses due to sugar beet yellows in Britain in 1957 were about 1,000,000 tons. The fungus-borne wheat mosaic and mite-borne wheat streak mosaic viruses have caused huge losses of wheat in some states in North America. Barley yellow dwarf has proved to be a potential menace to wheat, barley and oats in many temperate countries with large scale outbreaks under favourable conditions. Virus diseases have remained a major constraint to potato cultivation in various parts of the world. In India, the losses in potato yield have been estimated to be 20-25 percent and 40-85 percent due to potato leaf roll virus and potato virus Y, respectively. Tomato mosaic virus has been estimated to cause 15-20 percent loss in yield of the tomato crop.

IV. Structure and composition

A. Viral Structure

Morphologically, the plant viruses fall into two broad categories, anisometric (rigid or flexuous rods and bullet-shaped) and isometric (polyhedral). The isometric viruses have been referred as 'spherical'. High resolution electron microscope revealed many viruses to be symmetric polyhedra. X-ray crystallographic patterns has clearly shown the virus surface to be symmetrical. Crick and Watson (1956) suggested a paradoxical explanation that regular bonding of subunits from either of the three cubic symmetry, i.e. tetrahedral, octahedral or icosahedral which is likely to lead to isometric appearance. (Figure-1 & Figure-2)

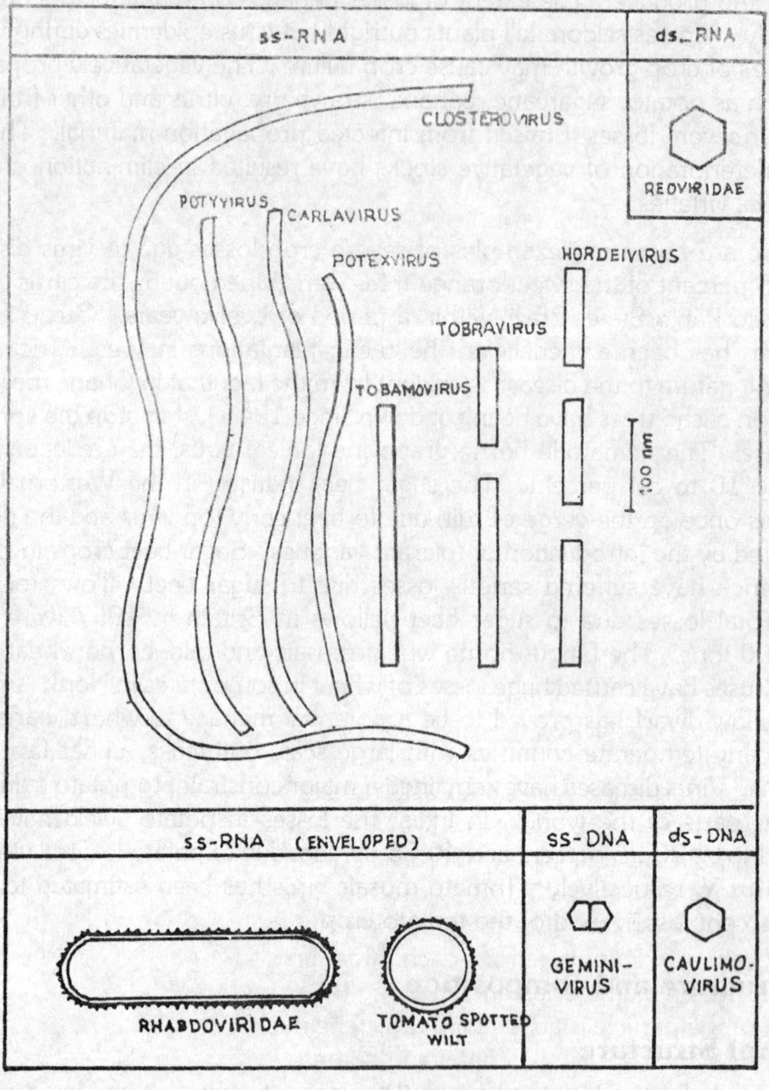
SS-RNA
ds-RNA
CLOSTEROVIRUS
REOVIRIDAE
POTYVIRUS
CARLAVIRUS
POTEXVIRUS
HORDEIVIRUS
TOBRAVIRUS
TOBAMOVIRUS
100 nm
SS-RNA (ENVELOPED)
SS-DNA
ds-DNA
GEMINI-VIRUS
CAULIMO-VIRUS
RHABDOVIRIDAE
TOMATO SPOTTED WILT

Fig. 1 : Diagram showing the particle shapes, relative sizes and genomic nature of some plant virus groups

a. *Rod -shaped :* All rods are helical in structure. The rod-shaped viruses may be rigid rods as are TMV particles, or flexuous like the potyviruses.

(i) *Rigid - rods :* TMV is most widely studied virus in this group. The entire TMV particle is very stable due to interaction between protein subunits and

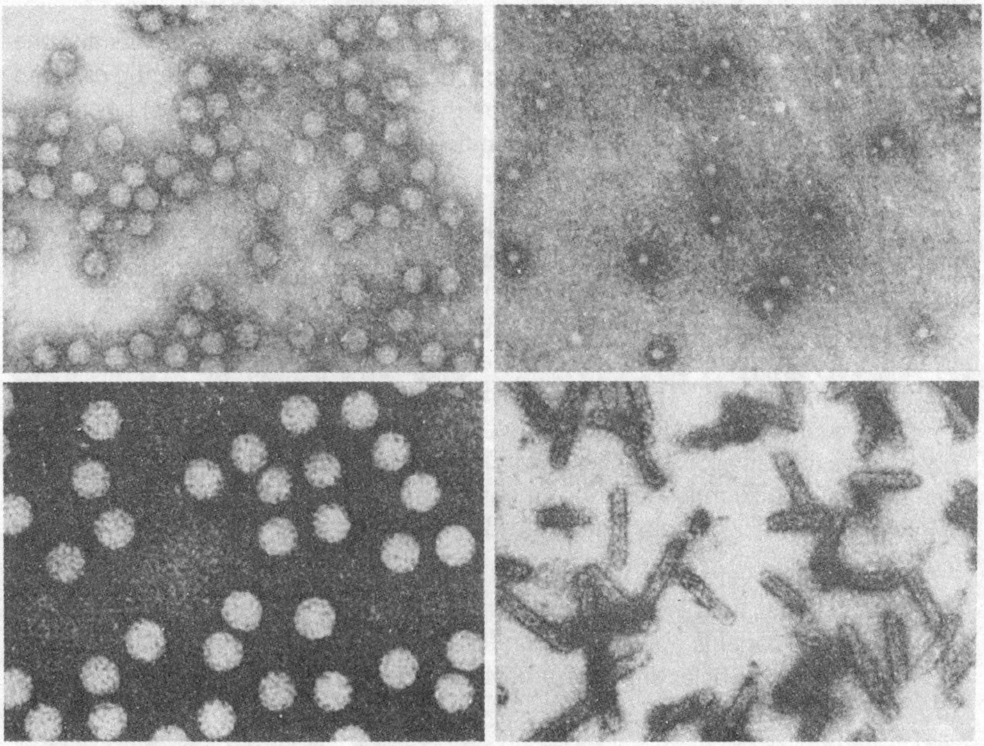

Figure-2: Different type of virus particle

RNA. The particles measure about 300 nm x 15 nm. The molecular weight of the particle is 39.4 x 10^6 daltons. It consists of about 2100 helically arranged identical protein subunits along with axial canal, each with molecular weight of 17500 daltons and consisting of 158 amino acid residues, whose sequence is known for some strains. The pitch of the helix is 2.3 nm the particle structure repeats every 6.0 nm of its length each three turns of the helix. The particle has a hole of 4.0 nm diameter extending along the axis and the cylindrically average diameter of the particle is about 15.0 nm. The nucleic acid strand has a molecular weight of 2 x 10^6 daltons containing some 6,400 nucleotides which follows the pitch of helix and is embedded between the protein subunits 4 nm from the particle axis.

Other viruses with straight tubular particles are similar to TMV in structure but show some differences. For example, tobacco rattle virus (TRV) has particles of two modal lenghts, the shorter ones measuring 45 - 114 nm against 195 nm particles of the longer ones, the width in both being 21-23 nm. All particles are constructed from the same protein subunits but differ in size of nucleic acid molecules. The protein subunit repeats every three turns of the basic helix which has a pitch of 2.5 nm and each turn of helix contains 25.3 protein subunits and the total of 76 subunits in three turns of helix. One end of particle is slightly rounded and another end is flattend. Barley stripe mosaic virus (hordeiviruses) has atleast three different types of modal length ranging from 110 nm to 150 nm with five axial turn repeats of helix.

(ii) *Flexuous rods :* Such particles are not straight and show more or less marked bandings. The particles of potexviruses (PVX, white clover mosaic virus etc.) have flexuous particles, 480 to 580 nm long and 11-13 nm wide with cross bandings. The particles have basic helix pitch of 3.3 nm and 3.6 nm, respectively. The particles of Carlaviruses have straight and rigid structure but having a slight band at one end. They measure 650 nm in length and 12 nm in width with bandings or lines of subunits parallel to the long axis. The particles of those of potyviruses (bean yellow mosaic; cowpea aphid-borne mosaic, sugarcane mosaic and others) are slightly flexuous, measuring 750 x 11 nm. The particles of potyviruses have basic helix pitch of 3.3 to 3.6 nm.

Viruses with very flexuous particles include beet yellows and other members of closteroviruses with length ranging from 1250 to 2000 nm and width of 12 nm. The axial canal is obscured and pitch of basic helix is obvious and of 3.4 to 3.7 nm. These particles show very obvious cross-banding at the intervals of 3.3 to 3.6 nm.

b. *Isometric particles :* The isometric or icosahedral virus particles are complex in their structrue and architecture. An icosahedron is a polyhedron with 20 equilateral triangular faces. It shows three fold symmetry when rotated around the axis through the centre of each triangular face. There are 12 vertices when the five triangle corners meet and thus 60 asymmetric units are required to construct an icosahedron. The surface of an icosahedron can be divided into a large number of smaller identical triangles which is called triangulation number. A simple icosahedron having three triangulation numbers and nine triangle faces may have 180 subunits. These subunits are clustered in different ways. The clustering pattern of protein subunits is characteristic of a virus or related viruses. However, the pattern of protein subunit clustering may differ when more than one type of protein are involved. Basicallly, three types of clustering patterns have been described which are :

(i) *Trimer capsomere:* Three subunits cluster at the centre of each triangle face.

(ii) *Pentamer capsomere:* Subunits cluster at a vertice where faces of five triangles meet.

(iii) *Hexamer capsomere:* Subunits cluster at a vertice where faces of six triangles meet.

c. Virus Particles with Mambrane

c. *Spherical particle :* Tomatto spootted wilt virion has rounded irregular shape, measuring about 70 nm diameter *in situ*, while in purfied preparation, the particles are larger and more irregular having tail-like structure which seems to be protuberance of outer membrane. In dip preparations from other plants such as *N. glutinosa,* the particles appear to be of various sizes and shapes including spheroidal, tubular and filamentous with helical strands extending from central cores. The membrane binding the particles is about 5 nm thick which is composed of protein subunits. It has been suggested that the tomato spotted wilt virus particle is *Pleomorphic myxovirus.* Another virus in this group is carrot mottle, having more or less rounded particles which measure about 50 nm in diameter. The particles have poorly defined cores with an outer membrane which might be acquired from host while passing through tonoplast membrane.

d. *Bacilliform or Bullet-shaped particles :* This group of plant viruses are of complex structure and architecture. The group includes plant viruses such as maize mosaic virus, lettuce necrotic yellows, broccoli necrotic yellows, potato yellow dwarf, wheat necrotic yellows, and sowthistle yellow-vein. The particles of these viruses are very fragile. Most are 300 to 400 nm long and 50 to 80 nm wide *in situ* but become shorter and often bullet shaped when purified. The particles are helically constructed with basic helix pitch of 4.5 nm and an axial channel. The membrane around the helical core is lipid.

e. *Geminate particles :* Gemini viruses are unique group of viruses having geminate or twin particles (dimer) constituted by two icosahedral monomers. The two monomers are joined at the discontinuities created by missing capsomeres to give the characteristic flattended appearance at the interface. The dimer particle shows groove at the centre of its interface. The ratio of dimers and monomers vary in purified preparations.

B. Composition

Each plant virus consists of at least a nucleic acid and a protein. Some viruses consist of more than one size of nucleic acid and proteins, and some of them contain enzymes or membrane lipids. The nucleic acid makes up 5 to 40 percent of the virus, protein making up the remaining 60 to 95 per cent. The lower nucleic acid percentages are found in the elongated viruses, whereas the spherical viruses contain higher percentages of nucleic acid. The total mass of the nucleoprotein of different virus particles varies"from 4.6 million to 73 million daltons (Da). The weight of nucleic acid alone, however, ranges only between 1 and 3 million ($1\text{-}3 \times 10^6$) Da per virus particle for most viruses, although some have up to 6×10^6 Da. All viral nucleic acid sizes are quite small.

a. Viral Protein

Viral proteins consist of amino acids. The sequence of amino acids within a protein, which is dictated by the sequence of nucleotides in the genetic material, determines the nature of protein.

The protein shells of plant viruses are composed of repeating subunits. The amino acid content and sequence for identical protein subunits of a given virus are constant but vary for different viruses and even for different strains of the same virus. Ofcourse, the amino acid content and sequence are different for different proteins of the same virus particle and even more so for sequences of amino acids are known for the proteins of many viruses. For example, the protein subunit of tobacco mosaic virus (TMV) consists of 158 amino acids in a constant sequence, and it has a mass of 17,600 Da.

In TMV, the protein subunits are arranged in a 'helix' containing 16 1/3 subunits per turn (or 49 subunits per three turns). The central hole of the virus particle down the axis has a diameter of 4 nm, whereas the maximum diameter of the particle is 18 nm. Each TMV particle consists of approximately 130 helix turns of protein subunits. The nucleic acid is packed tightly between the helices of protein subunits. In the rhabdoviruses the helical nucleoproteins are enveloped in a membrane. In the polyhedral plant viruses, the protein subunits are tightly packed in arrangements that produce 20 (or some multiple thereof) facets and form a shell. Within this shell the nucleic acid is folded or otherwise organized.

b. Viral Nucleic Acid

The nucleic acid of most plant viruses consists of RNA, but at least 80 viruses have been shown to contain DNA. Both RNA and DNA are long, chain like molecules consisting of thousands of units called nucleotides. Each nucleotide consists of a ring compound called the base attached to a five-carbon-sugar (Ribose in RNA, deoxyribose in DNA), which in turn is attached to phosphoric acid. The sugar of one nucleotide reacts with the phosphate of another nucleotide, and this is repeated many times, thus forming the RNA or DNA strand. In viral RNA, only one of the four bases, adenine, guanine,cytosine, and uracil, can be attached to each ribose molecule. The first two, adenine and guanine, are purines, and interact with the two uracil and cytosine, the pyrimidines. DNA is similar to RNA with two small, but very important differences: the oxygen of one sugar hydroxyl is missing, and the base uracil is replaced by the base methyl uracil, known as thymine. The size of both RNA and DNA is expressed either in daltons or as the number of bases [Kilobases (Kb) for single stranded RNA and DNA or Kilobase pairs (Kbp) for double stranded RNA and DNA].

The sequence and frequency of the bases on the RNA strand vary from one RNA to another, but they are fixed within a given RNA and determine its properties. Most plant viruses (about 540) contain single stranded RNA, but 40 contain double stranded RNA, 30 contain double stranded DNA, and about 50 contain single stranded DNA.

V. VIRUS INFECTION AND SYNTHESIS

Plant viruses enter host cells through wounds made mechanically or by vectors, or by deposition into an ovule by an infected pollen grain. In replication of an RNA virus, the nucleic acid is first freed from the protein coat. It then induces the cell to form the viral RNA polymerase. This enzyme utilizes the Viral RNA as a template and forms complementary RNA. The first new RNA produced is not the viral RNA a complementary copy (mirror image) of that RNA. As the complementary RNA is formed, it is temporarily connected to the viral strand. Thus, the two form a double stranded RNA that soon separates to produce the original virus RNA and the mirror image (-) strand, the latter then serving as a template for more virus (+strand) RNA synthesis (Agrios, 1996)

The replication of some viruses differ considerably from the one described above. In viruses with different RNA segments being present within two or more virus particles, all the particles must be present in the same cell for replication and infection. In single stranded RNA rhabdoviruses the RNA is not infectious because it is the (-) strand. This RNA must be transcribed by a virus carried enzyme called transcriptase into a (+) strand RNA in the host, and the latter RNA then replicates as above. In the double stranded RNA isometric viruses, the RNA is segmented within the same virus, is noninfectious, and depends for its replicaiton in the host on a transcriptase enzyme also carried within the virus.

On infection of a plant with a double stranded DNA (dsDNA) virus, the viral ds DNA enters the cell nucleus, where it appears to become twisted and supercoiled and forms a minichromosome. The latter is transcribed into two single stranded RNAs; the smaller RNA is transported to cytoplansm, where it is translated into virus coded proteins; the larger RNA is also transported to the same location in cytoplasm, but it becomes encapsidated by coat protein subunits and is used as a template for reverse transcription into a complete virion ds DNA. The single stranded (ss DNA) DNA replicates by forming a rolling circle that produces a multimeric (-) strand, which serves as a template for the production of multimeric(+) strands that are then cleaved to produce unit length (+) strands (Agrios, 1996)

As soon as new viral nucleic acid is produced, some of it is translated, that is, it induces the host cell to produce protein molecules codes by its nucleic acid. Protein synthesis depends on the presence of aminoacids and the cooperation of ribosomes, m-RNA and t-RNA. Each t-RNA is specific for one amino acid, which it carries toward the appropriate nucleotide sequence along the m-RNA, which is produced in the nucleus and reflects part of the DNA code, determines the kind of protein that will be produced by coding the sequence in which amino acids will be arranged. The ribosomes seem to travel along the m-RNA and to provide the energy for the bonding of the prearranged amino acids to form the protein.

For virus protein synthesis the part of the viral RNA coding for the viral protein plays the role of m-RNA. The virus utilizes the amino acids ribosomes, and t-RNA of the host; however, it becomes its own blue print (m-RNA), and the proteins formed are for exclusive use by the virus as a coat. When new virus nucleic acid and virus protein subunits have been produced, the nucleic acid organizes the protein subunits around it, and two are assembled together to form the complete virus particle (Virion).

VI. NOMENCLATURE AND CLASSIFICATION

A. NOMENCLATURE

Since a plant virus is first recognized by its manifestations or the symptoms of the disease it causes, virus names were originally common or vernacular names. For instance, the virus causing a mosaic of light and dark green areas in tobacco leaves, was called tobacco mosaic virus. The discovery of more and more viruses in a single plant species, the scientists proposed that a plant virus should be named after common names of the host in which it was initially found and be given a distinguishing number to denote serial order in which it was described on that host. Thus, tobacco mosaic virus became tobacco mosaic virus 1, being the first virus recorded on tobacco. The second virus on tabacco became tobacco virus 2 and so on. Later on this system was modified by replacing the common name of the host by its Latin generic name. Accordingly, tobacco virus 1 was transformed into *Nicotiana* virus 1. With rapid increase in the number of viruses described, the limitations of the numerical code were more and more exposed. Subsequently, the letter codes like X,Y,A, etc. were introduced for a number of potato viruses which are still in use.

A number of attempts were made to name viruses scientifically by the Latin binomial system. To mention a few which achieved temporary following, Holmes (1939) named genera of viruses on the basis of symptoms induced, while the names of the host plants were used for virus species. The tobacco mosaic virus became *Marmor tabaci,* the generic name denoting the mosaic symptoms caused by the virus. The system introduced by Hansen (1956) was rather complex but more descriptive. In this system the generic name was composed of symbols representing the independent characters of the virus viz., direct transmission, vector transmission and type of virus particle. Thus, potato virus X (PVX) was named *Minflexus solani*. In this symbol *M* denotes mechanical transmission, *in* denotes virus without specific arthropod vector and *flexus* denotes particle type. It was the first attempt to name and classify plant viruses based on their properties. All these exercises in Linnaean style of nomenclature added to unnecessary complexity and confusion. With the growing realization that there can be no acceptable binomial nomenclature without a stable system of virus classification, they fell out of favour. That is why plant virologists nowadays prefer to stick to the time honoured vernacular or common names such as tobacco mosaic virus or rice dwarf virus. To make the vernacular names more meaningful, Gibbs *et al.*(1966) suggested such names to be supplemented with cryptograms, composed of coded data on some properties of the virus.

The vernacular names of plant viruses are usually long and are, therefore, abbreviated. Thus, TMV stands for tobacco mosaic virus and BYDV is the abbreviated version of barley yellow dwarf virus. These abbreviations are yet to be standardized. The list of English common names of plant viruses was first published by the CMI, England, under the editorship of E.B. Martyn (1968, 1971), which is universally accepted.

B. Classification

All viruses belong to the kingdom VIRUSES. Within the kingdom, viruses are distinguished as RNA viruses and DNA viruses. They are further divided depending on whether they possess one or two strands of RNA or DNA of either postitive or negative sense, either filamentous or isometric. Within each of these groups there may be viruses replicating via a polymerase enzyme (+RNA or DNA viruses) or via a reverse transcriptase (-RNA or DNA viruses). Most viruses consist of nucleic acid surrounded by a coat protein, but some also have a membrane attached to them. Some viruses have all their genome in one particle (monopartite viruses), but the genome of other (multipartite) viruses is divided among two, three, or rarely, four particles. Other characterstics in classification include symmetry of helix, number and arrangement of protein subunits size and physicochemical and biological properties. Figure 1 : shows various groups of plant viruses. Classification of plant viruses is summarized below :-

CLASSIFICATION OF PLANT VIRUSES - A GLANCE

I. ELONGATED, HELICAL, ssRNA

A. RIGID

(a) Monopartite

1. Tobamoviruses (300 x 18 nm)

(b) Multipartite

1. Hordeiviruses (148 x 22 nm)
2. Tobraviruses (190 x 26 nm)
3*. Soil-borne wheat mosaic virus (250-300 x 18 nm)
4*. Peanut clump virus (245 x 14 nm).

B. FLEXUOUS - All monopartite

1. Potexviruses (480 - 580 x 13 nm)
2. Potyviruses (720-900 x 11 nm)
3. Carlaviruses (610-700 x 12 nm)
4. Closteroviruses (1250-1450 x 12 nm)

II. ISOMETRIC

A. SINGLE-STRANDED RNA

(a) Monopartite

1. Tymoviruses (28nm)
2. Tombusviruses (30nm)
3. Sobemoviruses (28 nm)
4*. Tobacco necrosis virus (26nm)
5. Luteoviruses (20-24 nm)
6. Machloviruses (30nm)

(b) Multipartite

1. Comoviruses (25-30nm)
2. Nepoviruses (28-30nm)
3*. Pea enation mosaic virus (28nm)
4. Dianthoviruses (31-34 nm)
5. Cucumoviruses (20-30nm)
6. Bromoviruses (26 nm)
7. Ilarviruses (26-35 nm)

(b1) With envelope

8*. Tomato spotted wilt virus (80-85 nm)

B. DOUBLE-STRANDED RNA

Reoviruses

1. Phytoreovirus (70 nm)
2. Fijivirus (71 nm)

C. SINGLE-STRANDED DNA

1. Geminiviruses (28 - 30 nm)

D. DOUBLE-STRANDED DNA

1. Caulimoviruses (50 nm)

****III. BACILLIFORM (BULLET - SHAPED)**

A. Without envelope

1*. Alfalfa mosaic virus (58 x 18 nm)

B. With envelope

1. Rhabdoviruses (200-350 x 75-95 nm)

* Monotypic group. * * Single - standed RNA.

VII. TRANSMISSION OF PLANT VIRUSES

A. MECHANICAL TRANSMISSION

Mechanical inoculation is basically the manual transfer and deposition of biologically active virus to suitable sites in the living cells by sublethal wounding or abrasion. The success of mechanical inoculation depends on the source of inoculum, method of preparing the inoculum, stability of the extracted virus and also on the test plants. The environmental conditions such as light intensity and temperature, prevailing before and after inoculation may also detemine the success. The method of preparation of inoculum is very important to have the maximum recovery of infectious virus particles. During the extraction of virus from diseased tissues, various host metabolites and cellular debris are released simultaneously with the virus. Some of these compounds may inactivate virus or inhibit infectivity. It is, therefore, necessary to grind the leaf in a suitable buffer and use other additives to regulate pH, remove inhibitors and inactivators and to overcome oxidase and phenolase activity.

A common procedure is to grind a freshly picked leaf tissue in 0.1 M potassium phosphate buffer at a pH of 7.0-7.5. Since most plant ribonucleases possess a pH optimum between 5 and 6, higher pH values usually minimize nuclease activity as well as phenolase activity. High pH values (8-9) have been used successfully for various unstable viruses. Loss of infectivity due to tannins can be prevented by grinding the leaves in alkaline buffer or in nicotine or caffeine solutions. The addition of reducing agents such as mercaptoethanol, thioglycollic acid, cysteine hydrocloride, or chellating agents such as sodium diethyldithiocarbamate to the extraction medium helps to prevent loss of infectivity due to oxidase activity. Bentonite clay is sometimes added to the extraction medium to adsorb the nucleases. Phenol is used as an additive to denature or dissolve the various inactivating agents. It is also used to extract nucleic acid from intact nucleoprotein virus particles.

There are numerous ways of inoculating plants with virus-containing liquids. The earlier transmission of TMV was done by injecting sap of infected plants into healthy ones with capillary glass needle. Since then various methods have been adopted including the sophisticated techniques like the use of microsyringes, ultrasonic devices, spray devices etc., out of which inoculation by rubbing has been the most consistent practice for its convenience. The rubbing may be done with a wet finger or with a pad of inoculum-soaked cheese cloth, cotton gauze or sponge, a glass spatula or a stiff brush.

Inoculation by rubbing should be gentle to cause just minor wounds for entry of the virus without killing the injured cells. Fine abrasive powders such as 300-600 mesh carborundum (silicon carbide) or celite (crushed diatomaceous earth) are commonly used to produce superficial wounds and increase the number of infectible sites. Carborundum powder is usually dusted onto leaves to be rubbed with inoculum while celite, which is lighter than carborundum, is mixed with inoculum. Washing of inoculated leaves with a jet of distilled water a few minutes after inoculation, has been found to enhance the infectivity of a few viruses. Infection will occur only when the virus is able to multiply in the epidermal cells of the inoculated leaf. The virus may then spread into the other cells.

B. GRAFT TRANSMISSION

A prerequisite for successful graft transmission is the perfect union of the cambial layers of the stock and scion. While interspecific and intergeneric graft are often possible, grafts between taxonomically distant species are much less likely to be successful.

The two commonest methods are called detached-scion grafting and approach grafting. In detached-scion grafting, the type of scion varies with the nature of plant and may be shoot, leaf, bud, tuber, bark, patch etc. For herbaceous plants, shoots with wedge-shaped ends are most commonly used which is called wedge graft. In woody plants shield buds or occasionally bark patches are used. In approach grafting, both the virus donor and recipient plants are rooted. Suitable portions of the two plants are sliced to expose the cambium and then bound together tightly to ensure union.

In all kinds of grafting, the partners must be firmly held together by grafting tape or other binding material till the union is complete. The grafted plants have to be kept under humid conditions to prevent water loss. The time required for the virus to get established in the healthy partner may vary from several days to a few weeks in herbaceous plants while woody plants usually take several months to express symptoms.

C Transmission through seeds and Pollens

Some viruses are carried on the seed coat (testa) as surface contaminant such as TMV in tomato seed. Many other viruses may also be detected on the surface of immature seeds but not after the seeds have matured and dried. Some highly stable viruses such as tobacco and tomato mosaic viruses and cucumber green mottle mosaic virus can survive dessication of the testa as the seed matures. On rare occasions they may occur also in the endosperm but they have never been detected in the embryo. Such viruses may enter the seedling tissues through cells injured during the transplanting process.

The large majority of seed-borne viruses are carried in the embryo which is regarded as true seed transmission (fig-3). Since the embryo and endosperm are formed within the embryo sac after fertilization having no direct vascular connection and cellular contact with the mother plant, embryo infection has essentially to occur before flowering and formation of female gametes. Thus early infection of the mother plant allows the virus sufficient time to invade the embryonic tissue before its cytoplasmic separation from the mother plant but many viruses still fail to do so. Again, some highly stable and infectious viruses like TMV cannot gain access to embryo and pollen while less stable ones like bean common mosaic and soybean mosaic viruses can. Important viruses transmitted through seed and pollen are given in Table 1.

D Transmission through vectors

Amongst insects, the dominance of the homopteran vectors (aphids, whiteflies, coccids, psyllids, leafhoppers and planthoppers) is very much evident since most of the insect-borne plant viruses and virus-like diseases are transmitted by allowing inoculation of pathogens into specific vascular or non-vascular tissues without destroying the inoculated cells-a condition that is essential for the virus to establish itself. This may largely explain why suctorial insects are efficient vectors but it fails to explain why only a very small percentage of homopterans are vectors while others are not.

a. Aphids

The aphids comprise a small homopterous group of about 4,000 species. While many aphids are specific to a particular host family, genus or even species, a few species have exceptionally wide host rang. Most of the highly polyphagous species enjoy a very wide distribution and are more or less common in India. For instance, *Aphis citricola, A. craccivora, A. gossypii and Myzus persicae* are

Table 1 : Examples of plant viruses transmitted through seeds and pollens

Virus	Host species	*Seed transm-ission (%)	Pollen transmission
1	2	3	4
Abutilon mosaic	*Abutilon spp.*	14-24	-
Alfalfa mosaic	*Medicago sativa*	10-50	+
Barley stripe	*Hordeum vulgare*	15-100	+
Bean common mosaic	*Phaseolus vulgaris*	1-93	+
Bean yellow mosaic (Filamentous)	*Pisum sativum*	10-30	-
Black gram mottle	*Vigna mungo*	8	-
Cowpea mosaic (common bean)	*Vigna unguiculata*	25-40	
Cowpea mosaic (Banding)	*V. unguiculata*	2.7-19	-
Cowpea mosaic (Chavali)	*V. unguiculata*	17-23	-
Cucumber mosaic	*Stellaria media* (in many more hosts with varying percent)	21-40	-
Elm mosaic	*Ulmus americana*	1-3	+
Lettuce mosaic	*Lactuca sativa*	3-10	+
Lychnis ringspot	*Lychnis divaricata*	58	+
Mungbean and Urdbean mosaic	*Vigna radiata*	20	-
Pea early browning	*P. sativum*	37	-
Pea enation mosaic	*P. sativum*	1.5	-
Pea seed-borne mosaic	*P. sativum*	0-100	-
Peanut (groundnut) Indian clump	*Arachis hypogea*	12	-
Peanut mottle	*A. hypogaea*	9	-
Peanut stunt	*A.hypogaea*	0.1	-
Prune dwarf	*Prunus cerasus*	3-30	+
Raspberry ringspot	*Fragaria spp.*	50	+
Southern bean mosaic	*V. unguiculata*	3-7	+
Soybean mild mosaic	*Glycine max*	22-70	-
Soybean mosaic	*G.max*	30	+
Soybean stunt	*G.max*	39	+
Tobacco ringspot	*G.max*	78-82	+
Tomato black ring	*Rubus spp.*	5-19	+
Urdbean leaf crinkle	*Vigna mungo*	18	-
Vicia cryptic	*Vicia faba*	varying	+
White clover mosaic	*Trifolium repens*	6	-

* True seed transmission, the viruses borne in embryo.

widely prevalent in the plains as well as in the hills. The other highly polyphagous aphids like *Aulacorthum solani, Neomyzus circumflexus, Myzus ascalonicus, M. ornatus and Toxoptera aurantii* are virtually confined to the hills. Most of the polyphagous species show host plant preferences. For example, *A. craccivora* shows marked preference for Leguminosae; *A. citricola* for Compositae and Rosaceae; *M. persicae* for Solanaceae and Cruciferae and *T. aurantii* for Rutaceae and Theaceae. Beans are often heavily infested by *A. craccivora*. Sporadic heavy infestations of cowpea and groundnut by this species occur in many places. *Aphis gossypii* is apparently the commonest aphid in India and occurs on more hosts than any other species.

The graminaceous crops are infested by several aphids. The more important in India are *Rhopalosiphum maidis,* which is quite common on maize, wheat and barley. *Rhopalosiphum padi* has more restricted occurrence in the plains, infesting wheat, oat and barley. *Sitobion miscanthi* colonises the earheads of wheat, oat and barley. *Longiunguis sacchari* infests sorghum much more severely than sugarcane. While *Toxoptera citricidus* is almost restricted to Rutaceae, the banana aphid, *Pentalonia nigronervosa*, also infests *Colocasia* and cardamom.

b. White flies

The whiteflies belong to the family Aleyrodidae in the order Homoptera and are placed in Sternorrhyncha along with aphids, coccids and psyllids. The family is rather small, comprising 1156 species in 126 genera. *Bemisia tobaci* is a highly opportunistic insect, capable of colonising numerous cultivated plants and weeds and building up huge populations in a short time under favourable condi-

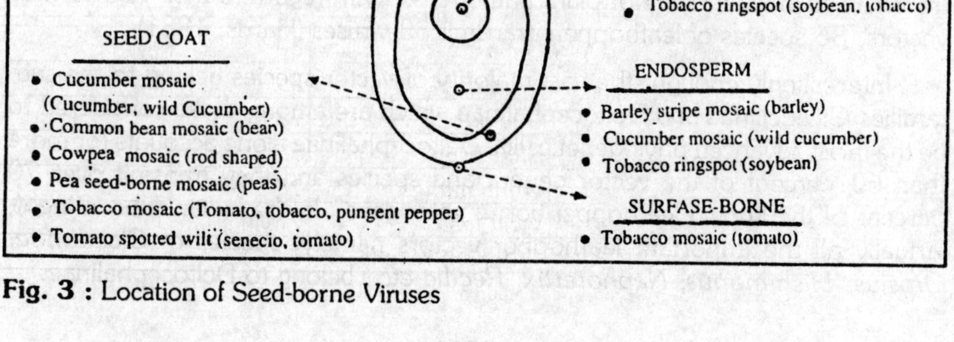
EMBRYO
• Barley stripe mosaic (barley, wheat)
• Common bean mosaic (bean)
• Cherry necrotic ring spot (cherry)
• Cowpea mosaic (spherical)
• Tobacco ringspot (soybean, tobacco)
SEED COAT
• Cucumber mosaic (Cucumber, wild Cucumber)
• Common bean mosaic (bean)
• Cowpea mosaic (rod shaped)
• Pea seed-borne mosaic (peas)
• Tobacco mosaic (Tomato, tobacco, pungent pepper)
• Tomato spotted wilt (senecio, tomato)
ENDOSPERM
• Barley stripe mosaic (barley)
• Cucumber mosaic (wild cucumber)
• Tobacco ringspot (soybean)
SURFASE-BORNE
• Tobacco mosaic (tomato)

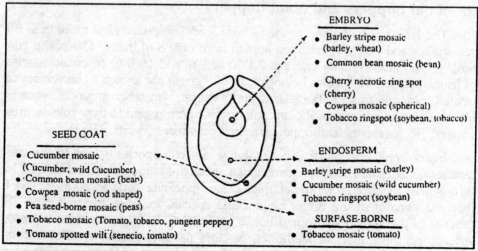

Fig. 3 : Location of Seed-borne Viruses

tions. Warm, humid conditions during *Kharif* season (July-October) have been found to be conducive for rapid multiplication of the whitefly on pulses and soybean. Duffus (1987) tentatively categorized the whitefly-transmitted viruses into seven groups based on the particle type, external symptoms and cytology, and methods of transmission. Besides listing thirteen diseases as caused by geminiviruses, he listed twenty others to be possibly caused by the same virus group. The six other virus groups possibly associated with whitefly-transmitted diseases (their numbers in parentheses) were as follows : Carlaviruse (3), closteroviruses(5), potyvirus(1), luteoviruses (3) (without evidence of particle type), rod-shaped DNA virus (1) and nepovirus(1).

c. Mealybugs

The mealybugs belong to Pseudococcidae, the only family in the superfamily Coccoidea having vector species. The mealybugs owe their name to the mealy appearance as they are dusted all over with white powdery material. An important morphological character of the family is the one- segmented tarsi with a single claw. Eighteen species of mealybugs have been reported as vectors, transmitting five viruses (Harris, 1981). The four viruses affecting cacao plantations are of great importance, especially the cacao swollen shoot virus. *Planococcoides njalensis* and *P. citri* are important vector species, transmitting most viruses in this group. The nymphs are more efficient vectors than adults. Starved mealybugs acquire virus more efficiently than the non-starved ones. Acquisition feeding periods of 48-72 hours and inoculation feeding for 3-4 hours result in maximum transmission. Cacao swollen shoot virus is retained by *P. njalensis* for 3-4 days.

d. Leaf hoppers and plant hoppers

The leafhoppers belong to the family Cicadellidae which has more than 60 subfamilies and vector species are known from only 8 of them. Out of the current conservative estimate of about 2,000 genera and 15,000 described species of leafhoppers, 151 species belonging to 65 genera are known to be vectors of viruses, phytoplasmas, spiroplasmas and bacteria. No other group of vectors transmits such wide range of plant pathogens. With regard to their role as virus vectors, 36 species of leafhoppers transmit 33 viruses (Harris, 1981).

Interestingly enough, the vast majority of vector species belong to the subfamilies Cicadellinae and Deltocephalinae which are among those considered to be the most advanced phylogenetically. Deltocephalinae alone accounts for more than 60 percent of the vector genera and species and they transmit over 70 percent of the known leafhopper-borne pathogens. In the Indian subcontinent, virtually all the important leafhopper vectors namely, species of *Cicadulina, Orosius, Hishimonus, Nephotettix, Recilia* etc., belong to Deltocephalinae.

The planthoppers are not as dominant as the leafhoppers in terms of the number of vector species and the number of plant pathogens transmitted. Out of approximately 9200 planthoppers described, 22 species of Delphacidae and 3 species of Cixiidae are known to serve as vectors. Delphacids are virus vectors, transmitting 23 viruses while the cixiids are vectors of phytoplasmas including tomato big bud and palm lethal yellowing.

e. Thrips

The importance of thrips as vectors lies in their well established role in field spread of the tomato spotted wilt virus (TSWV) and its strains. So far five species of thrips namely, *Thrips tabaci, Frankliniella schultzei, F.fusca, F.occidentalis* and *Scirtothrips dorsalis* are highly polyphagous and enjoy a wider distribution than the other vector species occurring all over tropical and temperate regions. Both *T. tabaci* and *A. schultzei* are known to transmit every strain of TSWV all over the world. *Scirtothrips dorsalis,* commonly known as the chilli thrips occurs in India while *F. fusca* and *F. occidentalis* are common in the eastern and western U.S.A., respectively.

f. Beetles

The beetles transmit 45 plant viruses, most of which can be placed in one of the four major groups namely, the comovirus, tymovirus, bromovirus and sobemovirus. A wide varity of host plants are affected, especially in Leguminosae, Cruciferae, Cucurbitaceae and Solanaceae. The comoviruses (cowpea mosaic, squash mosaic, radish mosaic, been pod mottle etc.) are efficiently transmitted by beetles.

g. Mites

The mites belong to the class Arachnida and are the only non-insect arthropod vectors of plant viruses. The ability to transmit virus is restricted to a few species of eriophyid mites (family Eriophyidae).

The eriophyid vectors are worm-like and very small, the adults measuring upto about 0.2 mm in length, possessing only 4 legs. The mouthparts are modified for piercing and sucking. Since the stylets are minute, feeding seems to be restricted to one or two layers of cells from the surface. They feed on leaves, buds and other tender parts of the plant. The mites are very delicate and dessicate readily. Special care has, therefore, to be taken to handle them for experiments. Due to very small size they are easily blown by wind which is the main means of their dispersal.

Only 3 plant viruses namely, wheat streak mosaic (WSMV), *Agropyron* mosaic

(AMV) and ryegrass mosaic(RMV) are definitely known to be transmitted by mites. These viruses are pathogenic to Gramineae only and are ecologically adapted to thrive in the northern temperate climates. They are all flexous rods of about 700 nm length, located abundantly in the mesophyll and parenchyma tissues and induce formation of pinwheel inclusion in host plant cells. In addition, at least three viruses are suspected to be mite-borne.

e. Nematodes

Since the first report of transmission of a plant virus by a nematode (Hewitt *et al.*, 1985), quite a number of plant viruses have been found out to be borne by this group of vectors. The nematode vectors are soil-inhabiting ectoparasites which feed on the roots, especially the root tips. The known vectors belong to 4 genera and can be divided into two distinct groups on the basis of their taxonomic affinities and viruses they transmit. Members of the genera *longidorus and Xiphinema* (family Longidoridae) transmit the polyhedral nepoviruses such as the grapevine fanleaf, *Arabis* mosaic, tomato ringspot. These vectors are larger in size (adults 2-12 mm long) than the *Trichodorus* and *Paratrichodorus spp*. (family Trichodoridae), which measure upto 2 mm in length and transmit the straight tubular tobraviruses such as tobacco rattle and pea early browning. Members of the four vector genera enjoy wide distribution although *Longidorus spp.* are more prevalent in temperate regions and *Xiphinema spp.* in the tropics.

f. Fungal vectors

The role of fungi as vectors of plant viruses was first discovered in 1960 (Teakle, 1960). The fungi which have been confirmed as vectors, belong to *chytrids* (fam. *Olpideaceae)* or the *plasmodiophorids* (fam. *Plasmodiophoraceae*). These fungi are obligate plant parasites, infesting roots in cool, wet soil. The fungus-borne viruses are generally more prevalent in temperate regions. At least three isometric viruses are known to be transmitted by the chytrid vectors, *Olpidium brassicae* (Tobacco necrosis virus (TNV), TNV satellite), and *O. radicale* (cucumber necrosis virus). *Olpidium brassicae* also transmits lettuce big vein and tobacco stunt diseases.

The three species of *plasmodiophorid* vectors transmit a number of straight tubular or filamentous particles including wheat mosaic (*Polymyxa graminins),* beet necrotic yellow vein (*P. betae)* and potato mop top (*Spongospora subterranea)* viruses.

g. Dodder Transmission

Out of about twenty species of dodders, tested for experimental transmission, *Cuscuta campestris* and *C. subinclusa* have been commonly used to transmit different viruses. They have wide host ranges, which add to their usefulness.

Transmission through dodder may be just passive transport of the virus since its multiplication in dodder is not essential for transmission. However, transmission is generally more efficient if the virus multiplies in the dodder itself. Dodder seeds may be germinated on the soil containing seedings of the host plant. The germinated dodder seed is placed in leaf axil of infected host and when it is established, its shoots are trained on to the test plant to be infected. The young apical leaves of the test plant show the initial symptoms of infection.

VIII. VIROIDS

A. Introduction

Prior to 1971, all infectious agent of plant diseases were believed to be limited to bacteria, viruses, fungi, protozoa, etc. It was during the 1960s, when potato spindle tuber disease of potato, which had been known to occur since 1920s in North America recaptured the attention of several laboratories. This was because, for the first time, the spindle tuber disease agent was transmitted to another host plant other than potato. During this period conflicting claims were made about the type of 'Virus' isolated from spindle tuber diseased plants. However by 1967-68 it became clear that the causal agent of the potato tuber disease was not a conventional virus, but a freely existing nucleic acid. Further attemps to characterize the spindle tuber agent led to the independent but simultaneous discovery of a low molecular weight Ribonucleic acid (Singh and clark, 1971) for which the name viroid was proposed (Diener, 1971).

It was conclusively shown that the potato spindle tuber viroid (PSTVd) is a single stranded covalently closed circular ribonucleic acid with unusal property. By 1978, Potato spindle tuber viroid (PSTVd) became the first plant pathogen whose complete nucleotide sequence became available. PSTVd is the prototype of the viroids.

Viroid is a single stranded covalently closed circular RNA of low molecular weight that can infect plant cells, replicate themselves and cause diseases.

B. Properties of Viroids differentiating them from Viruses

1. The pathogen (viroid) exist *in vivo* as an unencapsidated RNA; that is, no virion like particles are detectable in infected tissue.
2. The size of infectious RNA (viroid), has a low molecular weight.
3. Despite its small size, the infectious RNA is replicated autonomously in susceptible cells; that is, no helper virus is required.

C. Nature of viroids

Viroids are an independent class of plant pathogens. They are single stranded covalently closed circular RNA molecule with a nucleotide chain length of 240 to 375 (Table-2). Unlike viruses, viroids have no protein coat.

Table 2 : Chain lengths, sequence homology and common hosts with PSTV of different viroids

Viroids	Group	Isolates	Ribonucleotide chain length	% homology with PSTV	Common hosts with PSTV
ASBV[a]	1[b]	Australian	247	18	No
CCCV	1	RNA 1 fast	246	11	No
		RNA 1 slow	287	11	No
		RNA 2 fast	492	11	No
		RNA 2 slow	574	11	No
CEV	2	Californian	371	73	Yes
		Australian	371	59	Yes
		DE-25	371	--	Yes
		DE-26	371	--	Yes
CSV	2	English	354	73	Yes
		Australian	356	69	Yes
CPFV	3	Dutch	303	55	Yes
HSV	3	Japanese	297	55	Yes
PSTV	2	Severe	359	100	Yes
		Mild	359	99	Yes
TASV	2	Ivory Coast	360	73	Yes
TPMV	2	Mexico	360	83	Yes

[a]For full name of viroids see Table 3.[b]Group 1 with 11-18% sequence homology to PSTV and no common hosts; Group 2 with 59-83% sequence homology and many common hosts; and Group 3 with 55% sequence homology and few common hosts.

D. Occurrence and Importance

A review of the viroid diseases indentified in the last two decades (Table-3) shows that some of them were observed as early as 1917 (PSTV), 1927 (CCCV) or 1939 (ASBV). In the opinion of some ivestigators, viroid diseases are of recent origin and their appearance is attributed to the introduction of intensive agriculture, particularly the practice of monoculture. However, this reasoning applies equally to majority of plant diseases caused by viruses. It is likely that modern intensive agriculture contributed to the inadvertant spread of viroid diseases through the use of infected but symptomless plants for vegetative propagation. Accidental introduction of viroids from wild to cultivated plant also may explain the recent appearance of viroid diseases. This postulate assumes that viroids do not cause symptoms in native hosts or wild plants, but do so in cultivated crops.

The increase in number of reports (Table-3) of new viroid diseases in all parts of the world demonstrates an increase in detection and awareness of economic losses they cause for example, in the early 1920's PSTV was studied extensively because it infected 25 to 90% of the potato plants in some field of Maine, U.S.A.

Table 3 : Viroid diseases

Disease	Abbreviated names	Year first observed[1] (viroid etiology)	Distribution
Avocado sunblotch	ASBV	1939 (1971)	Australia, Israel, Peru, South Africa, U.S.A., Venezuela
Burdock stunt	BSV	1983(1983)	China
Carnation stunt	CarSV	1983(1983)	Italy, U.S.A.
Citrus exocortis	CEV	1943 (1972)	Argentina, Australia, Brazil, Corsica, Israel, Japan, Spain, South Africa, Taiwan, U.S.A.
Chrysanthemum chlorotic mottle	CCMV	1967(1975)	U.S.A.
Chrysanthemum stunt	CSV	1945(1973)	Australia, Canada, China, India, Japan, The Netherlands, United Kingdom, U.S.A.
Columnea viroid[2]	ColV	1978(1978)	U.S.A.
Coconut cadang-cadang	CCCV	1927(1975)	Philippines
Cucumber pale fruit	CPFV	1963(1974)	The Netherlands
Grapevine	GV	(1985)	Japan
Hop stunt	HSV	1952(1977)	Japan
Potato spindle tuber	PSTV	1917(1971)	Argentina, Australia[3], Brazil, Canada, China, Chile, Peru, Scotland[3], U.S.A. U.S.S.R. Venezuela
Tomato apical stunt	TASV	1981 (1981)	Ivory Cosat
Tomato planta macho	TPMV	1974(1982)	Mexico

[1]Dates mentioned inthe first report of a disease.
[2]No visible symptoms are observed on *Columnea erythrophae*, an ornamental plant.
[3]The viroid may be limited to potato germplasm collections only.

In the 1960's as many as 54% of the plants were infected in some provinces of USSR. PSTV is of concern to China because it is wide-spread there as well. PSTV has the potential of causing 25 to 64% yield loss. Cadang-cadang disease of coconut in the Phillipines has killed an estimated 20 million or more coconut trees since 1926. Similarly, stunt disease of Hops observed in 17% of the total acrage of hops in Japan in 1968, with some gardens having up to 60% of the plant infected.

E Symptoms and Host Range

The range of symptoms exhibited by viroid diseases is similar to that of plant viruses, except that most viroid infections induce'stunting' of some kind. Stunting of the entire plant is common in most herbaceous plants, accompanied by either smaller upper leaves, shortened internodes, or upright, dwarf apperance. Malformed or dwarf flowers are often observed in cucumber plants infected with CPFV or HSV. There is also a tendency of tubers (PSTV), cones (CCCV), fruits (CPFV) to be more "pointed" or "elongated" than normal. ASBV may cause streaks in the stems of avocados and CEV produces bark splitting in citrus. Motlling of leaves (CCCV, CCMV), of fruits (CPFV), and necrosis of leaves and stems (PSTV) may also occur.

The natural host range of viroids is limited to the species in which they were first reported. Although several viroids can cause disease symptoms in potato, only PSTV has been found naturally in potato. Viroids from grapevine, hop, and cucumber all cause similar symptoms in cucumber, although the viroids are not the same, and have different nucleotide sequences in their molecules.

The experimental host range of viroids shows a wide variation. Host plants for ASBV, CCCV and CCMV are confined to single families, e.g., Lauraceae, Palmaceae, and Compositae, respectively. Some plant families are susceptible to several viroids,e,.g., Compositae (CEV,CSV,CCMV,CPFV,PSTV, TASV and TPMV); Solanaceae (CEV,CSV,ColV, CPFV, HSV, PSTV, TASV, and TPMV); Cucurbitaceae (CEV,CSV,CPFV,HSV and GV); Scrophulariaceae (PSTV and TASV). In addition, PSTV can infect plants in families Boraginaceae, Companulaceae, Caryophyllaceae, Convolvulaceae, Dipaceae, Sapindaceae, and Valerianaceae. Similarly, CEV infects plants in the families Papilionaceae, Rutaceae, and Umbelliferae.

Some plant species are hosts of several different viroids. The plant symptoms vary with the viroid present. Cucumber (*Cucumis sativus*) is susceptible to a viroid from grapevine. HSV, CSV, CEV and CPFV, the potato (*Solanum tuberosum*) is susceptible to CEV, CSV, ColV, CPfV, PSTV,TASV and TPMV; *Scopolia sinensis* is susceptible to CEV,CSV, PSTV, TASV and ColV; *Solanum xberthaultii* is susceptible to CEV, CSV, PSTV and TASV and the tomato to : CEV, CSV, ColV, HSV, CPFV, TASV, PSTV, and TPMV. It is of some significance to note that those viroids which have many host plants in common also have higher nucleotide sequence homology with each other and belong to the same phylogenetic group. ASBV and CCCV with less sequence homology to PSTV, do not infect the same hosts, and belong to other viroid groups (Table 2).

F. Transmission and spread

Although viroids lack a protective protein coat, all of them can be transmitted mechanically. Rubbing viroid containing inoculum in high pH buffers onto carborundum dusted leaves is used experimentally in many viroid host combinations. In certain plants, e.g. in *G. aurantiaca* with CEV, in chrysanthemum with CSV, and in cucumber with CPFV, slashing or puncturing the stem with a scalpel or razor blade transmits the viroids more effectively. Viroids, like viruses, are also transmitted through ovule, pollen and seeds. However, unlike viruses, viroids are not transmitted by insect vectors at a high rate. Only low levels of experimental transmission have been achieved for PSTV and TASV by some species of aphids.

Viroids in field and orchards may be spread by contaminated tools and cultivation machinery . PSTV was transmitted to a large proportion of plants by cutting healthy seed tubers with a knife previously used to cut infected tubers. Eighty to 100% infection by PSTV was achieved by brushing actively growing, healthy plants with diseased foliage, or by excessive contact of large potato vines with cultivating hilling equipment contaminated with PSTV. Rapid transmission of CEV was achieved with contaminated budding knives; of CSV with contaminated tools, knives, and bare hands during cultural practices; of CPFV with pruning operations in greenhouses and of HSV with contaminated sickles or bare hands used to dress or pull shoots.

The useful indicater plants for different viroids are listed in Table-4.

G. Management of viroid diseases

Management of plant diseases is primarily aimed at reducing or eliminating losses to affected crops. It includes methods, which prevent the spread of pathogens from reservoirs, such as weeds to cultivated plants. There are different methods that might be recommended for viroid diseases. These methods have been based on measures used to control plant disease caused by other pathogens, such as viruses, bacteria and fungi. These methods include :

1. Use of viroid free seed

The use of healthy, viroid free seed or planting material is one of the best

Table 4 : The useful indicator plants for different viroids

Viroid	Indicator plants
Avocado sunblotch	*Persea americana* cvs. *'Haas'* and *'Collinson'*
Citrus exocortis	*Gynura aurantiaca*
Chrysanthemum stunt	*Chrysanthemum morifolium* cv. ' Mistletoe ' *G. aurantiaca*
Chrysanthemum chlorotic mottle	*C. morifolium* cv. 'Deep Ridge'
Coconut cadang-cadang	*Cocos mucifera*
Cucumber pale fruit	*Cucumis sativus* cv. *'Sporu'*
Grapevine	*C. sativus*
Hop stunt	*C. sativus* cv. *'Suuyou'*
Potato spindle tuber	*Lycopersicon esculentum* cvs. 'Sheyenne' 'Rutgers';, *Scopolia sinensis; Solanum xberthaultii*
Tomato apical stunt	*L. esculentum* cv. 'Rutgers"; *S. xberthaultii, Scopolia sinensis*
Tomato planta macho	*L. esculentum* cv. 'Rutgers'

means of preventing diseases that are seed borne. The most successful example is the control of potato spindle tuber viroid through the use of viroid free seed. The viroid disease citrus exocortis dispersed by using infected material as root stock. To avoid spread of CEVd, healthy stock should be used for grafting.

2. Selection and Breeding for Resistance

The best and reliable method available to control viroid diseases is the selection and breeding for disease resistance. Efforts are under way to find varieties resistant to viroids.

3. Decontamination of Tools and Hands

To maintain viroid- free status of vegetatively propogated material, it is necessary to disinfect tools, knives and other equipments with sodium hypochlorite or sodium hydroxide plus formalin.

4. Cross Protection

Cross protection is the interference in the expression of signs of viroid diseases by preinoculation with a 'mild' viroid strain. The mechanism of cross-protection in viroid is unknown. The occurence of cross-protection with viroids of potato spindle tuber has been discovered by Fernow (1967). He demonstrated that preinoculation of tomatoes with a mild strain of PSTVd protected the plants from developing severe symptoms of the disease when inoculated with severe strain of PSTVd. Although no symptoms of severe strain infection were observed in these plants, the severe strain could be isolated from inoculated "Protected" plants.

5. Chemical Protectants

No chemical protectants have been found to control viroid diseases. The antibiotics tetracycline and penicillin failed to cure cadang-cadang of coconut (Randles et al 1977). No data are available on antiviroid factors.

6. Replanting

Replanting is not a management method, but it has been recommended as a stop-gap method in Philippines. Replanting enabled the continuous production of coconut in areas where the disease was rampant. It helps in preventing the losses.

7. Quarantine

Strict quarantine, prohibiting the movement of viroids present on infected plant part into new viroid free areas would seem to provide an efficient control measure.

8. Tissue culture

Viroids are warm climate pathogens, and replicate at higher rates under high temperature conditions. Application of thermotherapy and meristem culture techniques has yielded a very low proportion of PSTVd free potatoes. However, a combination of cold treatment and meristem culture has eradicated several viroids.

Shoot tip culture has been successfully used to obtain CEVd-free citrus and tomato plants. Viroids from charysanthemum and grapevines have also been eliminated using meristem shoot tip culture.

Procedures which are effective for one viroid may not be applicable to other viroids, or to the same viroid in other host plants. For example, low temperature did not enhance the efficiency for producing HSVd free or HLVd free hop plants and shoot tip culture has failed to eliminate grapeving viroids.

IX Virusoids

In addition to viroids, there is a second group of law molecular weight ssRNAs associated with plant diseases; they are the plant virus satellite RNAs. Although some of these RNAs also do not show mRNA activity, they differ from viroids in two significant aspects: 1. they are dependent on a helper virus for their replication; and 2. they are encapsidated by either viral-coded or satellite RNA-coded coat protein. Amongst the satellite RNAs, the virusoids show most similarity to viroids in being covalently closed circular RNA molecules in the same size range, varying from 324 to 388 nucleotides. Virusoids are sigle stranded, circular, low molecular weight, viroid-like RNA (Mr. 1.1 to 1.3 x10^5). These satellite RNAs are dependent on a helper virus for replication and are encapsidated in the circular form by that virus. They exist in solution as a rod-like structures with secondary structures similar to viroids. The list of virusoids is given in Table-5.

Table 5 : List of virusoids

Virusoid	Abbreviation	No. of nucleotides
Lucern transient streak virus RNA 2	vLSTV	324
Solanum nodiflorum mottle virus RNA 2	vSNMV	378
Subterranean clover mottle virus RNA 2	v(388)SCMoV	388
-do- RNA2	v(332)SCMoV	332
Valvet tobacco mottle virus RNA 2	vVTMoV	366,367

X. Prusiner's Prions

Sometimes in 1972, an unusual patient was admitted in "School of Medicine-San Francisco, U.S.A. The patient had a peculiar disease, with symptoms of childish behaviour, short memory, drowsiness, and incoordination of movement. The cause could not be known. The patient died for want of cure. One young man (Dr. Stanley B. Prusiner) decided to study the unknown cause, and in 1982, he reported a completely new form of disease-causing agent — "proteinaceous infectious particles" which he called PRIONS (preeons). This propounder of the PRION THEORY was awarded the 1997 Noble-prize in medicine for discovery of a new biological principle of infection. Disease caused by prions are given in Table 6.

Table 6 : Prion Diseases in Different Species

Host	Disease	Geographic distribution	First clinical observation
Sheep	Scrapie	Worldwide except Australia, New Zealand and some European countries	1730
Goat	Scrapie	-	-
Man	Kuru.	Papua New Guinea	1990
	Creutzfeldt-Jakob disease	Worldwide	1920
	Gerstmann-Straussler-Schlenker syndrome	Worldwide	1927
	Fatal famillial Insomnia	-	-
Mink	Transmissible Mink Encephalolpathy	North America, Europe	1947
Mule deer	Chronic Wasting disease	North America	1967
Cattle	Bovine Spongiform Encephalopathy	U.K., Ireland and some other European countries	1985

Although prions have so far been found only in animals, it is not unreasonable to expect that they will soon be shown to affect plant also.

CHAPTER X

NEMATODES

I. GENERAL CHARACTERSTICS AND MORPHOLOGY

With some exceptions, adult plant parasitic nematodes are elongated worms ranging in length from about 0.30mm to over 5.0mm. The anterior end tapers to a rounded or truncated lip region, the body proper is more or less cylindrical, and the posterior end tapers to a terminus which may be pointed or hemispherical (Figs. 1,2). Proportions of the elongated body vary greatly, some species being more than 50 times longer than wide, others being only about 10 times longer than wide (Fig. 3). Females have greatly expanded bodies, sometimes nearly spherical, but always with a distinct neck (Fig. 4). The adult males are always slender worms (Fig. 2). Plant parasitic nematodes have no appendages. The mouth of a nematode is at the ANTERIOR end, and the terminus is at the POSTERIOR end. The excretory pore, vulva, and anus are on the VENTRAL side; and the opposite side is called DORSAL (Fig. 1). The right and left sides are called LATERAL. The cuticle is attached to several other layers of tissue which are separated laterally, dorsally, and ventrally by chords. These contain nerves, excretory organs, etc., and separate four bands of muscles which move the body (Fig. 9).

A. Alimentary Canal - The alimentary canal starts at the mouth (Fig. 5) and ends at the anus (Fig. 1, 2). It includes the oesophagus, intestine, and rectum.

B. Stylet - In plant parasitic nematodes of the "Tylenchida" group, the mouth contains a stylet or mouth spear, a hardened, hollow, culticular structure similar to a hypodermic needle (Fig. 5). Muscles are attached to three knobs at the posterior end of slylet and extend forward (Fig. 6). They are used to pull the stylet forward so that it projects from the mouth opening and can be used to pierce plant cells. All the food of the nematode is taken through the stylet.

C. Oesophagus - A slender tube is attacheב to the posterior end of the

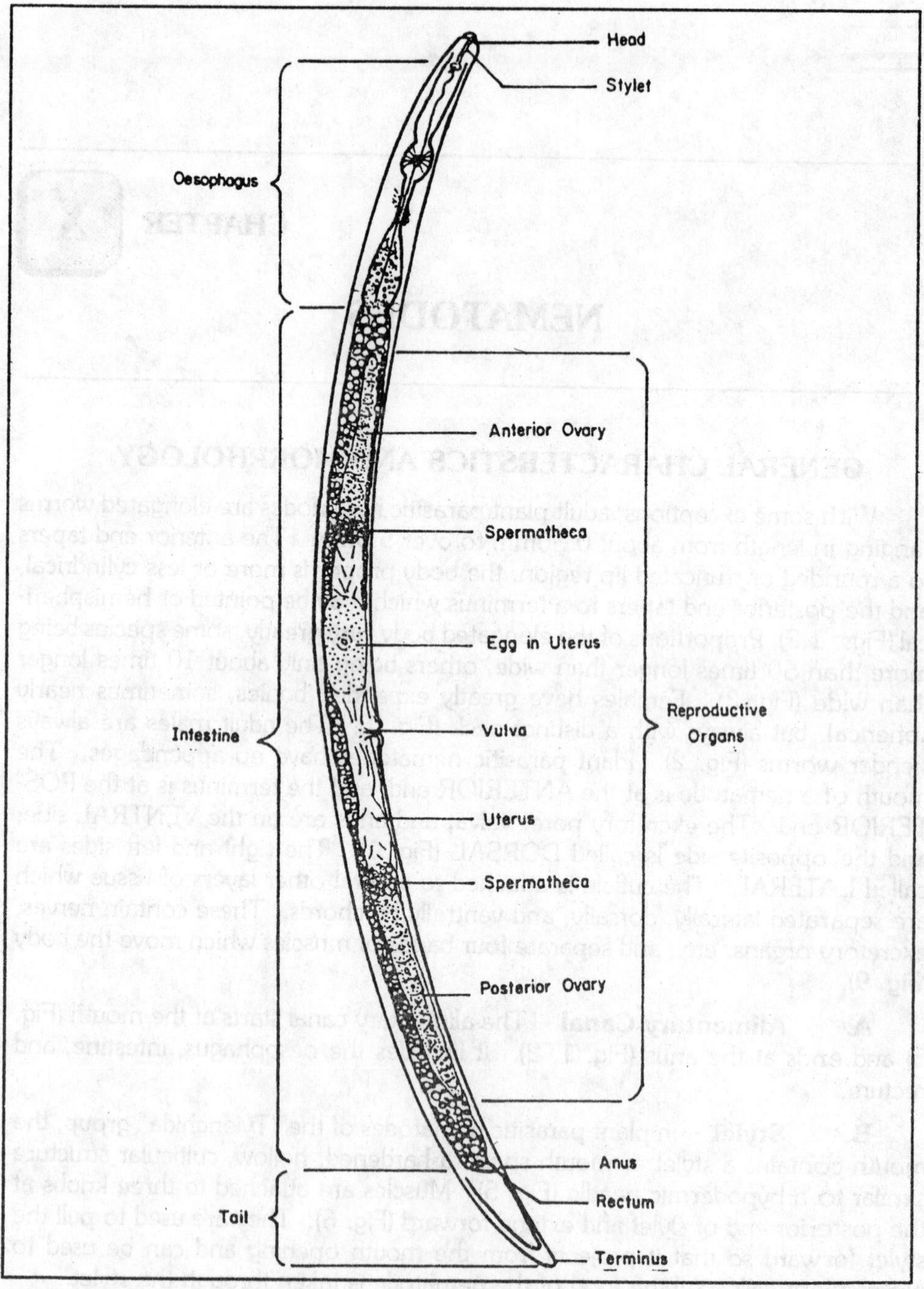
Head
Stylet
Oesophagus
Anterior Ovary
Spermatheca
Egg in Uterus
Reproductive Organs
Intestine
Vulva
Uterus
Spermatheca
Posterior Ovary
Anus
Rectum
Tail
Terminus

Fig. 1 : Anatomy of a typical female plant-parasitic nematode

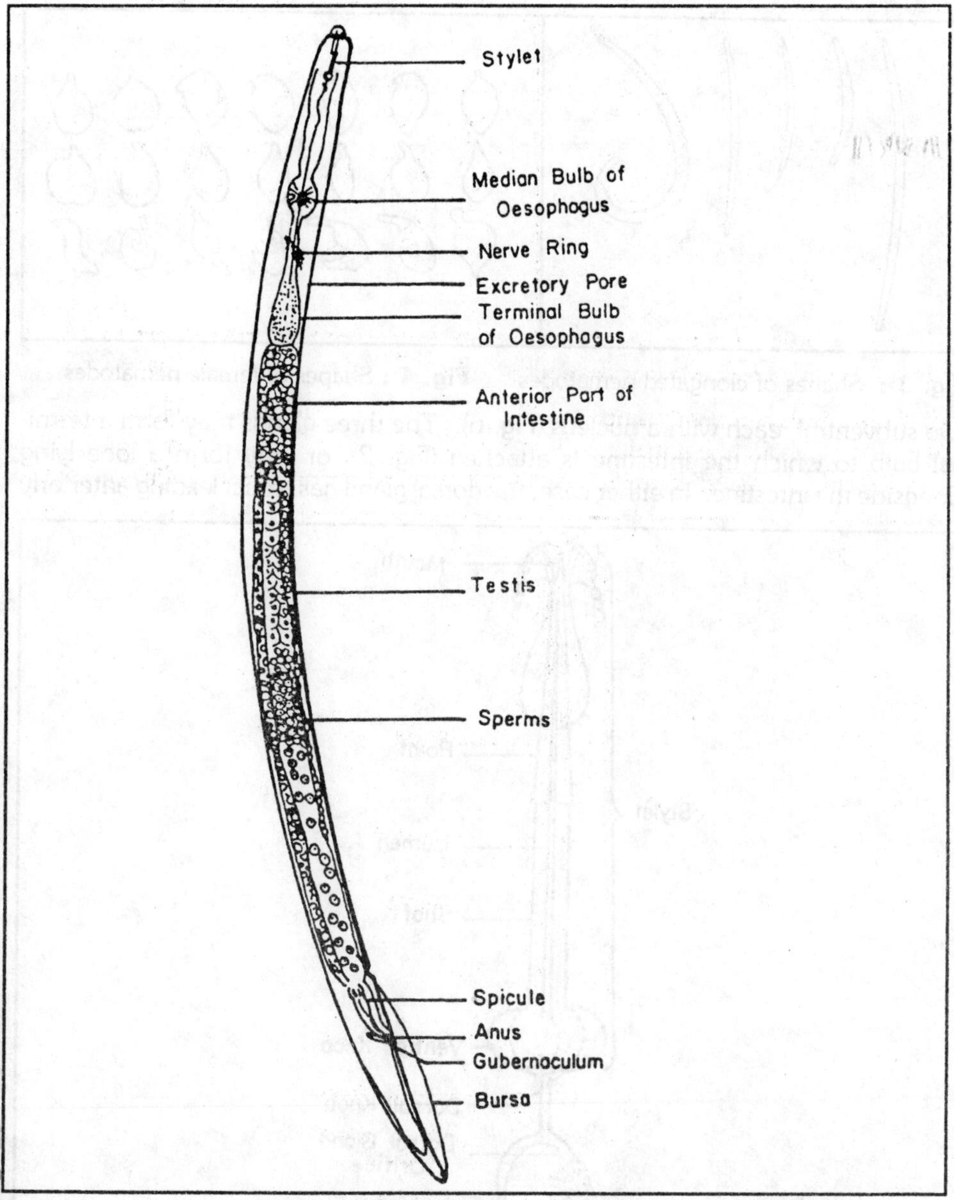
Stylet
Median Bulb of Oesophagus
Nerve Ring
Excretory Pore
Terminal Bulb of Oesophagus
Anterior Part of Intestine
Testis
Sperms
Spicule
Anus
Gubernaculum
Bursa

Fig. 2 : Anatomy of a typical male plant-parasitic nematode

stylet. This is the oesophageal tube (Fig. 5) leading to the median bulb which in turn is attached by means of another slender tube to the intestine (Fig. 6). Posterior to the median bulb, the oesophagus contains three glands, one dorsal and

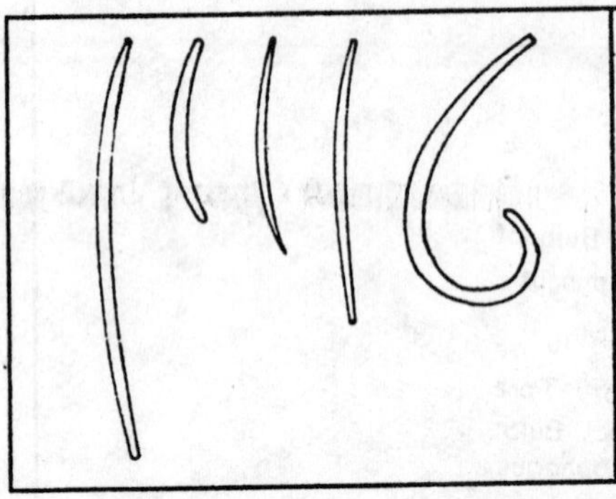

Fig. 3 : Shapes of elongated nematodes

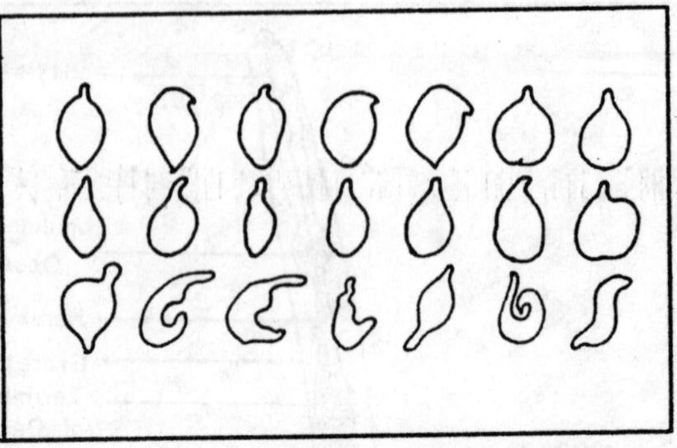

Fig. 4 : Shapes of female nematodes

two subventral, each with a nucleus (Fig. 6). The three glands may form a terminal bulb to which the intestine is attached (Fig. 2), or may form a lobe lying alongside the intestine. In either case, the dorsal gland has a duct leading anteriorly

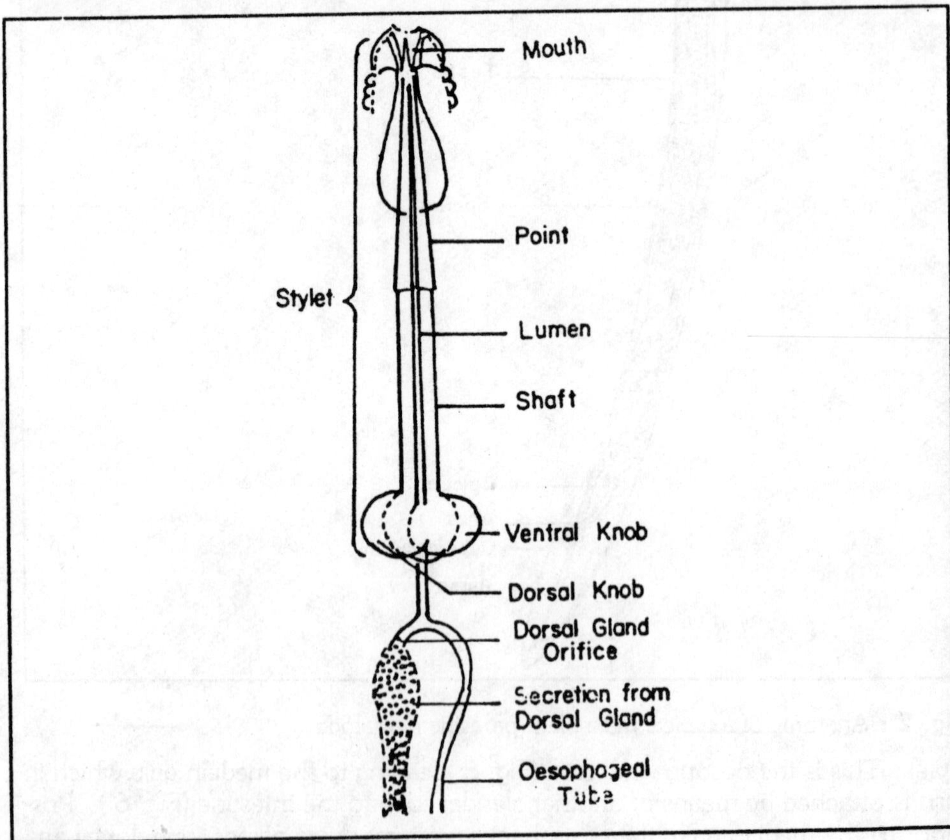
Mouth
Point
Stylet
Lumen
Shaft
Ventral Knob
Dorsal Knob
Dorsal Gland Orifice
Secretion from Dorsal Gland
Oesophageal Tube

Fig. 5 : Stylet and associated structures

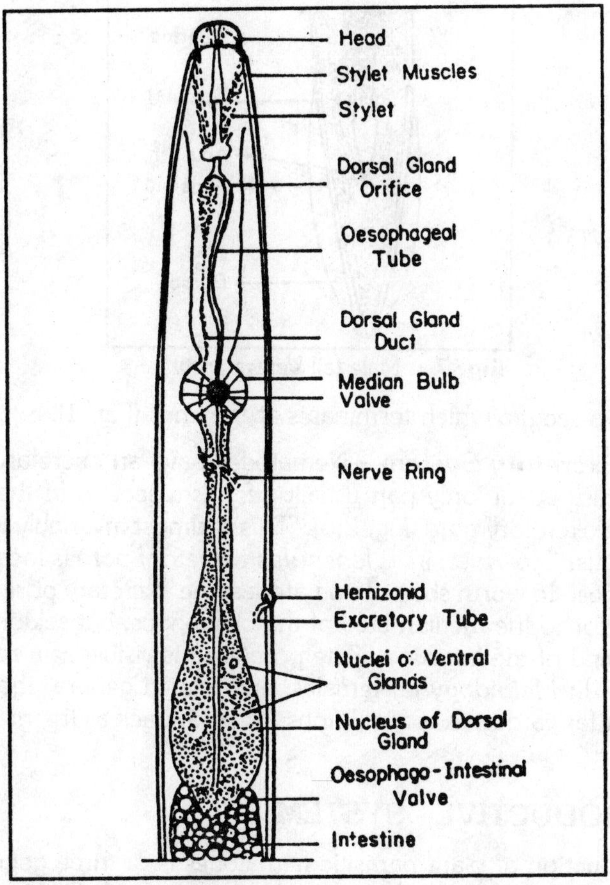
Head
Stylet Muscles
Stylet
Dorsal Gland Orifice
Oesophageal Tube
Dorsal Gland Duct
Median Bulb
Valve
Nerve Ring
Hemizonid
Excretory Tube
Nuclei of Ventral Glands
Nucleus of Dorsal Gland
Oesophago-Intestinal Valve
Intestine

Fig. 6 : Anterior end

through the median bulb and connecting with the oesophageal tube (Fig. 6). The connection is called the dorsal gland orifice.

D. Dorsal Gland Orifice - This in most species of plant parasitic nematodes is located behind the stylet at a distance seldom exceeding the stylet length and generally much closer (Fig. 5,6). At this point there is an opening into the oesophageal tube and often an abrupt bend in it.

E. Median bulb - The median bulb contains a "valve" to which muscle fibres are attached (Fig. 6). In cross-section, this structure is tri-radiate (Fig. 7). When activated by muscles, it functions as a pump, sucking food through the stylet and forcing it into intestine.

F. Intestine - It is a simple tube with walls one cell thick. It functions as a storage organ and is usually filled with globules of fatty substances. Posteriorly

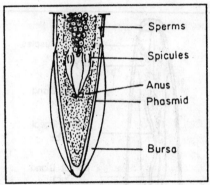

Sperms
Spicules
Anus
Phasmid
Bursa
Fig. 7 : Male tail-Ventral view

it narrows to a rectum which terminates at the anus (Fig. 1).

G. Excretory System - Nematodes have an excretary system, but in the plant parasites, the only part usually seen is a section of the excretory tube leading to the excretory pore (Fig. 2,6). This is almost invariably located opposite the oesophagus. However, in *Tylenchulus* excretory pore is located posterior to the oesophagus. In worm shaped nomatodes, the excretary pore is usually opposite or posterior to the median bulb of the oesophgus, but seldom as far back as the anterior end of the intestine. The pore itself is visible as a round opening in ventral view. In *Meloidogyne* females, and related genera, the excretary pore may be as far farward as the stylet knobs, or as far back as the median bulb of the oesophagus.

II. REPRODUCTIVE - SYSTEM

Reproduction of plant parasitic nematodes is of three general types, varying with species. In *bisexual* species the female is fertilized by the male. In *hermaphroditic* species both eggs and sperms are reproduced by the female. In *parthenogenetic* species the eggs develop without fertilization. The sex ratio varies according to these three reproduction types. It ranges from about equal numbers of males and females in species which are bisexual to very few or no males which are hermaphroditic or parthenogenetic. Some bisexual species may also reproduce parthenogenetically.

The female reproductive organs consist of ovaries and associated structures in which eggs are formed. There may be one or two ovaries. If there is only one ovary, the vulva is located on the posterior or the anterior quarter of the body and the ovary extends either forward or backward respectively. If there are two ovaries, the vulva is usually near the middle of the body, and one ovary extends forward and the other backward (Fig. 1). The ovary is a tube with very thin walls. At the end is a cell which actively divides to produce *oocytes*. These move down the ovary, increasing in size as they progress. They may be fertilized by sperms stored in a *Spermatheca* (Fig. 1), by sperms produced by a special apparatus near the end of the ovary, or may not be fertilized at all, depending on the species of nemetodes. In any case, they pass into the *uterus* and acquire a thin, flexible shell before being deposited through the Vulva. Eggs are usually, but not always, deposited in the one-cell stage. An important exception are the eggs of the cyst-forming nematodes of the genus *Heterodera*. In *Meloidogyne* and some other genera, eggs are deposited with a gelatinous substance and form an egg mass, other plant parasitic nematodes deposit their eggs in the soil or in plant tissues. The eggs develop by repeated cell division, and larvae are formed. In its life time, a female *Meloidogyne* might produce upward of 2,000 eggs, but the average is probable closer to 500. Cysts of *Heterodera* may contain 600 eggs.

The male reproductive organs consist of one or two testis and associated structures, two *spicules* and a *gubernaculum*. In addition, some species have a *bursa* (Fig. 2,7). The testis is similar to the ovary in that it has very thin walls and a cell which continually divides, producing *spermatocytes*. The spermatocytes move down the testis, increasing in size, then divide twice to form four sperms. Sperms are only a few microns in diameter and often globular. The posterior ends of the two spicules are close together at the anus, and the anterior ends spread apart. In copulation, these project through the anus and are used to open the vulva (Fig. 7).

The gubernaculum, which is located just behind the spicules, acts as a guide when the spicules are pushed out through the anus. The bursa is a thin membrane (Figs. 2,7) used in holding the female during copulation.

III. NERVOUS SYSTEM

Plant parasitic nematodes have a highly developed nervous system, though very little of it can be seen except for the *nerve ring* and *hemizonid* (Fig. 6). The nervous ring surrounds the oesophagus just behind the median bulb. The hemizonid is visible in the vicinity of the excretory pore. The nematodes have tactile sense organs called *papillae* located at the anterior end and else where on the body. They also have chemical sense organs called *amphids* located at the anterior end. On plant parasitic nematodes these are very dificult to see (Fig. 10). Most species of plant parasitic nematodes have *phasmids* (Fig. 8). The excretory pore, vulva, and anus of nematode are always on the ventral side. Amphids and phasmids (Fig. 7,8,10) are instead lateral.

IV. CLASSIFICATION

Plant parasitic nematodes are distinguished from non-parasitic forms by the presence of a stylet. For convenience, nematode parasites of plants are sometimes classified as ENDOPARSITES or ECTOPARASITES, and as SEDENTARY or MIGRATORY. The nematode species which move through the plant tissue

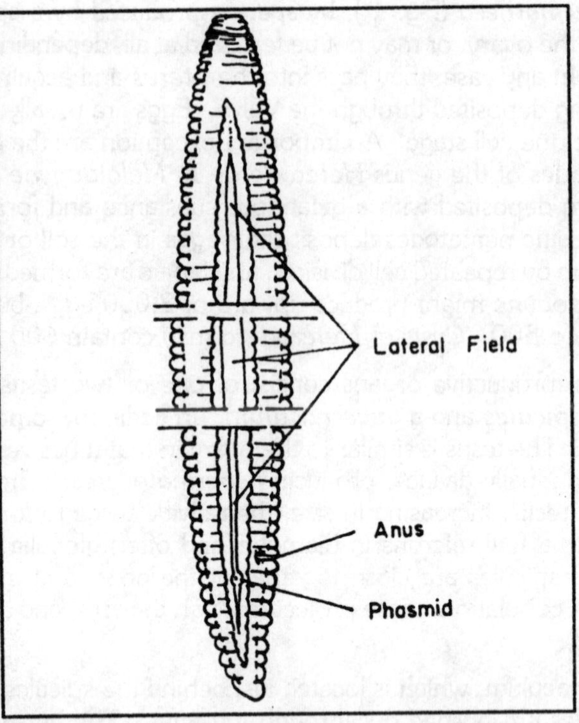
Lateral Field
Anus
Phasmid

Fig. 8 : Cuticle at anterior end, middle and posterior end, with annules and lateral fiel

are called *migratory endoparasites.* If the females are attached permanently t the roots, they are called *sedentary endoparasites,* even though part of the bod is outside the root, as with females of *Heterodera* or *Tylenchulus*. Ectoparasite feed on the outside of the plant and never have more than the anterior portion temporarily embedded in the root. They are always migratory.

Taxonomically, plant parasitic nematodes are classified in two large groups About 1000 of the approximately 1100 described species belong to the orde ***Tylenchida***. The reminder belong to ***Dorylaimida***. The differences betwee

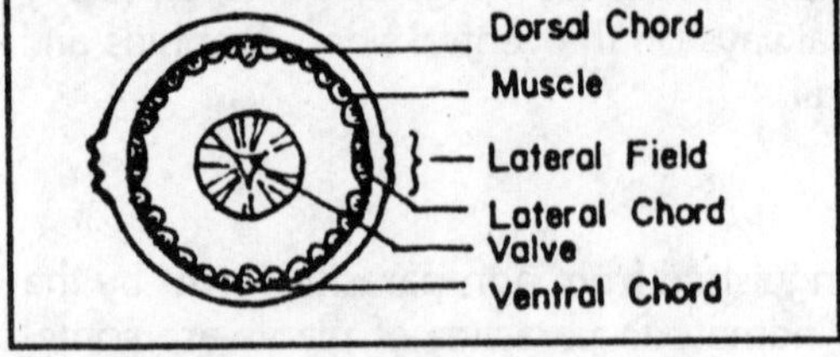
Dorsal Chord
Muscle
Lateral Field
Lateral Chord
Valve
Ventral Chord

Fig. 9 : Cross-section

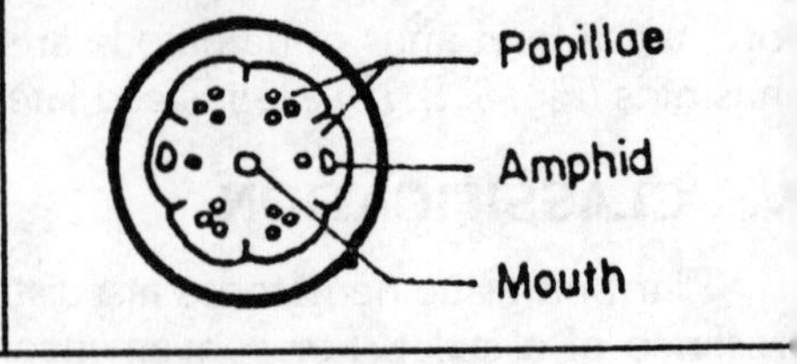
Papillae
Amphid
Mouth

Fig. 10 : Head of nematode with mout papillae and amphids the two is based essentially on the structure of stylet and of the oesophagus. though there are many other differences for example, Tylenchida always have annules, *Dorylaimida* do not. Most *Tylenchida* have stylets with knobs. An oesophageal tube is attached to the stylet, and they have a median bulb with a valve. While there is great variation in the size and shape of the stylet, the oesophageal tube, and the median bulb, they are nearly always present and easy to see. They are absent or very difficult to see in males of some genera, such as *Radopholus Pratylenchus, Hemicycliophora,* and Criconemoides.

The plant parasitic nematodes of the dorylaimoida group have an entirely different type of oesophagus. It consists of a narrow anterior portion and a thicker posterior portion, and is without valve. The four plant parasitic genera of this group are *Xiphinema, Longidorus, Paralongidorus*, and *Trichodorus*. The stylets of the first three are very long. The stylet of *xiphinema* has flattened flange-like knobs, while those of *longidorus* and *Paralongidorus* have no knobs. *Trichodorus* is very different by having a spear which is typically curved. This stylet does not actually have a lumen, and is essentially an elongated tooth. The food of *Trichodorus* in fact does not pass through the stylet. A broad classification with pertinent examples is given below:

Phylum : NEMATODA
ORDERS

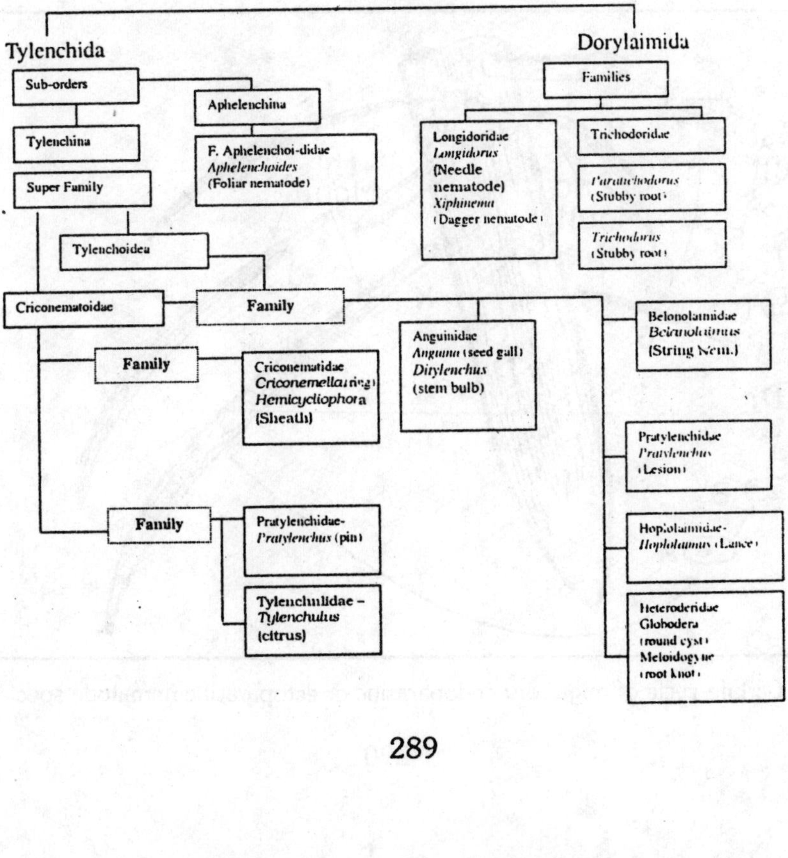
Tylenchida
Sub-orders
Aphelenchina
Tylenchina
F. Aphelenchoi-didae
Aphelenchoides
(Foliar nematode)
Super Family
Tylenchoidea
Criconematoidae
Family
Family
Criconematidae
Hemicycliophora
(Sheath)
Anguinidae
Anguina (seed gall)
Ditylenchus
(stem bulb)
Family
Pratylenchidae-
Pratylenchus (pin)
Tylenchulidae -
Tylenchulus
(citrus)
Dorylaimida
Families
Longidoridae
Longidorus
(Needle
nematode)
Xiphinema
(Dagger nematode)
Trichodoridae
(Stubby root)
Trichodorus
(Stubby root)
Belonolaimidae
(String Nem.)
Pratylenchidae
Pratylenchus
(Lesion)
Hoplolaimidae-
Hoplolaimus (Lance)
Heteroderidae
Globodera
(round cyst)
Meloidogyne
(root knot)

V. BIOLOGY

A. **Life-Histories :** The life-history of plant parasitic nematodes is usually very simple, with five distinct stages, the first four of which end in a molt. Nematodes molt by forming a new cuticle, after which the old cuticle may or may not be shed. Molting nematodes can be distinguished by the fact that the culticle is not closely attached at the head end and that the anterior portion of the stylet is attached to the molted cuticle. The life-history differs according to groups or species of nematodes. Nematodes which feed on outer layers of root, or feed inside the root without becoming permanently attached (migratory parasites) have simple life histories (Fig. 11). In bisexual species, the female is fertilized by the male and produces eggs which are deposited in soil. The one celled egg undergoes a series of cell divisions, forming the first stage larvae. The first molt takes place in the egg, and the second stage larvae emerges from the egg. The nematode often remains in the second stage until it finds a source of food, usually the roots of a living plant. After it starts to feed, it passes through three additional molts. Between the third and fourth molts, the sexual organs begin to develop; and at the fourth molt, there is evidence that males of some species do not feed after reaching the adult stage.

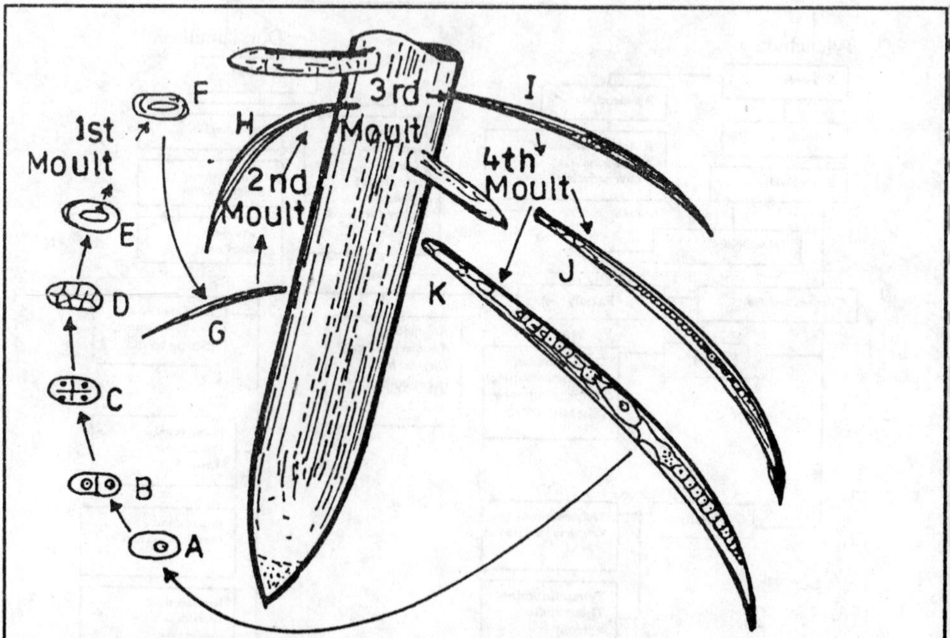
1st Moult
2nd Moult
3rd Moult
4th Moult
A
B
C
D
E
F
G
H
I
J
K

Fig. 11 : Life-cycle of migratory endoparasitic or ectoparasitic nematode species

B. Meloidogyne (Root-knot nematodes)

Parasitism and Habitat :

Females as well as third and fourth stage larvae are sedentary parasites on many plants. Males and second stage larvae are migratory and can be found in soil.

Morphological Characters :

Body : Elongated in larvae (0.5mm) and males (1.0 - 2.0 mm); typically saccate, spheroid, with a distinct neck in females (0.8 mm long and 0.5mm wide).

Stylet : Strong with rounded knobs in males; in females more slender than in males or larvae but with strong basal knobs.

Oesophagus : With very large median bulb followed by a short isthmus.

Excretory Pore : Often seen with part of excretory tube in the area between posterior stylet knobs and opposite median bulb.

Vulva and Anus : Of females, typically opposite to neck and surrounded by a pattern of fine lines.

Spicules : Very near the terminus of males.

Important species : *M. Javanica, M. arenaria*

C. Life cycle (Fig. 12)

A Eggs deposited by the adult female in an egg mass develop and larvae are formed. The larvae molt once in the egg, then hatch and enter the soil. In the soil, the larvae may enter a root immediately, or only after several months.

B The second stage larvae enter roots near the root tip and start to feed.

C After the second molt, the larvae are in the third stage, the root gall has started to develop, and giant cells have been formed.

D After the third molt, the larvae are in the fourth stage.

E Near the end of the fourth stage, the reproductive organs of the female begin to develop.

F The male becomes a long, slender worm coiled in the larval cuticle.

G The male leaves the root and moves through soil to fertilize the female.

H The adult female starts to produce eggs which are deposited in the egg mass. The egg mass may be outside the root if the posterior end of the female is at the root surface. It the female is completely embedded in the root the egg mass is inside the root.

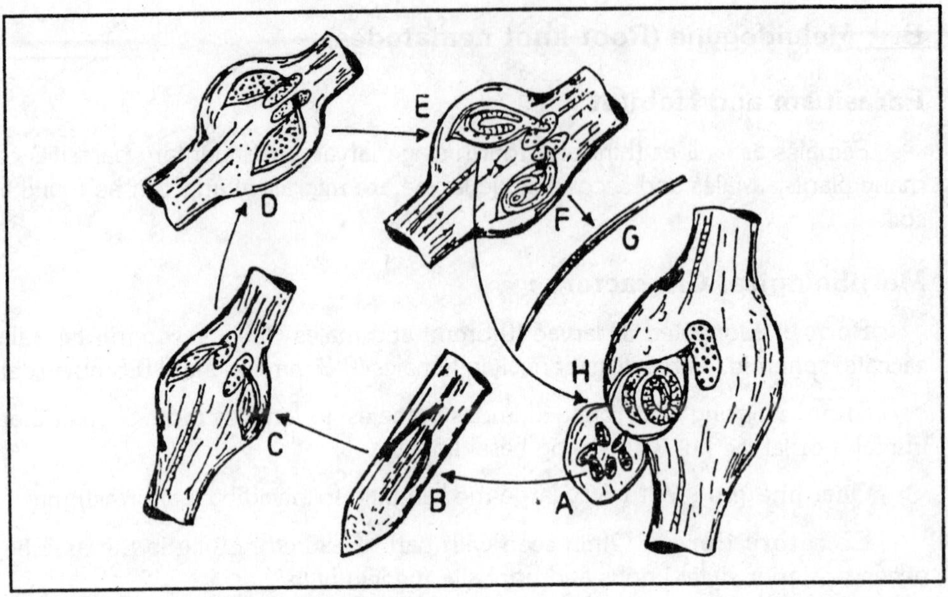
E
D
F
G
H
C
B
A

Fig. 12 : Life-cycle-*Meloidogyne* spp.

D. Heterodera (Cyst nematodes)

Parasitism and Habitat : Parasitic on many plants notably sugarbeet, oats and other grains, clover, soybean and various cruciferous plants. Adult females with neck embedded in plant roots and with body exposed larvae, males and cysts found in soil.

Morphological Characters :

Body : Slender in males (1.0 to 2.0 mm) and larvae (0.3 to 0.6 mm); in females, typically swollen, lemon (0.5 to 0.8mm in length), white or yellow in colour, cysts dark brown, lemon shaped (0.8 mm long and 0.5 mm wide) or nearly the same shape as the *Meloidogyne* female.

Stylet : Short in males with rounded basal knobs; in larvae, more than 0.02 mm long.

Oesophagus : With well developed median bulb and lobe extending back and overlapping the intestine.

Spicules : Near the posterior end of males.

Important species : *H. major*, *H. cajani*, *H. avenae*, *H. schachtii*

E. Life-cycle of cyst nematode, *Heterodera* species (Figure-13)

A. Cyst in soil is the body of a female filled with eggs containing second stage larvae.

B,C. Some of the larvae emerge from the cyst and enter the young roots of growing plants near the root tip.

D,E. The larvae feed and grow, passing through the second molt and becoming third stage larvae. Reproductive organs begin to develop.

F,G. After the third molt, the larvae reach the fourth stage. The female on the left is breaking through the surface of the root. The male on the right has become long and slender, but is still enclosed in the fourth stage cuticle.

H. The adult female remains attached to the root with the body mostly outside. It has begun to deposit egg in an egg mass. Later, eggs accumulate in the body, and at the end of its life, the female culticle becomes a cyst filled with eggs.

I. The adult male has left the root to move through the soil and fertilize the female.

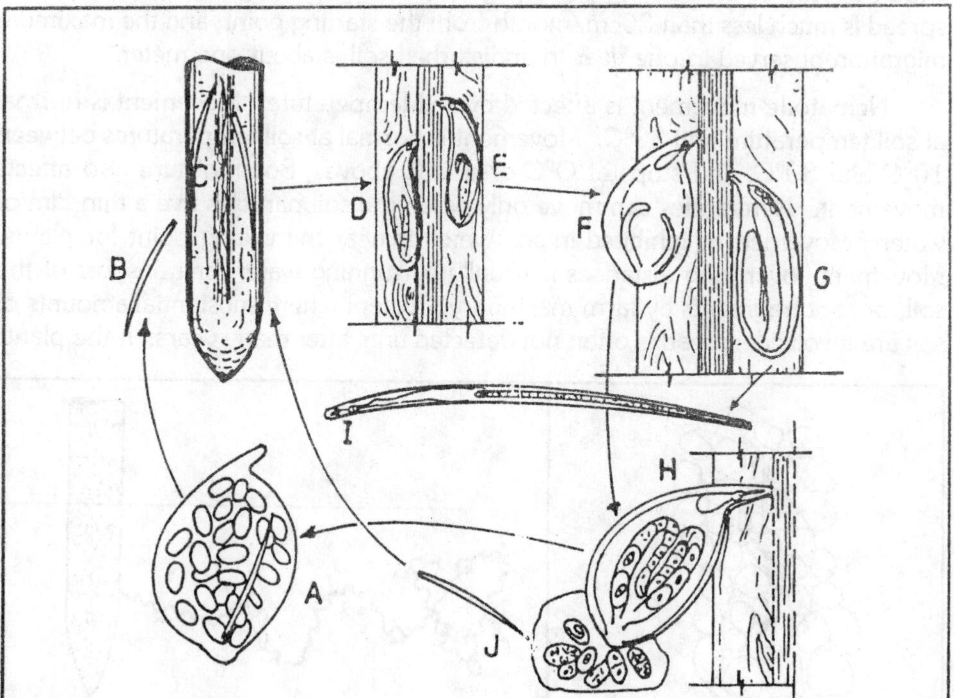
C
B
D
E
F
G
I
H
A
J

Fig. 13 : Life-cycle- *Heterodera* spp.

J. The eggs in the egg mass develop and larvae hatch. These also move to infect the young roots.

VI. Movement

Plant parasitic nematodes, with few exceptions, progress by undulatory motion in a darso-ventral plane (Fig. 14) rather than in a lateral plane. A living nematode seen on a flat surface under microscope is probably lying on its side, since their ability to force a passage by moving soil particles is limited. Nematodes must make their way through the pore spaces of the soil. Movement seems to be mostly random until the nematode comes near a root (Fig. 15). It then moves towards the root, apparently in response to root excretions which are detected by the amphids. Root excretions are attractive only over a distance of 2 or 3 centimeters; but even so, attraction increases the chances of a nematode finding the roots. In experiments where large number of nematodes have been placed at a starting point in sterilized field soil, maximum movement has averaged about 30cms/month. This is the maximum; most nematodes in fact, will be found within 20cms of the starting point (Fig. 16). In succeeding months, the rate of spread is much less than 30cm/month from the starting point; and the maximum migration observed in one year in undisturbed soil is about one meter.

Nematode movement is affected by soil temperature. Movement is normal at soil temperature and 30°C. Movement is normal at soil temperatures between 10°C and 30°C, and stops at O°C or a little above. Soil moisture also affects movement. Nematodes can move only when the soil particles hve a thin film of water. Movement is inhibited in soil which is near the wilting point for plants. Movement over long distances is usually by running water, or transport of the soil, or root fragments by farm machinary. Except where substantial amounts of soil are involved, spread is often not detected until after many years. If the plants

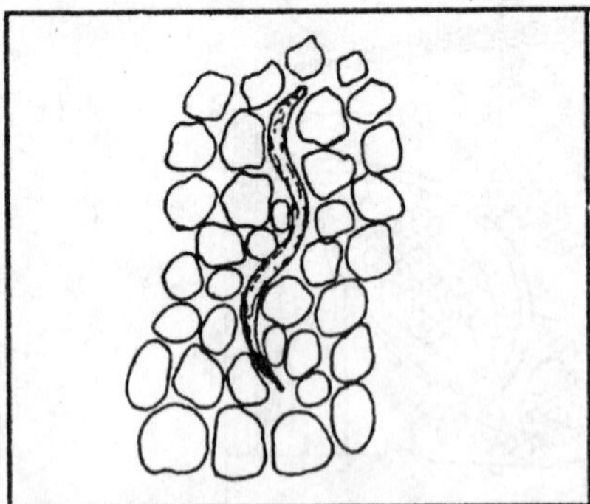

Fig. 14 : A nematode moving through soil

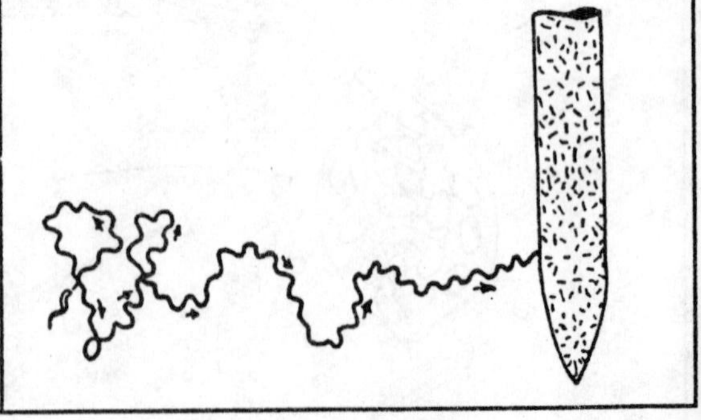

Fig. 15 : Movement of nematode through soil

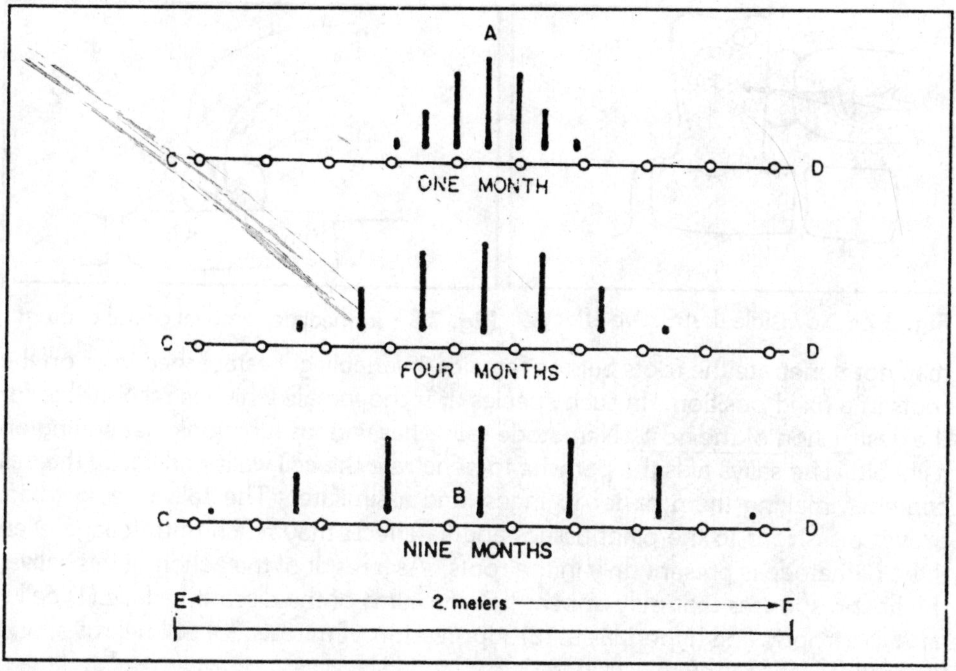
A
C
D
ONE MONTH
C
D
FOUR MONTHS
B
C
D
NINE MONTHS
E
2. meters
F

Fig. 16 : Distance covered by plant parasitic Nematodes in Soil

are propagated by sprouts with roots taken from the parent plant, the nematodes affecting the parent plant almost surely will be carried alongwith the sprout. Rooting of plants, or bulbs, corms, and rhizomes from plants grown in nematode-infested soil can spread nematodes to new locations. Seeds, stems, and leaves, when contaminated/infested by nematodes which are able to withstand desiccation, can also spread over long distances.

VII. Feeding and Damage to plants

The mechanical injury directly inflicted upon plants by the nematodes during feeding is slight. Most of the damage seems to be caused by secretion injected into plants while the nematodes are feeding. This secretion called 'Saliva', is produced in three glands from which it flows forward into the oesophagus and then injected through the stylet. Some nematodes are such rapid feeders that in a matter of few minutes they pierce a cell, inject saliva, withdraw the cell contents, and move on. Others feed more leisurly and may remain at the same juncture for hours or days.

The sedentary forms (*Heterodera and Meloidogyne*) remain attached to one point in the tissue throughout their lives. The males of the species may or

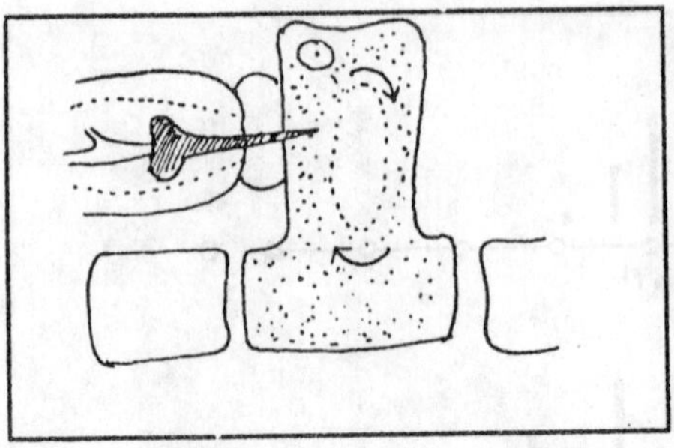

Fig. 17 : No visible destructive effect

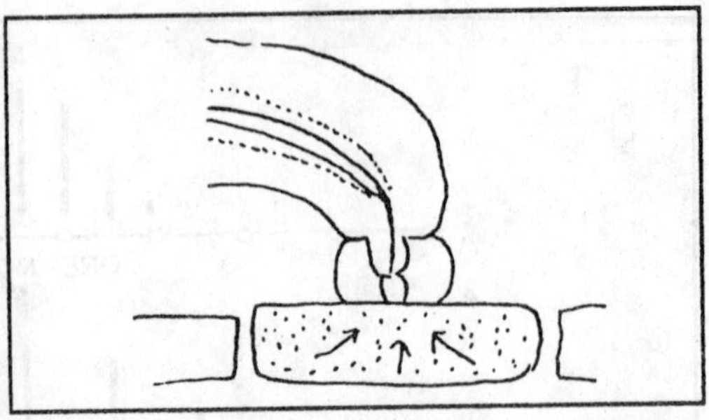

Fig. 18 : Immediate removal of cell content

may not penetrate the roots but the females invariably get established in or on the roots in a fixed position. In such species, it is the female which is responsible for the destruction of the host. Nematode saliva has various functions, depending on its habit. The saliva aids the parasite to penetrate the cell walls and liquify the cell contents, making them easier to ingest and assimilate. The saliva being toxic, proves disastrous to the plant tissues and its effects may reach upto foliage even if the nematode is present only in the roots. As a result of the action of this saliva, the host responses that may appear as symptoms of the disease include (1) cellular hypertrophy and hyperplasia, (2) suppression of mitosis, (3) cell necrosis, and (4) stimulation of growth.

The feeding of some nematodes produces only slight trauma in host cells. When the nematode *Tylenchorhynchus dubius* feeds upon the root hairs of *Lolium perenne* a spherical mass is formed at the tip of the stylet within the host cells. For about 30 seconds, the nematode remains quite, after which its bulb begins to pulsate and the spherical mass diminishes and finally disappears after a minute or so (Figure-17). In some associations, the stylet acts simply as penetrating organ and suction tube. The nematode *Paraphlenchus acontiodes* penetrate the host cells and withdraws their contents within 2 to 3 seconds (Fig. 18). Many plant parasitic nematodes affect only those cells upon wihich they feed or a limited number of cells in the immediate vicinity of the feeding site. Cells of infected tissues separate, undergo hypertrophy and loose chlosoplasts (Fig. 19). In certain host parasite combinations the pathogen stimulates changes in host cells, bringing about alterations in metabolism vital to the growth and development of the pathogen but do not bring about any cell destruction. The nematode *Tylenchulus semipenetrans* brings about an increases in the size of nuclei, and nucleoli, the cytoplasm of the host cells becomes dense their wall thickens and the vacuoles of the infected cells disappear (Fig. 20). *Anguina* and *Nothanguina* species stimulate gall formation in leaves and flower parts of their hosts. The galls develop due to hypertrophy and hyperplasia of parenchymatous tissues with a central cavity harbouring the nematode. No syncytia develop.

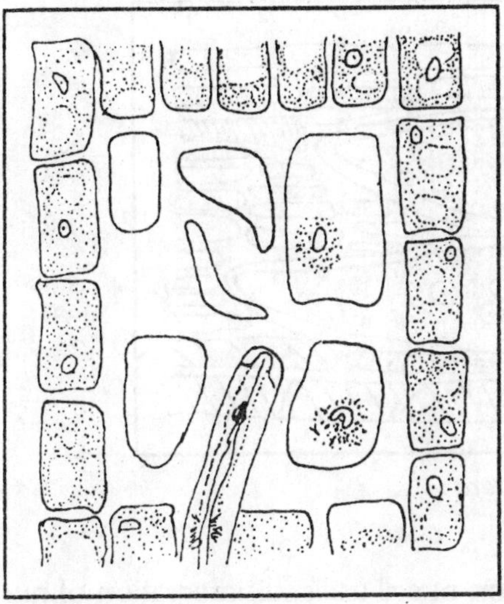

Fig. 19 : Cell Lysis

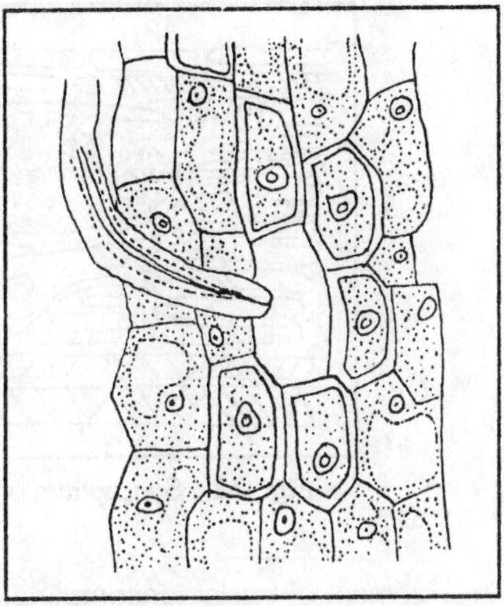

Fig. 20 : Nurse Cells of *Tylenchulus*

Mambers of family *Heteroderidae*, show the greatest morphological adaptation to parasitism. At the feeding site, a group of cells develop into charactersitic syncytia around the head of the parasite. The cytoplasm becomes dense and the size of nuclei and nucleoli enlarges considerably. The cell walls are usually altered. Species of *Meloidogyne* induce hyperplasia in the pericycle which results in the formation of galls. *Heterodera* species do not induce excessive/extensive hyperplasia. *Nacobbus* sp. forms galls on sugarbeet roots due to hypertophy of cortex and epidermal cells.

The stele is not affected. *Synchytial* mass is formed by the alteration of cells immediately adjacent to the nematode's head and extends longitudinally along the root (Fig. 21). In *Heterodera* infection synchytium is formed as result of the

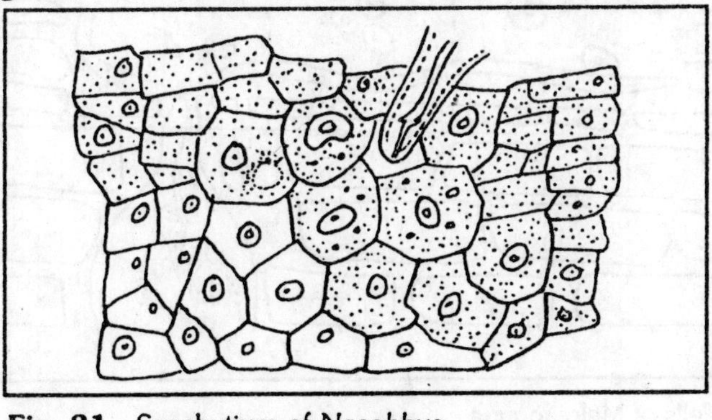

Fig. 21 : Synchytium of *Nacobbus*

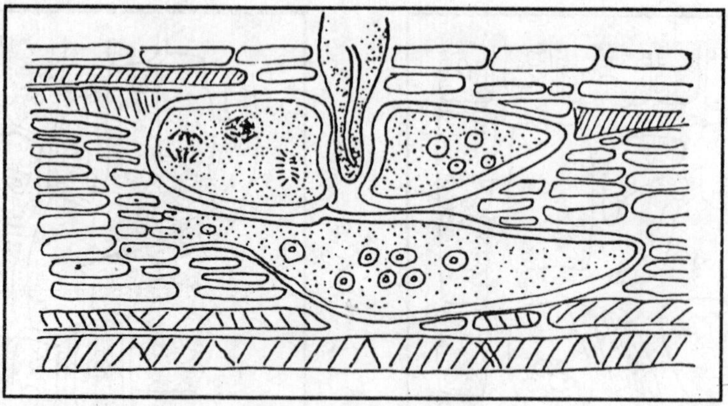

Fig. 22 : Synchytium of *Heterodera*

enlargement of nuclei, incomplete dissolution of cell walls, disintegration of nucleoli, thickening of cell walls and by the cytoplasm becoming dense and granular (Fig. 22). Giant cells of *Meloidogyne* are formed as a result of the enlargement of nuclei and the cells become polyploid and undergo synchronous mitosis. The cytoplasm of these cells become granular and new cells are incorporated by cell wall dissolution. The walls of the synchytium become thickened and the cells of pericycle divide repeatedly (Fig. 23).

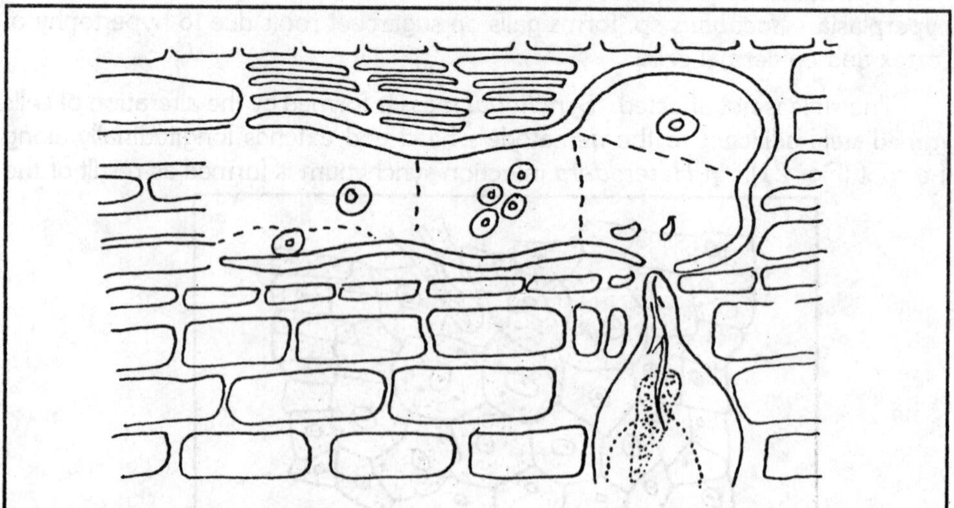

Fig. 23 : Giant Cells of Meloidogyne

CHAPTER XI

PHANEROGAMIC (ANGIOSPERMIC) PLANT PARASITES

I. INTRODUCTION

A number of flowering plants are parasitic on economic plants and cause considerable damage. Important plant families that have such parasites are *Loranthaceae, Convolvulaceae, Scrophulariaceae, Orobanchaceae, Lauraceae, Santalaceae*, and *Balanophoraceae*. Kwifit (1969) has described in detail the biology of parasitic flowering plants. In India, Kumar (1940) has described flowering plants that attack economic crops. Tarr (1972) has described several types of parasitism. They are:

A. The parasite is rooted in the soil and is attached to the roots of the host plant by haustorial discs through which the vascular tissues of both host and parasite become joined, such as in witchweed (*Striga*). In some cases green aerial shoots and flowers are produced.

B. The Parasite has perennial underground stems which are attached to the host roots and produce aerial flowering shoots each year, such as *Lathraea squamosa* (toothwart). The leaves are without chlorophyll.

C. The tissues of the parasite merge with those of the host root to form a swelling from which aerial flowering shoots arise. There may be varying numbers of rudementary or scalelike leaves or the host may be leafless, for example, broomrape, *Orobanche*.

D. The parasite invades the branches of the host, often a tree, and develops as a bushy perennial plant with green leaves but does not root in the soil, for example, mistletoe, *Viscum*.

E. The seeds of the parasite germinate in the soil and produce a hair - like shoot which makes circular movements until a suitable host is found. Thereafter, it twines around the host stem, obtaining nutrients through haustoria. The stem of the parasite dies at its lower end and thus contact with soil is lost. Chlorophyll is absent, although there may be small, rudimentary scale leaves. Examples are the dodders (*Cuscuta* spp).

F. There may be a more complete integration of the parasite and host, the former developing as a kind of hollow root or shoot. The flowers eventually burst through the cortex to the exterior. This situation is found in the *Rafflesiaceae*, for example *Rafflesia arnoldei*, where the vegetative organs of the parasite are reduced to a network ramifying in the host tissues.

The phanerogamic plant parasites belong to several widely separated botanical families and vary greatly in their dependence on their hosts. Some of them attack roots (root parasites) while others parasitize the stem(stem parasites). Some are devoid of chlorophyll and depend entirely upon their hosts (holoparasites) while others have chlorophyll and take only the mineral constituents of food from the hosts (semiparasites). The common parasitic flowering plants (Plate-I) may be grouped as follows :

1. DENDROPHTHAE (LORANTHUS, BANDA, GIANT MISTLETOE)

It is common parasite of fruits, waste lands, avenue and forest trees (plate-I). The most common is the mango tree, and in northern India 60 - 90% mango trees besides *Fiscus* spp. *Albizzia* spp., and *Delbergia* spp. are infected. In central part of India, *Madhuca latifolia* is most common host. It is semiparasitic on tree trunks and branches and can be easily spotted as a dense cluster of small twigs bearing smooth broad leaves and long, tubular, orange coloured flowers with red berries. *Dendrophthae falcata* is the most common species in India. It forms haustoria which penetrate deep into host tissues. Through these haustoria water and minerals flow from the food conducting system of the host to the parasite. The green leaves of the parasite manufacture sugar and starch. The parasite is disseminated mainly through its seeds, carried by birds.

2. CUSCUTA (DODDER, AMARBEL)

There are about 100 species of *Cuscuta* attacking trees, clover, alfalfa, flax, sugarbeet, onion but not cereals. *C. gronovei* is the most common dodder in India. *Cuscuta* spp. are non-chlorophyll, bearing leaflets, twining, parasitic seed plants. They attach their yellow, orange or pink, thread like stems to the stems or other parts of the host plant. When the contact is made, haustoria from the parasite penetrate the host and absorb nutrients. *Cuscuta* spp. produce flowers in clusters or racemes. The flowers are white, yellowish, or pinkish in colour. The calyx is shorter and coloured than the corolla. The corolla is tubular, with triangular lobes. The ovary has two locules and each cell is with two ovules. There are

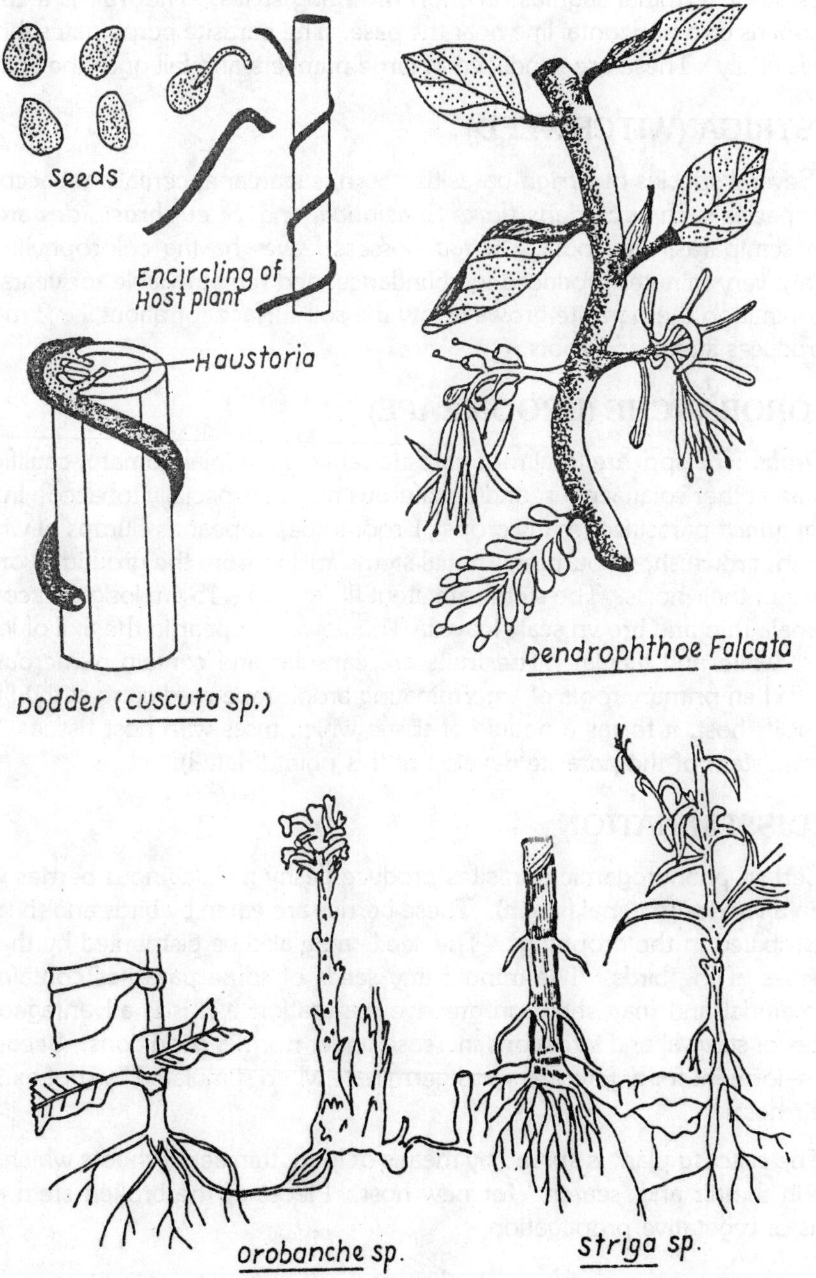
Seeds
Encircling of Host plant
Haustoria
Dodder (cuscuta sp.)
Dendrophthoe Falcata
orobanche sp.
Striga sp.

Plate-I : Phanerogamic Plant Parasites

two sessile or terminal stigmas on short diverging styles. The fruit is a capsule which opens by a horizontal line near the base. The parasite perpetuates through its seeds easily. These are produced in large numbers and fall onto the soil.

3. STRIGA (WITCHWEED)

Several species of *Striga* parasitize corn, sugarcane, cereals, tobacco, etc. Three species such as *S. densiflora*, *S. asiatica*, and *S. euphrasioides* are well known semiparasite in India. *Striga* possess leaves having chlorophyll. The seeds are very minute, produced in abundance, and remain viable for years. After germination the parasite grows below the soil surface for about 1 - 2 months and produces stems and roots.

4. OROBANCHE (BROOMRAPE)

Orobanche spp. are total root parasites affecting brinjal, tomato, cauliflower, turnip and other solanaceous, and cruciferous plants, especially tobacco. In India *O. aegyptiaca* parasitizes many crops. Broomrapes appear as clumps of whitish, yellowish, brownish, or purplish annual stems arising from the ground at or near the base of their hosts. The stems are stout fleshy, 15 - 45 cm long and covered with small, thin and brown scaly leaves. The flowers appear in the axil of leaves, and are white and tubular. The fruits are capsular and contain numerous tiny seeds. When primary roots of a germinating broomrape seed contact the fibrous roots of its host, it forms a nodule of tissue which fuses with host tissues. New roots and stem of the parasite develop at this point (Plate-I).

II. DISSEMINATION

Certain phanerogamic parasites produce fleshy mucilaginous berries which contain a viscous material (viscin). These berries are eaten by birds and the seeds are distributed in the droppings. The seeds may also be distributed by the feed and beaks of the birds. The minute tiny seeds of some parasites contain little food material and may show progressive maturation. This is advantageous as chances of survival and infections increase under normal conditions. Seeds may remain dormant for a few years and germinate when stimulated by root exudates of their hosts.

The *cuscuta* plant is spread by means of long, thin aerial shoots which wave about in the air and 'search ' for new host. Pieces of the broken stem act as organs of vegetative propagation.

III. MANAGEMENT

(1) Seeds and other planting materials must carry "phytosanitary" level. It should be free from infection / infestation.

(2) In parasites like *Loranthus* and *Cuscuta*, tree surgery is useful. Cut out and destroy infected plant plarts.

(3) Follow a suitable crop rotation in which crops grown should have properties to discourage build up of inoculum existing in soil.

(4) If available, trap crops must be grown to reduce inoculum potential.

(5) Herbicides which could selectively kill the parasites should also form part of an integrated approcah.

CHAPTER XII

PRINCIPLES AND PRACTICES OF PLANT DISEASE MANAGEMENT

I. INTRODUCATION

Information on etiology, symptoms, pathogenesis, and epidemiology of plant diseases are intellectually interesting and scientifically justified, but most important of all, they are useful as they help in formulation of methods for management of diseases and thereby increasing the quantity and improving the quality of plant and plant products. Practices employed for disease management vary considerably from one disease to another depeding upon the type of pathogens, the host, and the interaaction of the two under overall influence of the environment.

II. DISEASE CONTROL *VS* DISEASE MANAGEMENT

In the past, plant pathologists always aimed at the impossible task of destroying pathogens to control plant diseases. However, the diseases controlled are really few. No plant pathogen has ever been wiped out from the face of the earth. So long as the pathogen suvives and its host is cultivated, the chances of disease incidence will persist. It is a different matter that due to some direct or indirect efforts on the part of man, the disease causing agent has been subdued or has been reduced to an innocuous level. These efforts could better be called as "management practices" whereby the population of the pathogen or its disease causing potentialities have been kept under check so that it failed to cause noticeable loss to the crop.

Although disease control is an established and widely understood term, there is convincing rationale supporting the substitution of "management" for "Control". The word "control" evokes the notion of finality, the final disposal of the problem (Apple, 1977). How many plant diseases have been finally disposed-off ? "Management" conveys the concept of a continuous process rather than an event accomplished through application of an intrinsic factor. The fact that in almost all examples of recommendations for disease control, we recommend schedules that are to be follwed almost every year in the crop suggests that these disease control methods are part of a continuous process, and the disease has not been disposed-off. The diseases are inherent components of the agroecosystem that must be dealt with on a continuous, knowledgeable basis (Apple, 1977). Management is based on the principle of maintaining the damage or loss below an economic injury level. If this concept is conveyed to the farmer, he will not be frustrated if he finds a few diseased plants in his field even after adhering to recommendations made by the plant pathologists.

III. PRINCIPLES OF DISEASE MANAGEMENT

(A) Based on Practices

Whetzel(1929) was the first to classify methods for the control of plant diseases. His list included exclusion, eradication, protection, and immunization. As a consequence of advances in plant pathology and to accomodate control measures developed, two more principles - avoidance and therapy were included (NAS, 1968).

1. Avoidance of the Pathogen

Avoiding disease by planting at times when, or in areas where, inoculum is ineffective due to environmental conditions, or is rare or absent.

Many diseases can be managed by proper selection of field, choice of sowing time, selection of cultivars, seed and planting stocks, and modification of cultural practices. The aim of these measures is to enable the host to avoid contact with the pathogen or the susceptible stage of the plant and the fovourable condition should not coincide. The main principles under this group are selection of geographical area, selection of field, choice of sowing time, selection of planting materials, disease escaping varieties, modification of cultural practices, etc.

2. Exclusion of Inoculum

Preventing the inoculum from entering or establishing in the field or area where it does not exist. Seed certification, crop inspection, growing crops in regions unfavourable for the pathogens, and quarantine measures are some of the means of preventing the spread of the pathogens.

3. Eradication of the Pathogen

Reducing, inactivating eliminating, or destroying inoculum at the source, either from a region or from an individual plant in which it is already established.

This is attempted through the methods such as biological control, crop-rotation, sanitation, etc.

4. Protection

The inoculum of many fast spreading infectious diseases is brought by wind from neighbouring fields or any other distant place of survival . Protective measures are necessary to destroy or inactivate such inocula. It is possible by creating a chemical toxic barrier between the plant surface and the invading pathogen. Methods employed to achieve such results are chemical sprays, dusts, modifacation of environment and host nutrition, etc.

5. Disease Resistance

Preventing infection or reducing the effect of infection by managing the host through improvement of resistance in it by genetic manipulation or by chemotherapy. In any crop, resistance against a specific disease can be developed by selection or hybridization. This type of resistance is genetic. Biochemical resistance of nongenetic nature can be developed in plants by chemotherapy or modification of nutrition. This type of resistance is induced and temporary, lasting until the chemical or nutrient is effective in the plant.

6. Therapy

Reducing severity of a disease in an infected individual.

The first five of these principles are mainly preventive (prophylatic) and constitute the major procedure of plant disease management. They are applied to the population of plants, i.e., the crop. The last, therapy, is a curative procedure and is applied to individuals under the concept of "disease management". These principles have been classified into following five categories (Horsfall and cowling, 1977).

(a) Management of physical environment including cultural control,
(b) Management of associated microbiota which includes antagonism,
(c) Management of host genes,
(d) Management with chemicals, and
(e) Management with therapy, radiation, and meristem culture.

(B) Based on Epidemiology

Epidemiology is the science of epidemics, which are widespread out break of disease. As Van der Plank (1963,1968) pointed out, most control measures reduce either the initial inoculum from which an epidemic starts or reduce the rate at which infection builds up during the epidemic, one may, of course, also do the both (Ven der Plank, 1968.) crop sanitation, the growing of vertically resistant crop varieties. The sowing of disease free planting material, the destruction of pathogens on or in the planting material by chemical or heat treatments, and soil fumgation all reduce the initial inoculum. Measures which reduce the rate at which pathogens spread, include the growing of horizontally resistant crop varieties, and the application of protective fungicides; these measures reduce the number of diseased plants and the severity of infection so that the amount of inoculum at the end of season is also reduced. *Monocyclic* diseases are most efficiently suppressed by reducing the amount or efficacy of initial inoculum while *polycyclic* diseases are most efficiently suppressed by reducing large amounts of initial inoculum and/or by limiting potentially rapid rates of disease increase.

The six principles that characterize the modern concept of plant disease management should be viewed from three stand points:

(i) Whether they involve reduction in the initial inoculum or the rate of disease development.

(ii) Whether they primarily involve control of pathogen population, the cure and defense of the suscept or involve the environment as it relates to disease and,

(iii) Whether th ey involve interruption of survival, dispersal, inoculation, penetration, infection or the actual of disease (Robert and Boothroyd, 1972). These interactions originally proposed by Baker (1968) and Robert and Boothroyd(1972) and subsequently modified by us are illustrated in **figure-1**.

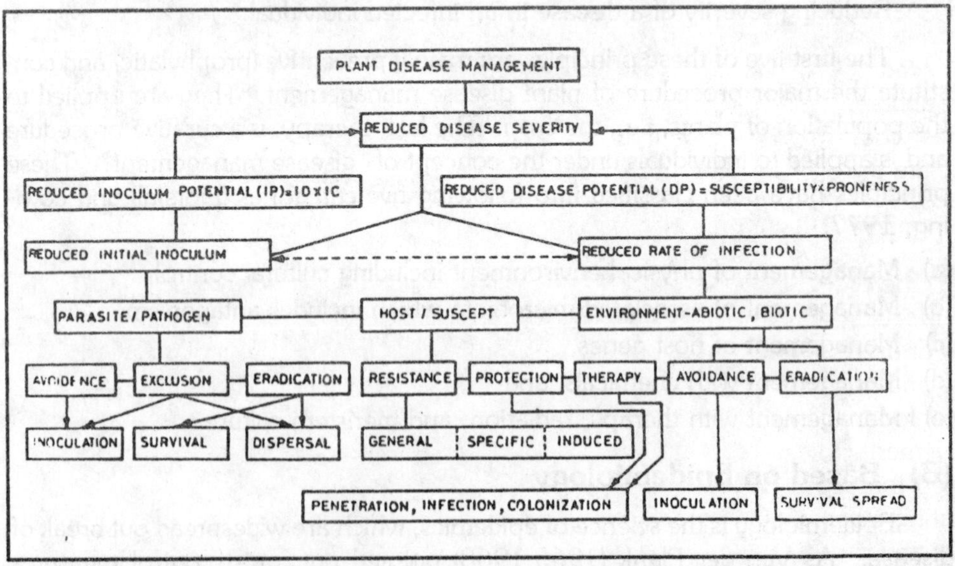
PLANT DISEASE MANAGEMENT
REDUCED DISEASE SEVERITY
REDUCED INOCULUM POTENTIAL (IP) = I D x I C
REDUCED DISEASE POTENTIAL (DP) = SUSCEPTIBILITY x PRONENESS
REDUCED INITIAL INOCULUM
REDUCED RATE OF INFECTION
PARASITE / PATHOGEN
HOST / SUSCEPT
ENVIRONMENT-ABIOTIC, BIOTIC
AVOIDANCE
EXCLUSION
ERADICATION
RESISTANCE
PROTECTION
THERAPY
AVOIDANCE
ERADICATION
INOCULATION
SURVIVAL
DISPERSAL
GENERAL
SPECIFIC
INDUCED
PENETRATION, INFECTION, COLONIZATION
INOCULATION
SURVIVAL SPREAD

Fig. 1 : The relationship between principles, practices, components of disease pyramid, and elements of pathogenesis.

CHAPTER XIII

REGULATORY METHODS

I. INTRODUCTION

Unrestricted movement of plants and plant products within and between countries has resulted in worldwide distribution of many plant pathogenic organisms. While it is not always possible for alien organisms to survive under new ecological and climatic conditions, many of them can adapt themselves to new environments to become eventually potentially dangerous to the crop plants of those areas. The international exchange of plants and their parts is now practiced widely to improve the crops of a country and their genetic base. In addition, shiploads of grains for consumption or large quantity of seeds for sowing is practiced in many countries. Even minute quantities of soil and plant debris contaminating true seeds can disseminate pathogens. Therefore, in order to prevent introduction and spread of plant pathogens and their races/ biotypes into the country or states where they are not known to exist, certain federal and state laws must regulate the exchange of such materials.

I. PRINCIPLES

One of the six principles of management of plant diseases is the exclusion of the inoculum by preventing the inoculum from entering or establishing into the field or area where it does not exist. In this commercialized world today, through easy and quick means of transportation, there is exchange of not only grains and other food stuffs between nations but also of germplasms or seeds for the improvement of crops especially in developing nations. Past experiences have revealed that no country or state can afford unrestricted movement of plant and planting materials. Thus, the knowledge and methodology of exclusion of the pests and pathogens are enforced by a legally constituted authority to prevent the entry and thus spread of injurious crop pests and pathogens in the public interest.

III. PRACTICES EMPLOYED FOR DISEASE MANAGEMENT

A. Plant Quarantine

1. Concept and Importance

Quarantine is derived from the Latin word *Quarantum* meaning 40. It refers to a 40-d period of detention of ships arriving from countries with bubonic plague and cholera in the middle ages. Plant quarantine may be defined as "the restriction imposed by duly constituted authorities on the production, movement and existence of plants or plant materials, or animals or animal products or any other article or material or normal activity of persons and is brought under regulation in order that the introduction or spread of a pest may be prevented or limited or in order that the pest already introduced may be controlled or to avoid losses that would otherwise occur through the damage done by the pest or through the continuing cost of their control." Thus, plant quarantines promulgated by a government or group of governments, restrict the entry of plant or plant products, soil, culture of living organisms, packing materials, and commodities as well as their containers and means of conveyance.

The adoption of "Quarantine Regulations and Acts" by different countries of the world has arisen out of the fact that extensive damages, often sudden in nature, have been caused not by indigenous organisms but by exotic ones which have been introduced along with plants, plant parts or seeds in the normal channel of trade or individual transit. Instances may be cited of the introduction of grape *Phylloxera (Phylloxera vitifolii)* from the U.S. to France which caused destruction of French vineyards; Mexican boll weevil *(Anthonomus grandis)* whose original home was in Mexico or Central America, about 1892 entered the U.S. and later to various countries of the world, causing extensive damage to cotton; European corn borer *(Ostrinta nubilalis)* which reached North America probably through broom corn from Italy or Hungary and has since become a major pest there. Pink boll worm *(Pectinophora gossypiella)* considered to be one of the six most destructive insects of the world probably a native of India, is now established as a highly destructive pest in almost all cotton growing areas of the world.

Some of the examples cited above thus, clearly reveal that through unrestricted movement of plants and planting materials, diseases and pests have been introduced in regions, where they were not known to exist. Some of the important plant pathogens introduced into different countries of the world and where they caused severe losses, are listed in Table 1.

Table 1 : Some Examples of Plant Pathogens Introduced into Some Other Countries

Disease and Pathogen	Introduced from	Introd-uced into	Year of intro-duction
American goose berry mildew *(Sphaerotheca morsuvae)*	N. America	England	1899
Bacterial canker of tomato *(C.michiganense pv. michiganense)*	U.S.	U.K.	1942
Bacterial leaf blight of paddy *(X.campestris pv oryzae)*	Phillipines	India	1959
Black rot of crucifers *(X. campestris pv. campestris)*	Java	India	1929
Black shank of tobacco (*Phytophthora nicotianae)*	Holland	India	1938
Blister rust of pines (*Cronartium ribicola)*	Europe	U.S.	1910
Bunchy top of banana(viral)	Sri Lanka	India	1940
Chestnut blight *(Endothia parasitica)*	Asia	U.S.	1904
Citrus canker *(X. campestris pv. citri)*	Asia	U.S.	1907
Powdery mildew of cucurbits *(Erysiphe cichoracearum)*	Sri Lanka	India	1910
Downy mildew of grapes *(Plasmopara viticola)*	U.S. Europe	France India	1878 191 0
Downy mildew of maize *(Sclerospora phillipinensis)*	Java	India	1912
Dutch elm (*C.ulmi)*	Holland	U.S.	1928-30
Flag smut of wheat *(Urocystis tritici)*	Australia	India	1906
Fire blight of apple *(E. amylovora)*	N. America	New Zealand	1919
Golden nematode of potato *(G. rostrochiensis)*	Europe	U.S. Mexico India	1881 1961

Disease and Pathogen	Introduced from	Introduced into	Year of introduction
Hairy root of apple(Viral)	England	India	1940
Late blight of potato *(Phytophthora infestans)*	S. America U.K.	Europe India	1830 1883
Leaf rust of coffee *(Haemelia Vastatrix)*	Sri Lanka Asia, Africa	India Brazil	1879 1970
Onion Smut *(Urocystis cepulae)*	Europe	India	1958
Paddy blast *(M. grisea)*	S.E. Asia	India	1918
Peanut rust *(Puccinia arachidis)*	Brunei U.S.	Brazil	
Powdery mildew of grape *(Uncinula necator)*	North America	England	1845
Powdery mildew of rubber *(Oidium heavea)*	Malaya	India	1938
Rye grass seed infection *(Gleotinia temulenta)*	New Zealand	Oregon	1940
Wart of potato *(Synchytrium endobioticum)*	Netherlands	India	1953
Wheat bunt *(Tilletia caries)*	Australia	California, U.S.	1854
Witches broom of cocoa *(Marasmius perni)*	Trinidad	South America	1974-75

2. Principles, Basis, and Justification

All the nations of the world are passing through a period of intensive agricultural development in an effort to accelerate food production. All available resources are being mobilized to step up food production which include introduction of high yielding exotic germplasm. However, associated with these introductions is the danger of introducing some serious pathogens or their biotypes or races unknown in the importing country. Sometimes grains imported for milling purposes find their way into the farmers field thereby introducing organisms. Thus, international trade with regards to plant and plant products has eventually put a great burden on quarantines all over the world.

Plant quarantines promulgated by a government or group of governments, restrict the entry of plants, plant products, soil, culture of living organisms, packing materials, and commodities as well as their containers and means of conveyance and thus, help to protect agriculture and environment from avoidable damages by hazardous organisms. However, quarantines are justified only if the organism has little or no chance of spreading naturally. Thus, before its enforcement, the nature of the pathogen, its mode of spread/transmission, host range, and natural barriers should be properly understood. In addition, socioeconomic and geopolitical factors likely to work against the interests of exporting and importing countries too should be taken into account.

Plant quarantine regulations in order to be effective have to be based on sound scientific principles. They are (1) the biology and ecology of the organism against which quarantine measure is proposed to be enforced should be known; only those organinsms which are supposed to pose threat to major crops and forests should be taken into consideration, (2) in the event of its introduction whether the organism is likely to be established and cause damage of any consequence, (3) formulated to prevent or control the entry of the organisms and not to hinder trade or attainment of other objectives, as quarantine measures are for the crop and not for trade protection, (4) derived from adequated legislation and operated solely under the law, (5) amended as conditions change or further facts become available, and (6) attended by trained and experienced workers and the public must cooperate on an international scale for effective operation of the quarantine regulations.

3. Some Important Quarantines

Strict and timely application of quarantine laws along with suitable eradication practice have resulted in control of several plant diseases. For example, rubber plantation in Southeast Asia. In Southeast Asia, rubber trees are susceptible to *Mycrocyclus ulei* (South American leaf blight). The environment is also congenial for the development of the pathogen and disease. The quarantine regulations, publicity, surveillance, and eradication of diseased trees prevented the pathgen from being established in Southeast Asia. Similar is the case with citrus canker (*X. campestris pv. citri*) in U.S. The disease was first observed in 1912. Early in 1914 the disease was confirmed in all Gulf Coast states and Florida and prohibited importation of citrus stock. Subsequently, the U.S. government supported eradicatory efforts. Burning the infected trees in nurseries and groves was undertaken. Subsequently, continued vigilance and eradication practices have prevented establishment of *X .campestris* pv. citri.

Wart of potato caused by fungal pathogen *Synchytrium endobioticum* was observed in 1919 in West Virginia. Eventually, 70 garden sites were found infested with the pathogen. In 1921 the area was quarantined. Only varieties of potato immune to *S. endobioticum* were allowed to be grown and movement of soil and root crop vegetables required a special permit. Persistent quarantine and eradication efforts resulted in extinction of the disease. By 1973, all infested areas were declared free of the pathogen. Similar is the story of this disease in India. *S. endobioticum* was introduced in the Darjeeling hills in 1913 from Netherlands. Quarantine and other eradicatory and preventive efforts have succeeded in restricting the disease only in Darjeeling hills.

Golden nematode of potato *(Globodera rostochiensis)* was first discovered in New York state (Long Island) in 1941, and subsequently in Western New York, Delaware, and New Jersey. Infestations in Delaware and New Jersey appear to have been contained. With active and coordinated efforts of state and federal agencies, growers and universities, the population of the nematode has been brought below levels at which dispersal is unlikely to occur. The effort has prevented crop loss as well. This pathogen was introduced into Nilgiri hills, India in 1961 from Europe. Quarantine regulations and continuous monitoring of the disease have resulted in restriction of the disease in Nilgiri areas so far.

4. Entrance and Establishment of Pathogens

An exotic species must first gain entry and then it must become established. It is an extremely difficult preposition to predict accurately whether an exotic organism will become established and, once established, become economically important. In a relatively few cases, the pathogeographical approach has led to prediction of the occurrence of pathogens based on knowledge of the life cycles, distribution of the pathogen, and ecological characteristics of the host and pathogen. The assessment revealed that two organisms, the potato wart fungus (*Synchytrium endobioticum*) and the pathogen of citrus scab *(Elsinoe fawcetti)*, could not survive unfavourable seasons. Thus, the potato wart fungus does not survive in soil where the temperature rises to 30^{0}C in any season. Consequently, it was recommended that quarantine regulations against this pathogen be relaxed in tropical countries.

Some factors which affect entry amd establishment include: (1) hitchhiking potential compared with natural dispersal; (2) ecological range of the pest as compared with ecological range of its host i.e. climate vs. life-cycle vs. natural enemies vs. population of susceptible hosts, etc; (3) weather; (4) ease of colonization including reproductive potential; and (5) agricultural practices including pest management.

5. Plant Quarantine Regulations

a. Brief History

In 1660, France promulgated first quarantine law against barberry. In 1873, an embargo was passed in Germany to prevent importation of plant and plant products from the U.S. to prevent the introduction of the Colorado potato beetle, and in 1877, the United Kingdom Destructive Pests Act prevented the introduction and spread of this beetle. In 1891, the first plant quarantine measure was initiated in the U.S. by setting up a seaport inspection station at San Padro, California, and the first U.S. quarantine law was passed in 1912. The Federal Plant Quarantine Service was established in Australia in 1909. In India, a Destructive Insect and Pests Act was passed in 1914. Since then most of the countries have formulated quarantine regulations.

On a global basis, the first International Plant Protection Convention (the *Phylloxera* Convention) was signed in 1881, with the objective of preventing the spread of severe pests.

This convention was amended in 1889, 1929 and 1951. The International Plant Protection Convention (IPPC or Rome Convention) under the FAO was established to prevent the introduction and spread of diseases and pests through legislation and organizations across international boundaries. This convention provided a model phytosanitary certificate (Rome certificate) to be adopted by member countries. Within this convention, ten regional plant protection organizations have been established on the basis of biogeographical areas.

These are European and Mediterranean Plant Protection organization (EPPO), Inter-African Phytosanitary Council (IAPSC), Organismo International Regional de-Sanidad Agropecuaria (OIRSA), Plant Protection Committee for the South East Asia and Pacific Region (SEAPPC), Near East Plant Protection Commission (NEPPC), Comite Interamericano de Protection Agricola (CIPA), Caribbean Plant Protection Commission (CPPC), North American Plant Protection Organization (NAPPO), Organismo Bolivariano de Sanidad Agropecuaria(OBSA), and ASEAN region grouping Indonesia, Malasia, Phillipines, Thailand and Singapore, The regional organizations are concerned with the coordination of legislation and regulations within their area, agreement on the quarantine objects, inspection procedures etc.

b. Plant Quarantine in India

The earlier activities concerned with the introduction of plant pests and diseases with plant material was in early 1900s. Fumigation of all imported cotton bales was required to prevent introduction of the Mexican boll weevil *(Anthonomus grandis).* In 1914, the Government of India passed the **"Destructive Insects and Pests Act"** prohibiting or restricting the import of plant and plant materials, insects, fungi, etc., to India from foreign countries. This comes under foreign quarantine. Rules and regulations have been made prohibiting or restricting the movement of certain diseased and pest-infested materials from one state to another in India. This comes under domestic quarantine.

The enforcement of plant quarantine regulations is carried out by the technical officers of the Directorate of Plant Protection and Quarantine, Ministry of Agriculture, Government of India, under the overall supervision of Plant Protection Advisor. There are eight quarantine stations at seaports, seven at airports and seven at land frontiers. Research material is examined by three agencies, the National Bureau of Plant Genetic Resources, New Delhi, for agricultural and horticultural crops; the Forest Research Institute, Dehradun, for forest plants and Botanical Survey of India, Calcutta, for all other plants of general economic significance.

c. International Organizations

In 1929, a first world Plant Protection Convention was signed in Rome. In 1945, the United Nations and its specialized agency, the Food and Agriculture Organization (FAO), were established. This provided the best basis for the development of plant protection on a global level. The real breakthrough in this respect occurred with the establishment of the 1951 FAO International Plant Protection Convention. For implementation of various tasks the signatory governments have to make provision for : (1) the establishment of an official plant protection service, mainly in charge of quarantine matters, including plant health certification, and surveillance of the phytosanitary situation within the country and controlling harmful organisms of major importance;(2) a system securing technology transfer; and (3) a research organization. At present there are eight regional plant protection organizations covering different areas of the World (Table-2).

The regional organisations listed in Table-2, differ from each other in scope and functions, some of them dealing with all protection technologies and other specifically with quarantine. However, they all serve as advisory and coordinating bodies to participating governments. They operate as a network to promote concerted action for combating pests of international importance and to propose quarantine measures for preventing the spread of these pests.

d. Guidelines for Import of Germplasm

1. Import from a country where the pathogen(s) is absent.

Table-2 : Regional Government Organizations

Region	Organisation	Member Governments	Establishment
Western Hemisphere	Central America (OIRSA)	7	1955
	FAO Caribbean Commission	12	1967
	South America Northern Part (OBSA)	3	1965
	Southern Part (CIPA)	6	1965
West and East Palaearctic	EPPO	35	1951
	FAO Near East Commission	16	1963
Africa	North and Central Africa (IAPSC)	41	1967
	South Africa (SARCCUS)	8	1950
Asia	FAO Committee (SEAPPC)	18	1956

2. Import from a country with an efficient plant quarantine service, so that inspection and treatment is done.
3. Obtain plant material from the safest known source within the selected country.
4. Obtain untreated seeds so that detection of seed borne pathogens is facilitated.
5. Obtain clean healthy looking seeds free from impurities.
6. Obtain an official certificate of freedom from pests and diseases from the exporting country.
7. Import the smallest possible amount of planting material; the smaller the amount the less the chance of its carrying infection. It will also simplify post entry inspection.
8. Inspect material carefully on arrival and treat.
9. If other precautions are not adequate, subject the material to intermediate or post entry quarantine.

e. Problems in Plant Quarantine

Quarantines serve as a filter against the introduction of hazardous pathogens but still pathogens are introduced. Possible reasons are that;

1. It is difficult to detect all types of infectious pathogens by conventional methods.
2. The methods may not be sensitive enough to reveal traces of infection.
3. The latent infections may pass undetected under post entry quarantine.
4. Destruction of all infected or suspected material, and
5. Lack of sensitive methods for testing fungicide treated seeds.

CHAPTER XIV

PHYSICAL METHODS

. PRINCIPLES

The scientific principle involved in thermotherapy is that the pathogens present n seed material are inactivated or eliminated at temperatures nonlethal for the ost tissues. The exact mechanism by which heat inactivates the pathogen is not ully understood. However, it is universally accepted that heat causes inactivation nd not immobilization of the pathogen (viruses) by heat.

There are two schools of thought regarding inactivation of pathogen (vi-uses) by heat. One school of thought holds the opinion that the heat treatment timulates enzymes that cause the degradation of virus. The other school pursues he idea that heat causes what is known as "loosening" of bonds both in nucleic cid and the protein components of the pathogen. In the nucleic acid, when the onds are disrupted the linear arrangement of nucleotides is disturbed and thus he virus looses infectivity. In proteins, the bonds holding the chains of amino cids together may be broken, or more likely, the architecture of the folding of he chain may be destroyed. Disruption of bonds causes denaturation of protein nolecules, which become less soluble in water, and finally leads to coagulation. It still a matter of conjecture whether breakage in the nucleic acid chain or protein enaturation is responsible for the inactivation of the pathogen in host tissues. he rate at which the pathogen is inactivated is determined by temperature, the igher the temperature, the faster is the inactivation. At constant temperature, he drop in the density of pathogenic inoculum is exponential.

I. METHODS

Different physical methods employed for reduction and/or elimination of rimary inoculum are as follows:

A. Hot-Water Treatment

Hot water is widely used for the control of seed-borne pathogens, especiall bacteria and viruses. A list of some important seed-borne diseases claimed t have been controlled by hot-water treatment is given in Table-1, 2 and 3.

Table-1: Control of Seed-Borne Pathogens through Hot- wate Treatment of Seed

Crop	Disease	Causal organism	Treatment
Brassica spp.	Black rot	*X. campestris pv. campestris*	50°C for 20 or 30 min
Clusterbean, Guar (Cyamopsis tetragonolcba)	Blight	*X. campestris pv. cyamopsidis*	56°C for 10 min
Cucumber (*Cucumis sativus*)	Seedling blight	*Ps. syringae pv. lachrymans*	50°C and 75% RH fc 3 days
Lettuce (*Lactuca sativa*)	Leaf spot	*X. campestris pv. vitians*	70°C for 1 to 4 d
Peanut (*Arachis hypogea*)	Testa nematode	*Aphlenchoides arachidis*	60°C for 5 min afte soaking for 15 min i cool water
Pearl millet (*Pennisetum typhoides*)	Downy mildew	*Sclerospora graminicola*	55°C for 10 min
Potato (*Solanum tuberosum*)	Potato phyllody	*MLO*	50°C for 10 min
Rice (*Oryza sativa*)	Udbatta	*Ephelis oryzae*	54°C for 10 min
	White tip	*Aphlenchoides besseyi*	51-53°C for 15 mi after dipping for 1 d i cool water
Safflower	Leaf spots	*Alternaria spp.*	50°C for 30 min
Teasel (*Dipsacus spp.*)	Stem nematode	*Ditylenchus dipsaci*	1 h at 50°C or 48° for 2 h
Tobacco (*Nicotiana tabacum*)	Hollow stalk	*E. carotovora pv. carotovora*	50°C for 12 min
Tomato (*Lycopersicon esculentum*)	Black speck	*Ps. syringae pv. tomato*	52°C for 1 h

B. Hot-Air Treatment

Hot-air treatment is less injurious to seed and easy to operate but also le effective than hot water. It has been used against several diseases of sugarcan

able-2 : Control of Seed-Borne Pathogens through Hot-water Treatment of Sugarcane Cuttings

Disease	Pathogen	Treatment
)owny mildew	*Perenosclerospora sacchari*	54°C for 1 h. dried at room temperature for 1 d and again treated in 52°C for 1 h
Grassy shoot	*MLO*	54°C for 2 h
.eaf scald	*Xanthomonas albilineans*	Soaking in cold water for 1 d and then treating the cutting at 50°C for 2-3 h
Aosaic	Virus (potato virus Y group)	20 min treatment each day on 3 successive days at 52, 57.3 and 57.3°C, respectively.
Ratoon stunting	*Clavibacter xyli* ssp. *xyli*	50°C for 3 h
Red rot	*Colletotrichum falcatum (Physalospora tucumanensis)*	54°C for 8 h
mut	*Ustilago scitaminea*	55 to 60°C for 10 min
pike	*Virus*	52°C for 1 h
Vhite leaf	*MLO*	54°C for 40 min
Vilt	*Acremonium sp. Fusarium moniliforme*	50°C for 2 h

Complete control of red rot in varieties Co527, CoS 510, B0 3, and B0 32 by ot -air treatment at 54°C for 8 h has been reported. It is used for treating ugarcane stalks on a commercial scale in Louisiana to control ratoon stunting isease(RSD). It is employed for treating canes which are soft and succulent. Hot ir treatment at 54°C for 8 h, effectively eliminates RSD pathogen without im-airing the germination of buds. Similarly, grassy shoot disease of sugarcane has lso been controlled by hot air at 54°C for 8 h.

:. Steam and Aerated Steam

The use of aerated steam is safer than hot water and more effective than hot ir in controlling seed-borne infections.

The heat capacity of water vapour is about half that of water and 2.5 that of ir, hence the temperature and time required may be higher than that of hot ater and lower than that of hot air. The advantages of this method include asier drying of seeds, low loss in germination, easy temperature control and no

Table-3 : Time and Temperature Recommendations for Hot-Wate Treatment for Denematizing Planting Stocks

Nematode	Planting stock	Time (min)	Temp. (°C)
Aphelenchoides ritzemabosi	Chrysanthemum stools	15	47.8
		30	43.0
A. fragariae	Easter lily bulbs	60	44.0
D:tylenchus dipsaci	Narcissus bulbs	240	43.0
D. destructor	Irish bulbs	180	43.0
Meloidogyne spp.	Cherry root stocks	5-10	50-51.0
	Sweet potatoes	65	45.7
	Peach root stocks	5-10	50-51.1
	Tuberose tubers	60	49.0
	Grapes rooted	10	50.0
	Cuttings	30	47.8
	Begonia	30	48.0
	Tubers	60	45.0
	Caladium tubers	30	50.0
	Yam tubers	30	51.0
	Ginger rhizomes	10	55.0
	Strawberry roots	5	52.8
	Rose roots	60	45.5

damage to seed coat of legumes. Besides its use in controlling sugarcane di eases, it has been used against citrus greening.

D. Solar Heat Treatment

Solar heat treatment is effective in controlling both, seed-borne, and sc borne diseases.

1. Seed-Borne Diseases

Solar heat treatment controls effectively the loose smut of wheat (*Ustilag segatum*). In this method the seed is soaked in water for 4 h (8 a.m. to 12 noo on a bright summer day. After this presoak, that seed is dried in the sun for 4 from 12 noon to 4 pm. Solar energy and sun-heated water methods are suitab for controlling the disease in north Bihar (India) conditions. Loose smut of barle (*Ustilago nuda*) has also been effectively controlled by solar heat.

Ascochyta rabiei, the causal organism of Ascochytosis of chickpea survive in seed. Chaube(1984) studied the effect of sun drying on the survival of the pathogen in chickpea seeds. The seeds were exposed to bright sunlight during the last week of May and the first week of June. The seeds were spread on cemented floor from 8 am. to 4 pm. daily for 15 days. Direct exposure of seeds on cemented floor reduced the recovery of the fungus from 31.5 to 16%. In seeds covered with polyethylene sheets on cemented floor the reduction in survival of *A. rabiei* was of higher magnitude. In both the exposure methods, there was no effect on germination of the seed.

2. Soil Solarization

a. Introduction

Solar heating or soil solarization or plastic/ polyethylene tarping or polyethylene or plastic mulching of soil is a soil disinfestation method which aims to reduce or eradicate the inoculum existing in soil. It was to the credit of Israel extension workers and growers who suggested that intensive heating that occurs in mulched soil might be used for the control of soil-borne diseases. Since then, this approach to control soil-borne pathogens and weeds has been widely used in Israel and other countries.

b. The Principles

Solar heating method for disease control is similar, in principle, to that of artificial soil heating by steam or other means, which are usually carried out at 60 to 100°C.

There are, however, biological and technological differences with soil solarization there is no need to transport heat from its source to the field. Solar heating is carried at relatively low temperatures, as compared to artificial heating; thus, its effects on living and nonliving components of soil is likely to be less drastic. Of the four components of disease severity, inoculum density is the most affected component by solarization, either through the direct effect of the heat or by microbial processes induced in the soil. The other components except the host susceptibility which is genetically controlled, might also be affected.

c. Solar Radiation and Soil Temperature

Absorption of solar radiation varies according to the colour, moisture, and texture of the soil. In general, the soil has a relatively high thermal capacity and is a poor heat conductor. This results in a very slow heat penetration. On an average, one square centimeter area outside the earth's atmosphere and parallel to its surface receives 2 cal/cm^2/min (solar constant) of energy in the form of solar radiation, but only about half of it finally reaches the ground. The heat that does penetrate the soil surface is stored in the soil and at night, when the thermal gradient is reversed, it is lost again, resulting in a cyclic reversal in the direction of heat flow.

d. Soil Solarization and Plant Disease Control

Soil solarization as a practice for management of soil-borne pathogens has been demonstrated in several cases. A list of pathogens controlled by solarization is given in Table-4.

Table-4 : Effect of Solarization on Soil-Borne Plant Pathogen

Pathogen	**Diseases and crop**
	Fungal Pathogen
Bipolaris sorokiniana	Crater disease of wheat
Didymella lycopersici	A tomato disease
F. oxysporum f.sp. melonis	Wilt of watermelon
F. oxysporum f.sp. vasinfectum	Wilt of cotton
F. oxysporum f.sp. ciceri	Wilt of chickpea
F. oxysporum f.sp. lentis	Wilt of lentil
Phytophthora cinnamomi	Root rot of several plants
Pyrenochaeta lycopersici	Brown root rot or corky root disease of tomato,
Pyrenochaeta terrestris	Pink root rot of onion
Pythium ultimum	Damping off and root rot of several crops
Rhizoctonia solani	Root rot of several crops
Sclerotinia minor	Root diseases of row crops
Sclerotinia sclerotiorum	
Sclerotium cepivorum	White rot of onion
Sclerotium oryzae	Stem rot of paddy
Sclerotium rolfsii	Root rot of several crops
Thielaviopsis basicola	Root rot of several crops
Verticillium dahliae	Wilt of cotton, tomato, potato and egg plant
	Nematodes
Ditylenchus dipsaci	Garlic bulb nematode
Globodera rostochiensis	Golden nematode of potato
Helicotylenchus digonicus	The spiral nematode
Heterodera trifolii	Clover cyst nematode on carnation
Meloidogyne hapla	Root knot nematode
Meloidogyne javanica	Root knot nematode
Pratylenchus thornei	*Pin nematode of potato and other crops*

e. Beneficial Side Effects

i. Weeds and Other Pests:

Researches conducted on soil solarization have revealed that polyethylene mulching of soil could be an effective method for weed control, lasting in cases for the whole year or even longer. Weed populations of both perennial grasses and broad leaf weeds are highly reduced by polyethylene mulching. This treatment continued to be effective until the end of the season.

In general, most of the annual and many perennial weeds such as *Amaranthus* spp., *Amsinckia douglasiana, Anagallis sp., Avena fatua, Chaenopodium spp., Convolvulus spp., Cynodon dactylon, Digitaria sanguinalis, Echinochloa crus-galli, Eleusine sp., Fumaria sp., Lactuca sp., Lamium amplexicaule, Mercuriales, Molucella, Montia, Notabaris sp., Phalaris spp., Poa sp., Portulaca oleraceae, Sisymbrium sp., Senecio vulgaris, Solanum halpense, Sorghum, Stellaria and Xanthium pensylvanicum* have been controlled. Many graminae are especially sensitive, while others like *Melilotus,* remain unaffected. *Cyperus rotundus* was only partially controlled and Orobanche was effectively controlled in a number of studies.

The possible mechanisms of weed control by solarization are: (1) thermal killing of seeds, (2) thermal kiling of seeds induced to germinate, (3) breaking seed dormancy and consequent killing of the germinating seed, and (4) biological control through weakening or other mechanisms. Volatiles that accumulate under the mulch may play a role in direct killing of weeds or may affect seed dormancy or germination.

The bulb mite *Rhizoglyphus robini,* which causes heavy damage to certain crops, are drastically reduced by soil solarization.

ii. Increase in Yield

The amount of yield increase is a function of disease reduction as well as of level of soil infection, and the damage caused to the crop by the concerned disease.

Some important examples of yield increase in solarized soil are given in Table-5.

iii. Solarization and Integrated Control

Being a nonchemical method, integrating soil solarization with chemical, biological and cultural practices of control, is both possible and promising. Findings of Chet et.al. (1982).and Elad et.al. (1980) who combined the antagonist

Table-5 : Selected Examples of Yield Increase by Soil Solarization

Crop	Pathogen	Increase over control (%)
Carrot	*Orobanche aegyptiaca*	ca. 70 ton/ha as compared to 0 in untreated control
Cotton	*F. oxysporum f.sp. vasinfectum*	40-70
	Verticillium dahliae	60
Egg Plant	*V. dahliae*	215
Onion	*Pyrenochaeta terrestris*	60-125
Peanut	*Sclerotium rolfsii*	42-64
Potato	*V. dahliae*	35
Tomato	*Pyrenochaeta lycopersici*	100-300
Safflower	*V. dahliae*	113

Trichoderma harzianum with heating or solarization in *Rhizoctonia solani* infested soil, to achieve improved disease and pathogen control and delayed inoculum build up, is an invitation to scientists to try to integrate solarization with biocontrol agents for long lasting disease control. Similarly, fungicide vapam and solarization when combined together had synergistic effect in controlling delimited shell spots of groundnut pods. A proven cultural practice, crop sequence when combined with solarization resulted in an improved and longer-lasting control of cotton wilt (*F. oxysporum f.sp. vasinfectum).*

f. Mechanisms of Disease Control

Solarized soils undergo changes in their temperature and moisture regimes, the inorganic and organic composition of their solid, liquid and gaseous phases and their physical structure, all of which in turn affect the biotic components. Thus, reduction in disease incidence occurring in such soils, results from effects exerted on each of the three living components, i.e., the host, pathogen and microorganisms alongwith physical and chemical environment of the soil (Figure 1). Although these processes occur primarily during solarization, they may continue to various extents and in different ways, after the removal of the polyethylene sheets and planting.

Of the four components of disease severity, inoculum density is the one most affected by solarization, either through the direct physical effect of heat or by microbial processes induced in the soil. The other components, however, (except for susceptibility which is genetically controlled), might also be affected.

E. Burning

Controlled burning may alter the environment and affect plant disease re-

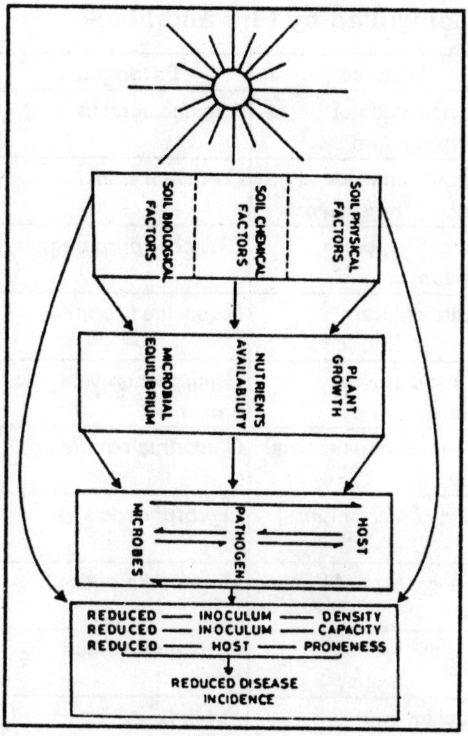

SOIL BIOLOGICAL FACTORS
SOIL CHEMICAL FACTORS
SOIL PHYSICAL FACTORS
MICROBIAL EQUILIBRIUM
NUTRIENTS AVAILABILITY
PLANT GROWTH
MICROBES
PATHOGEN
HOST
REDUCED — INOCULUM — DENSITY
REDUCED — INOCULUM — CAPACITY
REDUCED — HOST — PRONENESS
REDUCED DISEASE INCIDENCE

Figure-1: Schematic diagram of the mechanisms of disease management through soil solarization

sponse, providing both a temperature effect and a means of destroying the pathogen. Burning is a single most important practice in grass seed production in the Pacific Northwest. It was initiated there to control the blind seed disease of perennial rye grass caused by *Gloeotinia temulenta*. It also effectively controlled *Claviceps purpurea* (ergot of rye), *Anguina agrostis* (seed nematode) and silver top. Most of the crops to which fire can be applied are cereals in which inoculum can be destroyed after harvest, or pasture grasses which can periodically be freed from inoculum before they make new growth. In addition, where residues are dry and inflammable enough to be burn without fuel, this can be a cheap and fairly cost effective. It has also been realize that the increasing success of nontillage in some crops, and the resultant problems of debris management, make burning an attractive and effective proposition for reduction of inoculum, provided the cost involved is reasonably low.

Diseases that have been successfully managed by burning or flaming crop residue are given in Table-6.

Table-6 : Diseases Controlled by Fire and Flame

Pathogen	Disease	Pathogen	Disease
Anguina sp.	Seed nematode of *Lolium rigidum*	*Anguina agrostis*	Seed nematode of *Festuca rubra*
Anguillulina tumefaciens	Leaf gall nematode of *Cynodon transvalensis*	*Corticium sasaki*	Sheath blight of rice
Claviceps paspali	Ergot of *Paspalum dilatatum*	*Claviceps purpurea*	Ergot of *Lolium perenne*
Cuscuta spp.	Parasite of lucern	*Diaporthe vaccinii*	Die back of low bush blueberry
Drechslera poae	Leaf mold of *Poa pratensis*	*Gaeumannomyces graminis*	Take all of wheat
Gerlachia nivalis	Snow mold of wheat and barley	*Glueotinia temulenta*	Blind seed of *Lolium perenne*
Godronia cassandrae	Canker of *Vaccinium* spp.	*Leptosphaeria* spp.	Leaf blight of sugarcane
Phleospora indahoensis	Stress eye spot of Festuca rubra	*Pleiochaeta setosa*	Brown spot of Lupin
Pseudocercosporella herpotrichoides	Eye spot of wheat	*Puccinia menthae*	Rust of peppermint
Puccinia asparagi	Asparagus rust	*Puccinia poaenemoralis*	Leaf rust of *Poa pratensis*
Puccinia graminis	Stem rust of *Poa pratensis*	*Puccinia striiformis*	Stripe rust of *Poa pratensis*
Rhynchosporium secalis	Eye blotch of barley	*Sclerotium oryzae*	Stem rot of rice
Septoria avenae	Leaf blotch of oat	*Septoria nodorum, Septoria tritici*	Leaf blotch of wheat
Selenophoma bromigera	Leaf spot of *Bromus inertis*	*Urocystis agropyri*	Flag smut of wheat
Verticillium dahliae	Wilt of potato and pepermint		

F. Flooding

Flooding fields and orchards to reduce or eliminate soil-borne inoculum of plant pathogens is an ancient practice. Flooding has been recognized to be one of the key factors for the low incidence of soil-borne diseases in present day chinese agriculture.

Several explanations of the harmful effects of prolonged flooding on soil-borne pathogens have been suggested. Lack of oxygen may be involved in some cases or, more often perhaps, accumulation of CO_2 in the soil. The survival of *F. oxysporum f. sp. cubense* in soil after 2 weeks depends on formation of chlamydospores, since the conidia are not apparently long lived in soil CO_2 and flooded soil both largely inhibited chlamydospore formation, whereas they at first stimulated the production of conidia. Consequently the fungus, although able to survive in banana plantation soil containing organic matter, is likely to die out in a fallow, flooded field where organic matter is in short supply. The main factor in the elimination of the fungus by flooding is a high CO_2 content in the flooded soil combined with a decreased availability of colonizable substrate. In flooded soil CO_2 stimulates germination of conidia, presumably by overcoming the fungistatic factor present in soil, but prevents the formation of chlamydospores so that the fungus dies out when the organic matter is exhausted. A similar situation perhaps holds for other soil borne fungi, but few cases have been investigated.

A list of plant diseases that have been successfully controlled by flooding fields and orchards is given in Table-7.

Table-7 : Plant Diseases Controlled by Flooding

Pathogen	**Disease**	**Pathogen**	**Disease**
Alternaria porri f. spp. solani	Alternaria blight of tomato and potato	*Alternaria dauci*	Blight of carrot
Aphelenchoides besseyi	White tip of rice	*F. oxysporum f.sp. cubense*	Wilt of banana
Meloidogyne spp.	Root knot of celery	*Orobanche spp.*	Phanerogamic plant parasite of several crops
Phytophthora parasitica var. nicotianae	Black shank of tobacco	*Pyrenophora teres*	Canker and blight of barley
Radopholus spp.	Burrowing nematode of banana	*Sclerotinia sclerotiorum*	White mold of vegetables
Trichodorus Similis	Stubby root nematode of celery	*Tylenchorhynchus* spp.	Stunt nematode of celery
Verticillium dahliae	Wilt of cotton		

CHAPTER XV

BIOLOGICAL CONTROL

I. INTRODUCTION

"Biological control is the reduction of inoculum density or disease producing activities of a pathogen or parasite in its active or dormant state, by one or more organisms accomplished naturally or through manipulation of the environment, host or antagonist, or by mass intorduction of one or more antagonists" (Baker and Cook 1974).

II. THEORIES AND MECHANISMS

Management of associated microbiota is a major form of biological control Cook (1977) has discussed five elements in the theoritical base.

A. Reduction of Inoculum Density

Inoculum density can be reduced by destroying propagules or by preventing their formation. Crop rotation adds chemically different plant residues to soil and, therefore, helps in complexities of soil microbiota. It starves the pathogen due to absence of host and weakens to the extent that it is more rapidly destroyed by the microflora. Dormant sclerotia are killed off by *Trichoderma spp.*, *Fusarium roseum*, *Coniothyrium minitans*, and other fungi and bacteria. Microbial activities helping in decay of organic residues and release of several chemicals, stimulate germination of resting structures like sclerotia and reduce their resistance so that they are easily colonized by antagonists and destroyed. The germtube that comes out is lysed without forming fresh structrues. Other treatments that predispose resting structures to microbial decay are wetting and drying, flooding, and sublethal fumigation. Destruction of resting structures occurs in soil by direct activities of bacteria and actinomycetes. There is evidence that perforations appear in the walls of resting propagules due to activities of some microbes. These microbes subsequently enter through these perforations and destroy them.

B. Displacing the Pathogen from host Residues

This applies to those pathogens that depend for survival on occupancy of the host remains during host free period. The pathogens use the residues both as shelter and as a food base. This gives them the advantage of pioneer colonization. The system of residue possession by root pathogens is either PASSIVE, or ACTIVE or both. *Pythium spp.* that attack succulent roots exemplify passive possession. They invade thoroughly., digest extensively, store the surplus food in their resting bodies (oospores) and then abandon the fragile exhausted host remains to other saprophytes (Bruchl, 1975). In active possession the organisms invade the residue, usually as parasite while the tissues are still alive and active, become established in some tissues in which they persist and are metabolically active within the dead host remains. Utilization of the substrate is slow and the pathogen persistently defends the substrate against saprophytes. Normally the active possessor does not retreat into a dormant structure. *Cephalosporium gramineum* is a typical example. *G. graminis* and *F. graminearum* have some characterstics of active possessors. It is for these pathogens that efforts are to be made for their displacement or for nullifying their pioneer colonization. *F. oxysporum, F. solani*, and *F. culmorum* come under the category of combination possessors, that is, they are both active as well as passive possessers. They also need displacement for control through cultural practices.

The hold of *C. gramineum* on wheat straw is weakened in alkaline soil or by reducing moisture in the soil. These conditions permit entry of *Penicillium spp.* Nitrogen deficiency reduces the hold of the take-all fungus which is displaced by other saprophytes. *Armillaria mellea* on citrus wood is weakened by application of CS_2 and looses its capacity to produce antibiotics. This enables its antagonist *T. viride* to displace it from the wood.

C. Suppression of Germination and Growth of the Pathogen

This form of biological control has two aspects : 1. reduction or prevention of germination (soil fungistasis) and 2. Slowing down of growth of germlings due to starvation, antibiotics, bacteriocins, mycoviruses, etc.

D. Protection of Infection court

This approach aims at encouraging the soil microbiota in or on the infection court which slow or prevent infection by the particular pathogen. Such protection mainly includes conditions where a weak pathogen or non pathogenic organisms take possession of the sites of infection on the host. The mechanisms by which these precolonizers may protect the infection court include 1. prior use of essential nutrients or oxygen needed by the pathogen, 2. modification of rhizosphere pH, redox potential, and other abiotic factors that place the pathogen at a competitive disadvantage, 3. production of antibiotics, 4. hyperparasitism or exploitation of the pathogen, and 5. modification of the host resistance.

E. Stimulation of Host's resistance Response

This includes cross protection provided by a weak strain of the same pathogen or by another pathogen. A tomato variety resistant to *F. oxysporum f. sp. lycopersici*, if inoculated with that pathogen, becomes resistant to *Verticillium dahliae*. Mint is resistant to *V. dahliae* if inoculated first with *V. nigrescens*.

III. MECHANISMS AND PROCESS OF DECLINE

A. Antagonism — Antagonism is one main sub-division of microbial associations in soil. It implies that in any association of two or more species, at least one of the interacting species is harmed due to the activities of one or more of the rest. The mechanisms of antagonism operate through:

B. Antibiosis — It is defined as the condition in which one or more metabolites excreted by an organism have harmful effects on one or more other organisms. In such antagonistic relationship species A produces a chemical substance that is harmful or inimical to species B without species A deriving any direct benefit. However, the species A may have an indirect benefit in having a better competitive ability, thereby getting an advantage over species B for substrate colonization.

C. Competition — It is the indirect rivalry between two species for some feature of the environment that is in short supply. Broadly speaking competition could involve all kinds of interplay between organisms in which one is favoured at the expense of the other. But in strict sense, if we keep antibiosis or even exploitation restricted to their specific mode of action and result, competition has been defined as " a more or less active demand in excess of the immediate supply of material or condition on the part of two or more organisms.

D. Exploitation — When species A inflicts harm by the direct use of species B for its own benefit, it is exploitation (parasitism and predation). The two terms *parasitism* and *predation* have basically same effect, i.e., destruction of the host or prey. From a plant pathologist's view point, predation is a form of parasitism. However, the mode of operation makes the two terms some what distinct. In parasitism, some sort of etiological relationship between the parasite and the host is established and the host is not rapidly eliminated. A predator physically eliminates its prey by direct feeding on it without establishing any etiological relationship.

Mycoparasitism, hyperparasitism, direct parasitism, or interfungus parasitism are terms interchangeably used to refer a phenomenon in which one fungus

Table 1 : Some Example of Mycoparasitism

Mycoparasite	Host	Type of Parasitism
Cephalosporium sp.	*P. megasperma var. sojae*	Produce chlamydospores within oospores and hyphae extend out from oospores
Coniothyrium minitans	*S. sclerotiorum*	Invasion of sclerotia
Corticium sp.	*R . solani*	Hyphal invasion
Dactylella spermatophaga	*P. megasperma var sojae*	Parasitism of oospores
Didymella exitialis	*G. graminis*	celi wall penetration, grows into thallus, break down of cell wall by chitinase
Fusarium oxysporum	*R. Solani*	Coiling and penetration
Fusarium roseum	*C. purpurea*	Parasitism of sclerotia
Humicola fuscoatra	*P. megasperma var sojae*	Destruction of oospores
Penicillium Vermiculatum	*R . solani*	Hyphal penetration
Pythium sp	*P. megasperma var. sojae*	oospore destruction
R . solani	*R . stolonifer*	coiling and penetration of hyphae

is parasitic on another through a nutritional relationship established during the life of the host. Almost all taxonomic groups of fungi are involved in this phenomenon and often species within the same genus (e.g. *Pythium*) interact as host and parasite.

IV. MYCOPARASITISM

Mycoparasitism is an act where one fungus parasitizes the other one. Barnett and Binder (1973) classified mycoparasitism into two main groups, i.e. necrotrophic and biotrophic, on the basis of nutritional relationship of parasite with the host. The necrotrophic (destructive) parasite makes contact with its host, excretes toxic substances which kills the host cells and utilizes the nutrients that are released. The mycoparasitism is of common occurrence and examples can be found among all groups of fungi from chytrids to the higher basidiomycetes. Some examples of mycoparasitism are given in Table 1. The mycoparasitism includes different kinds of interactions, viz. coiling of hyphae, penetration, production of haustoria and lysis of hyphae (plate I). Hyphae of majority of *Trichoderma* coil around hyphae of different host fungi. Metabolites produced from *T. harzianum* inhibit growth and sporulation of host fungi. Table 2 lists the mechanism of parasitism of *Trichoderma spp.* on different plant pathogenic fungi.

V Biological control by Introduced Antagonists

From numerous experiments that have been conducted all over the world, a common fact emerges: reduction in the activity of a pathogen is often correlated

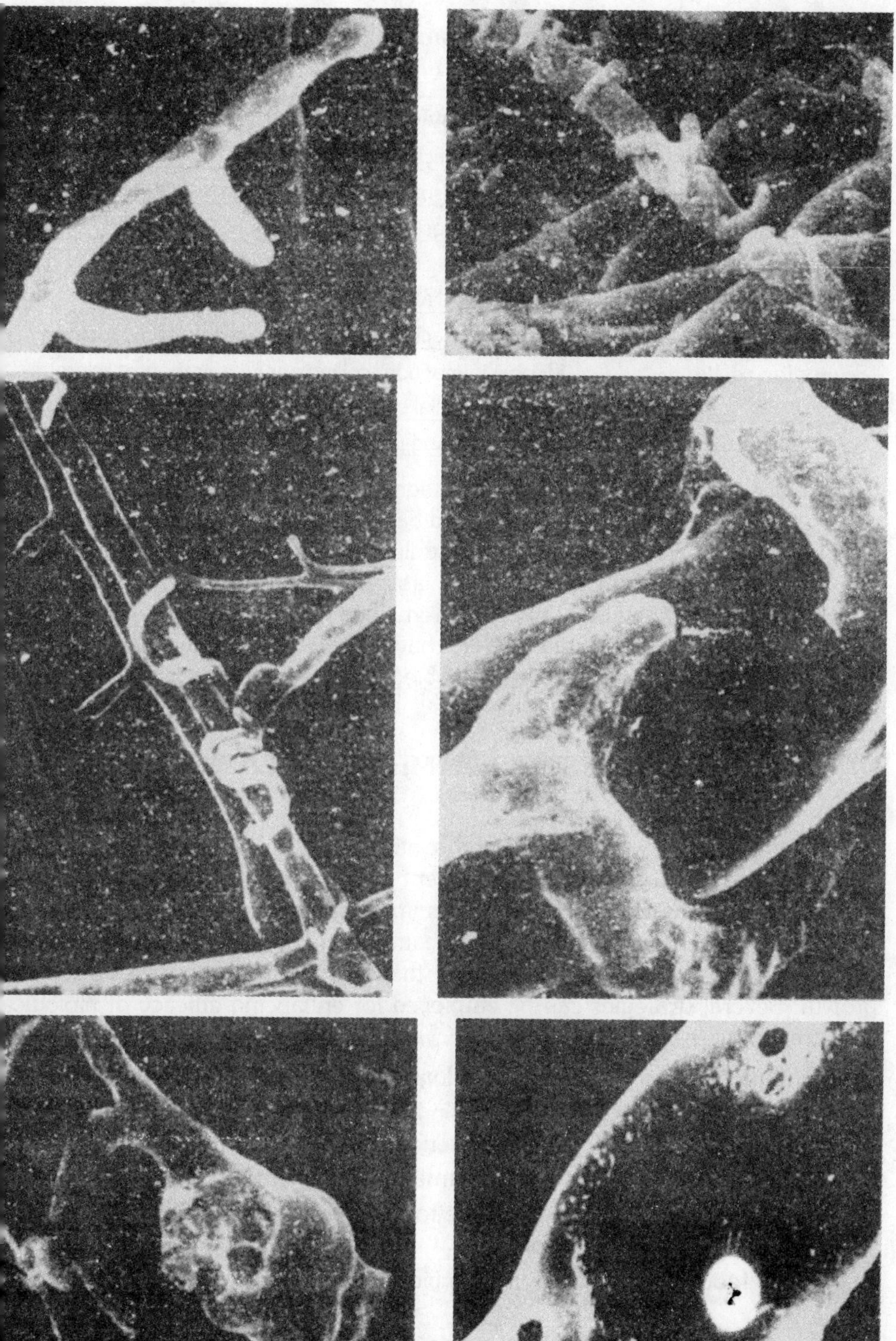

e-I: Scanning electron micrograph showing interaction of *Trichoderma harzianum* with *R. solani* and *S. rolfsii*.

Table 2 : Mechanism of Parasitism of Trichoderma spp. on some parasitic Fungi

Plant Pathogen	Mode of antagonism
Botrytis cinerea	Hyphal interaction
Fusarium spp.	Mycoparasitism
F. oxysporum	Hyphal interaction
Pythium aphanidermatum	Hyphal interaction, coiling, penetration, lysis of hyphal cells antibiosis
Rhizoctonia solani	Hyphal interaction, penetration, cell wall lytic enzyme
Sclerotium rolfsii	Hyphal interaction. coiling and penetration of sclerotia, lysis of mycelial cells, etc.
Sclerotinia sclerotiorum	Plasmolysis and penetration of sclerotia

with an increase in the populations of antagonists as assessed in the soil. The use of selected antagonists is a direct method based on the theory that these antagonists when introduced in the soil, can act directly on the behaviour of the pathogen. However, with few exceptions, the biological control of plant pathogens by augmenting soils with antagonists has remained restricted to research studies Nevertheless continued scientific efforts have resulted in identification of several antagonists which have given successful biocontrol in the experimental plots and field when augmented in soil (Table 3).

VI BIOLOGICAL SEED TREATMENT - A FEASIBLE DELIVERY SYSTEM

A more economical, effective and relatively non polluting delivery system of the antagonists is biological seed treatment because only small amount of material is applied per unit area and comes in immediate contact with the target site There are two general classes of seed treatment: those that alleviate stress associated with soil environment, and those that directly improve or increase plant growth. Several strategies can be employed for enhancing efficacy of biological seed treatment, as:

1. Permitting the bioprotectant to colonize the seed surface before planting,
2. Applying the bioprotectant in a thin layer that permits it to proliferate for a few hours after planting in the absence of competitive microflora, and
3. Inclusion of adjuvants in seed treatment, which helps to,

a) control pH to a level favourable to the biocontrol agent but unfavourable to competitive microflora,

b) add food base selectively to the bioprotectant,

Table 3 : Examples of Biological control by Introduced Antagonists

HOST DISEASE	PATHOGEN	ANTAGONISTS
Carnation disease	*F. oxysporum f. sp. dianthi, F. roseum 'culmorum', R. solani*	*Bacillus subtilis, Pseudomonas fluorescens, Thrichoderma spp.*
Chickpea root rot/wilt complex	*F. oxysporum f. sp. ciceri, R. solani, S. rolfsii*	*Trichoderma harzianum, G. virens*
Cotton root rot	*R. Solani*	*T. harzianum*
cotton wilt	*F. oxysporum f. sp. vasinfectum*	*T. harzianum* *G. virens* *Ps. fluorescens*
Crown gall of woody trees	*Agrobacterium tumifaciens*	*Agrobacterium radiobacter*
Damping off of crops/vegetables	*Pythium spp.* *R. solani* *S. rolfsii* *Fusaruim spp.*	*B. subtilis* *Ps. fluorescens* *Trichoderma spp.* *G. virens*
lentil root rot wilt complex	*F. oxysporum f. sp. lentis* *S. rolfsii, R. solani*	*Trichoderma spp.* *G. virens* *Ps. fluorescens*
Potato scab	*Streptomyces scabies*	*B. subtilis*
Take all of cereals	*G. graminis var tritici*	*Ps. fluorescens*

c) use selective toxicants which do not restrict the growth of antagonist, and

d) after the timing i.e. the period when antagonist is active in relation to the pathogen. Table-4 enlists some examples of biological seed treatment with *Gliocladium* and *Trichoderma spp.*

VII ORGANIC AMENDMENTS - MODIFICATION OF SOIL ENVIROMENT - BIOLOGICAL CONTROL

Modification of soil environment is one of the methods to achieve biological control. Amendment of soil with decomposable organic mattter is recognized as an effective method of changing soil and rhizosphere environment. Such changes adversely affect the pathogens and empower plants to resist infection through better vigour and/or altered root physiology. Such methods bring multiple pathogen suppression, lasting effects, less cost, no hazards, and improvement of soil

Table 4 : Biological seed Treatment

Host	Pathogen	Antagonist
Bean	*Pythium spp.* *S. Sclerotiorum* *B. cinerea* *R. solani, Fusarium spp*	*Trichoderma spp.* *T. Koningii* *G. catenulatum*
Chickpea	*S. rolfsii, R. solani* *F. oxysporum,* *M. phaseolina*	*T. harzianum* *T. viride* *G. roseum* *G. virens*
Cauliflower	*P. aphanidermatum*	*T. harzianum*
Cucumber	*Pythium ultimum* *S. sclerotiorum*	*Trichoderma spp.*
Cotton	*R. solani, Fusarium spp.*	*T. hamatum* *T. harzianum* *G. virens*
Ginger	*F. oxysporum*	-do-
Maize	*R. solani*	*T. harzianum*
Mungbean	*Fusarium spp.*	*G. virens*
Pea	*R. solani, P. ultimum*	*T. harzianum* *T. hamatum*
Radish	*Pythium spp., R. solani*	*Trichoderma spp.*
Soybean	*R. solani* *M. phaseolina* *Fusarium spp.*	*T. pseudokoninghii* *T. harzianum* *G. virens*
Sugarbeet	*S. rolfsii, R. solani*	*T. harzianum*
Sunflower	*Sclerotinia spp.* *R. solani* *S. rolfsii* *Fusarium spp.*	*Trichoderma spp.* *G. virens*

fertility and nutrients uptake by plants. Crop residue decomposition encourages microbial activity both quantitatively and qualitatively. The enhanced microbial activity has two effects; it increases the variety of complex organic compounds in soil and promotes the population of antagonists in soil. Together, these two factors increase the biological buffering capacity of soil. This capacity of the soil helps in disease prevention particularly in areas where the pathogen is not yet established.

Amendments act against diseases through 1. direct effect on the active pathogen on the root or in the rhizoshere, 2. direct effect on the pathogen during its survival in the absence of the host, and 3. indirect effect on the pathogenesis through the host . In the latter are included all those cosequences of amendment

PARASITISM		ANTAGONISM					HOST NUTRITION	CHANGES IN PHYSICAL ENVIRON-MENT
		ANTIBIOSIS			PATHOGEN NUTRITION			
		PLANT FUNGI-STASIS	MICROBIAL FUNGISTASIS		INDIRECT STARVATION	DIRECT STARVATION		
DIRECT	INDIRECT LYSIS AND DECOMPOSITION	TOXIC COMPO-NENTS OF ORIGIN	SPECIFIC FUNGI-STASIS (ANTI-BIOTIC)	GENERAL NON-SPECIFIC FUNGI-STASIS	NUTRIENTS ADEQUATE BUT POOR COMPETI-TIVE OR ANTIBIOSIS PREVENT UTILIZATION	LOW LEVEL OR ABSENCE OF ESSENTIAL NUTRIENTS CAUSES REDUCTION OR CESSATION OF ACTIVITY	RESISTANCE OR TOLERANCE OF HOST INCREASED BY IMPROVED NUTRITION THROUGH (A) MORE VIGOROUS HEALTHIER ROOTS (B) NUTRIENTS IN AMEND-MENTS	pH CO_2

Figure-1: Mechanism of disease control through organic amendments (after R.H. Stover, 1962).

which result from uptake of organic compounds by the plant roots, changed host physiology and possible dvelopment of resistance.. Therefore, the amendments can reduce inoculum density (ID), inoculum capacity (IC), host proneness and can also increase host resistance; the net result being reduction in disease severity. The list of some soil borne diseases managed by organic amendments of soil are

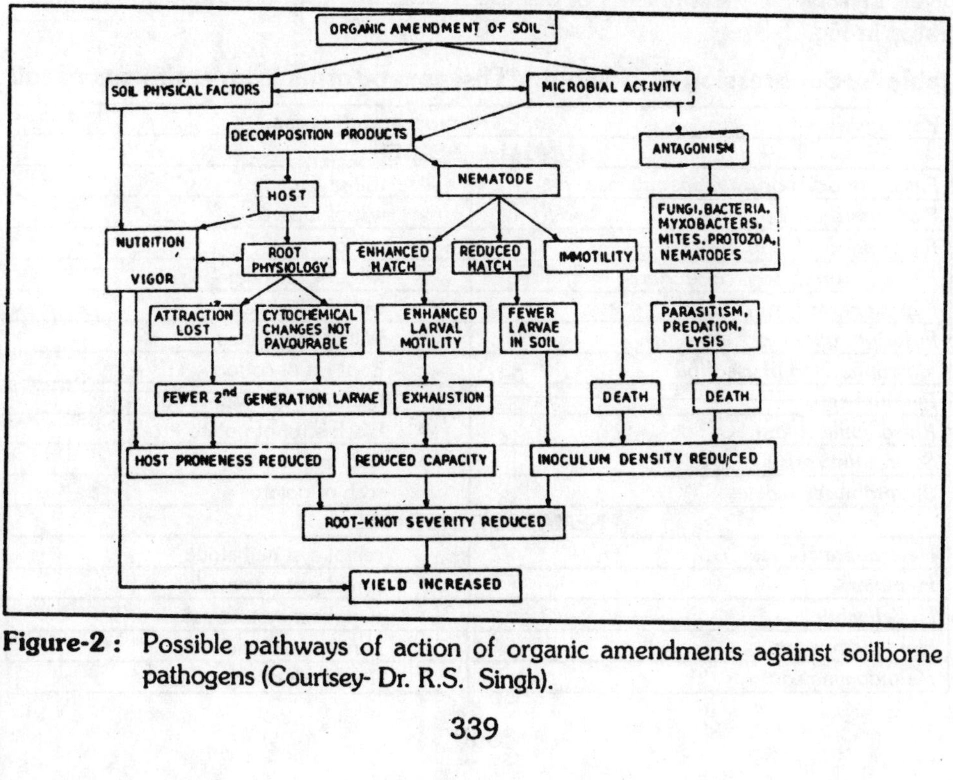
ORGANIC AMENDMENT OF SOIL
SOIL PHYSICAL FACTORS
MICROBIAL ACTIVITY
DECOMPOSITION PRODUCTS
ANTAGONISM
NEMATODE
HOST
FUNGI, BACTERIA, MYXOBACTERS, MITES, PROTOZOA, NEMATODES
NUTRITION VIGOR
ROOT PHYSIOLOGY
ENHANCED HATCH
REDUCED HATCH
IMMOTILITY
ATTRACTION LOST
CYTOCHEMICAL CHANGES NOT FAVOURABLE
ENHANCED LARVAL MOTILITY
FEWER LARVAE IN SOIL
PARASITISM, PREDATION, LYSIS
FEWER 2nd GENERATION LARVAE
EXHAUSTION
DEATH
DEATH
HOST PRONENESS REDUCED
REDUCED CAPACITY
INOCULUM DENSITY REDUCED
ROOT-KNOT SEVERITY REDUCED
YIELD INCREASED

Figure-2 : Possible pathways of action of organic amendments against soilborne pathogens (Courtsey- Dr. R.S. Singh).

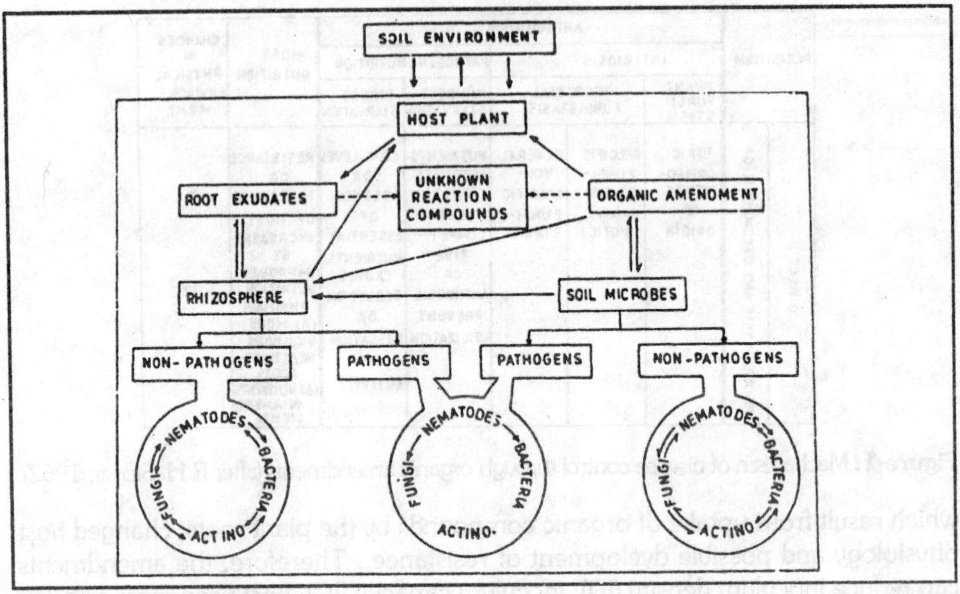
SOIL ENVIRONMENT
HOST PLANT
UNKNOWN REACTION COMPOUNDS
ROOT EXUDATES
ORGANIC AMENDMENT
RHIZOSPHERE
SOIL MICROBES
NON-PATHOGENS
PATHOGENS
PATHOGENS
NON-PATHOGENS
NEMATODES
BACTERIA
ACTINO
FUNGI

Figure-3 : Interaction of amendments, rhizosphere and soil microflora (Courtsey- Dr. R.S. Singh).

given in Table 5. Mechanisms of disease management in amended soil are illustrated in Fig. 1-3.

Table-5 : Suppression of Soil borne Diseases by organic amendments of soil

Pathogen	**crop and desease**
FUNGAL DISEASES	
Fusarium oxysporum f. sp. cubense	wilt of banana
F oxysporum f. spp. *lini*	wilt of Linseed
F oxysporum f. spp. *pisi*	wilt of Pea
F oxysporum f. spp. *coriander*i	wilt of Coriander
F oxysporum f. spp. *ciceri*	wilt of Chickpea
Fusarium udum	Wilt of pigeonpea
Macrophomina phaseolina	Root rot of cotton
Pythium spp.	Damping off, root rot of crops
Rhizoctonia solani	Black scurf of potato
Sclerotium rolfsii	Root rot of crops
Streptomyces scabies	scab of potato
NEMATODE DISEASES	
Heterodera avenae	cereal cyst nematode
H. major	cereal cyst nematode
H. Schachtii	sugarbeet cyst nematode
Hoploliamus spp.	Lance nematode
Meloidogyne spp.	Root knot of vegetables

CHAPTER XVI

CULTURAL PRACTICES

I. INTRODUCTION

Cultural practices which include manipulation and/or adjustment of crop production techniques have been as old as possibly agriculture itself. In early stages of agriculture development, the growers through their experiences and observations had known that repeated cultivation of a particular crop species or variety on a piece of land oftenly resulted in crop sickness. By proper crop rotations they had been avoiding such sickness. As a matter of fact, in the present day agriculture, cultural practices are being considered as essential backup methods for management of plant diseases. Adequate adjustment in crop production techniques can modify the environment in such a manner that it becomes unfavorable for the pathogen and pathogenesis. Based on this the disease control affected by cultural practices are preventive. These practices aim at reducing the activity and density of inoculum.

II. METHODS

Procedures for disease control through cultural practices are discussed under the following three heads:

1. Production and use of pathogen-free planting material
2. Adjustment of crop culture to minimize disease
3. Sanitation

A. Production and Use of Pathogen Free Planting Material

Many plant pathogens are transmitted by establishing themselves on or in the seed or other vegetative propagating materials or as contaminants. For successful disease control this source of primary inoculum must be destroyed. The following methods are followed to produce and use pathogen-free seed material.

1. Seed Production Areas

Seed should be produced in areas where the pathogens of major concern are unable to establish or maintain themselves at critical levels during periods of seed development. Areas with low rainfall and low relative humidity are favorable for production of high quality seeds. Some examples are anthracnose of beans and cucurbits, ascochytosis of pea and chickpea, bacterial blight of legumes, etc. Such crops can be grown in dry areas with the help of irrigation.

2. Inspection of Seed Production Plots

Periodical inspection of crops raised for seed production is an important procedure in the production of clean and healthy seeds. Destruction of diseased plants/organs at the time of inspection helps in reducing inoculum in the field and thus, the percentage of healthy seeds in the produce is increased. If disease incidence is very high, the entire crop may be rejected for seed.

3. Drying and Ageing of Seed

Germinability of seeds, not properly dried before storage, is reduced. Such seeds harbour several types of pathogens. Prolonged storage of seeds also reduces many pathogens. *Fusarium solani* f. sp. *cucurbitae*, infecting cucurbits, is eliminated if the seeds are stored for 2 years before sowing. Similar eradication of pathogen has been achieved in anthracnose of cotton.

4. Cleaning of Seed

In many cases the pathogen is present in plant parts mixed with the seed. When such seeds are used pathogen easily gets into the field. Thus, proper cleaning of seeds before sowing is essential. Common examples of diseases disseminated in this manner are ergot and smut of pearl millet, ear cockle of wheat, white rust of crucifers, ascochyta blight of chickpea, etc.

5. Harvesting Time

Appropriate timing of harvest from the point of view of crop disease, means essentially the attempt to escape disease; the intention is to harvest either before environmental conditions become very favourable to the pathogen, or before the crop becomes highly susceptible to age related pathogens, or both. A number of crops are helped by early harvest to escape pathogens. For example, potato (*Globodera rostochiensis,* potato virus X, Y, and leaf roll) in U.K., France and Germany; tomato (*Erwinia amyl vora*) in US; groundnut (*Pythium, Fusarium, Diplodia, Rhizoctonia*) in Libya and Israel; lucern (*Leveilula taurica, Uromyces* sp.) in Israel; red clover (*Rhizoctonia* sp., *Phoma trifolii*) in U.S.; sugarcane (*Glomerella tucumanensis*) in Australia, etc.

Disease incidence can also be minimized by reducing spread of inoculum through adjustment of harvesting time and practice. Potato harvested when tops are still green may easily get contaminated by *P. infestans* present on the leaves. One of the practices to avoid tuber infection or contamination is to first remove the tops and let them dry before digging the tubers.

B. Adjustment of Crop Culture to Minimize Disease

1. Crop Sequence and Crop Rotation

Different degrees of disease control through biological buffering of the soil against plant pathogens can be achieved through cultivation of crops in rotation, monoculture or through growing a mixture of two or more crops of different genetic composition simultaneously in the same field. Crop sequence implies cultivation of crops in monoculture or in sequence of different crops without any fixed cycle. On the other hand, crop rotation implies repeated growing of the same crop in at cycle at regular intervals.

The crops grown on a field determine its microbial makeup. Each crop species selects for a specific saprophytic or parasitic microbiota on the basis of nutrients it supplies. Monoculture of a crop on a field will favour the prepetuation of a specific and eventually a stable flora and fauna. This may ultimately give rise to supressive soil that supports heavy incidence of a disease in the first year of cultivation but when the same crop is grown season after season without interference by any other crop, it eliminates the pathogen through development of an antagonistic flora which becomes stable in the soil unless disturbed by variation in cropping system. However, when a different crop is grown each season, the soil microflora remain in a state of unstability and result in different dominant organisms each season. Thus, it becomes a rotation of not only the crops but biota also.

(a) Crop Rotation

Since ancient time farmers have known that soil becomes sick and unsuitable for cultivation of a crop when the same crop is grown continuously on the same land. This can be due to (1) exhaustion of a particular essential nutrient due to excessive use by same type of crop, (2) accumulation of organic acids and other toxic substances released by the crop, and (3) easy survival of the pathogen due to regular presence of the susceptible host.

To overcome these problems, crop rotation is necessary in crop management. It aims at (1) better use of nutrients, (2) desirable effects on soil texture, with deep rooted crops alternating with shallow rooted crops, (3) water economy, in particular conservation of water in years of fallow, (4) weed control, and (5) suppression of soil-borne plant pathogens.

i. Basis for Disease Control

Basis for control by crop rotation could be (1) release of toxic substances by the crop effective against the pathogen, (2) starvation of the pathogen and its consequences, (3) variation in microbial activity that increase chances of antagonism, and (4) change in the physico-chemical environment of the soil. Individually or in combinations these factors primarily reduce inoculum density of a pathogen. Nutritional factor (starvation of the pathogen and nutrition of saprophytic flora) remains, however directly or indirectly, the most important factor. Major root pathogens, like wilt causing fungi, are wholly or partly active possessors of host residues. After harvest of the crop the fungus remains active as saprophyte with reduced rate of metabolism. To displace such pathogens from the residue, weakening and starvation is necessary so that either they are lysed or are overcome by the saprophytic flora waiting outside.

Many plant species release toxic compounds in their root exudates. HCN is an example. These toxic compounds directly suppress the pathogen, reduce biological competition and permit activity of some microflora tolerant to them, to colonize the area and suppress the pathogen. The root exudates of non-host crops in the rotation, being non-specific, can stimulate germination or release dormant propagules from fungistasis at a wrong time i.e. in the absence of the host. Starvation and lysis may follow or the germlings may be destroyed by antagonists. Use of resistant cultivars of the same crop in the rotation may do the job efficiently if the pathogen is responsive to specific root exudates.

ii. Types of Pathogen Affected, Period of Rotation, and Selection of crop

Crop rotation as a practice for disease control is most effective in case of specialized root pathogens or soil invaders having low competitive capacity, having narrow host range, and requiring continued colonization of dead root tissues. It is not effective against diseases caused by weak, unspecialized parasites or soil inhabitants and the pathogens which have a large host range among crop and weed plants.

The length of rotation for disease control will depend on how quick the pathogen can be displaced from colonized tissues and exposed to antagonistic effects. This is determined by biotic and abiotic characteristics of the soil, treatments given to the soil, crops included in the rotation, and how far chances of recontamination can be avoided. Maintenance of high organic matter content of the soil, tillage to expose the residues to weathering, inclusion of the crops which do not permit high stimulation and subsequent growth of the pathogen, crops that are likely to produce toxins, are some of the criteria to be considered at the time of preparing crop rotation. Normally, legumes should be followed by cereals and vice-versa. The pigeon pea wilt fungus, *F. udum* is specific on pigeon pea. However, many other legumes such as cowpea, soybean, etc. stimulate its germination and growth in soil although they are nonhosts. If pigeon pea is avoided and replaced with one of these legumes, the chalamydospores will germinate abundantly. If the soil is rich in organic matter and microbial activity is intense, the germlings may be destroyed by antagonism before forming secondary chlamydospores. But in a soil with low biological activity, the fungus may continue to grow and subsequently produce more chlamydospores. Cereals like sorghum, maize, etc. also stimulate germination of chlamydospores but not to the same extent as legumes. Furthermore, the root exudates have some toxic materials (HCN) that may kill the germlings before they can form spores.

Larvae of *Meloidogyne*, an obligate root parasite, are hatched with limited food reserves and can not infect the host when this is exhausted in their search for the host. The nematode population in a field which contains no host plants will become noninfective and die sooner or later of starvation. However, the root knot nematodes have a large host range. To overcome this limitation, it has been proposed that noneconomic catch or trap crops may be included in the rotation, or only resistant or immune commercial crops should be grown for some time. Tomato is highly susceptible to all species and races of *Meloidogyne*. Groundnut is immune to all known races of *M. incognita, M. javanica* and to race 2 of *M. arenaria*. It is not immune to *M. hapla*. Cotton is highly resistant to all of the common root knot nematodes except races 3 and 4 of *M. incognita*. Antagonistic plants like *Tagetes* spp. *Chrysanthemum* spp. and *Ricinus communis* can be used in rotation in heavily infested soils. Larvae enter these plants but fail to reproduce. The plants are known to contain antinemic toxins. An extensive list of diseases in which rotation can play a significant role in disease management is given in Table-1.

(b) Monoculture

Monoculture means cultivation of single or closely allied species in annual or seasonal succession, with interruption only by fallow or intermittent growing of green mannure, or by application of soil amendments, not necessarily after each crop. From the general biological aspect, monoculture is clearly dangerous because it does not permit diversity. Serious crop losses from diseases are much less likely to occur in crop rotations than in monoculture. Monoculture may also exert selection pressure on pathogens, resulting in the emergence of new

Table-1 : Crop Diseases in which Crop Rotation Has Had a Major Effect in Disease Control

Crop	Disease	Pathogen	Crop interval (year)	Remarks
Onion	Bloat	*Ditylenchus dipsaci*	2	Rotate with beets, carrots, crucifers, lettuce, spinach
	Nematode	*Belonolaimus gracilis*	5-10	Use watermelon or tobacco and eradicate weeds
Pea	White rot	*Sclerotium cepivorum*	8-10	Rotate with nonhost crop
	Anthracnose	*Colletotrichum pisi*	2+	
	Bacterial blight	*Pseudomonas pisi*	2+	
	Leaf spot	*Septoria pisi*	4+	Use more than half crop sequence
	Root rot	*Aphanomyces euteiches*	6-10	in corn, grains, vegetables, not forage
	Seedling blight	*Rhizoctonia solani*	1+	Rotate with crops such as cereals, corn
Pepper	Anthracnose	*Colletotrichum piperatum C. capsici*	1	Avoid solanaceous crops in rotation
	Bacterial spot.	*Xanthomonas campestris pv. vesicatoria*	1	Control solanaceous weeds
Potato	Corky ring spot	*Tobacco rattle virus*	1	Barley in rotation
	Golden nematode	*Globodera rostochiensis*	8	Exclude potato or tomato in rotation; in short rotations use beans, corn, red clover, rye-grass
	Root knot	*Meloidogyne hapla*	1	Use corn or *Poa pratensis* in between crops
	Root rot	*Rhizoctonia solani*	3-6	Use alfalfa before potato
	Scab	*Streptomyces sp.*	1+	Soybean as green manure before potato
	Wilt	*Fusarium oxysporum*	3-6	Use alfalfa before potato
Sorghum	Root knot	*Meloidogyne naasi*	1	Rotate with root crops
	Stalk rot	*Fusarium moniliforme*	2	Winter wheat-grain sorghum-fallow with no tillage
Soybean	Brown spot	*Septoria glycinea*	1	Use nonsusceptible crops
	Brown stem rot	*Cephalosporium gregatum*	5	5 years corn before soybean or corn-soybean-oats-clover
	Downy mildew	*Peronospora manshurica*	1	Use nonhost crops
	Cyst nematode	*Herterodera glycinea*	1-5	Use nonhost crops
	Leaf spot	*Cercospora sojina*	1	Use nonhost crops
		Phyllosticta sojaeicola	1	Rotate and plow residues
		Pseudomonas glycinea	1	Rotate and plow residues

Crop	Disease	Pathogen	Crop interval (year)	Remarks
Spinach	Downy mildew	*Peronospora spinacice*	3	
Squash	Root and stem rot	*Fusarium solani*	2-3	
Sweet potato	Black rot	*Ceratocystis fimbriata*	2+	Use any other crop
Tobacco	Black shank	*Phytophthora parasitica*	4	Rotate with cotton, peanut, soybean, oats, rye or wheat but not legumes before tobacco
	Black root rot	*Thielaviopsis basicola*	5	Many crops in rotation but use small grains just before tobacco
	Wilt	*Pseudomonas solanacearum*	3-5	Rotate with corn, cotton, cowpea, soybean or small grains.
	Root knot	*Heterodera marioni*	2-3	Corn-oats-tobacco, cotton-peanut-tobacco, peanuts-oats-tobacco
Tomato	Anthracnose	*Collectotrichum phomoides*	2	Rotate only if susceptible crops not nearby
	Bacterial canker	*Corynebacterium michiganense*	1	Rotate with other crops and control solanaceous weeds
	Black speck	*Pseudomonas tomato*	1	
	Bacterial spot	*X. campestris* pv. *vesicatoria*	1	
	Blight and fruit rot	*Helminthosporium lycopersici*	1	Rotate and plow disease debris
	Septoria blight	*Septoria lycopersici*	1	Rotate other crops and control horse nettle
	Soil rot	*R. solani*	1	Rotate with pangola grass
Vegetable	Nematodes	*Pratylenchus* spp.	1	Vegetable crops after oats and peanuts then after corn or lupines.
Wheat (Spring)	Common root rot	*Helminthosporium sativum*	4	Use noncereal crops
		Fusarium roseum		
	Eye spot	*Cercosporella herpotrichoides*	2-3	Rotate with other crops but not wheat and barley
	Flag smut	*Urocystis tritici*	1-2	Any crop but susceptible wheat
	Foot rot	*H. sativum*	1	Use non-grass crops
	Seed gall	*Anguina tritci*	1-2	Rotate with non cereals
	Septoria leaf and glume blotch.	*Septoria avenae, S. nodorum, S. tritici*	2	Rotate with non cereals
Wheat (Winter)	Cephalospor-ium stripe	*Cephalosporium gramineum*	1-2	Rotate with corn or legumes
	Snow mold	*F. nivale, Sclerotinia borealis, Typhula spp.*	1	Rotate with spring cereals or legumes
	Take-all	*Gaeumannomyces graminis*	2-3	Rotate with non cereals

pathotypes. Thus, in general the plant pathologists view monoculture as a disease perpetuating system.

Two types of disease and pathogen development are associated with monoculture: irreversible increase in disease incidence until a certain level is reached and maintained, and the reversible phenomenon of disease first rising, but subsequently declining to a more or less fixed level.

The pathogens in the irreversible disease pattern group are *Verticillium albo-atrum, V. dahliae* (wilt of cotton), *Heterodera schachtii* on sugarbeet, *Globodera rostochiensis* on potato, *H. glycines* on soybean and *Meloidogyne* spp. on several crops. The examples of pathogens in the group showing reversible disease pattern in monoculture are those fungi which have low ability for saprophytic survival. The group includes *Phymatotrichum omnivorum, Streptomyces scabies, Gaeumannomyces graminis* and eye spot fungus *Pseudocercosporella herpotrichoides.*

(c) Mixed Cropping

Mixed crops such as wheat-barley, wheat-chickpea, pigeonpea-sorghum, etc. reduce the economic loss from diseases. Since same pathogen does not attack both the crops in the mixture, at least one crop is saved if the other is badly damaged by the pathogen. Mixed crops also reduce spread of disease.

If two crops are to be sown together, different types of plant species are usually chosen for the purpose. There are only few foliage pathogens, notably some powdery mildews, that will attack crops belonging to different botanical families. Each of the two crops thus, to some extent, shields plants of the other crop from the impact of air borne pathogens, as well as from vectors. Soil borne diseases may also be reduced by root exudates of one crop, eg. onions, being detrimental to root pathogens of the other crops. The reduction in disease incidence in a mixed crop can be attributed to one or more of the following:

1. Due to reduced number of host plants there is sufficient spacing between them and chances of contact between foliage or roots of diseased and healthy plants are greatly reduced.
2. The roots of nonhost plants may act as a physical barrier obstructing the movement of pathogens in soil. They may also release toxic substances in their root exudates which suppress the growth of pathogens attacking the main crop. HCN in root exudates of sorghum is toxic to *F. udum* attacking pigeon pea in the mixture.
3. Due to reduced number of host plants in a mixed crop the susceptible area for an air-borne foliar pathogen is decreased. Therefore, there is less primary infection and less production of secondary inoculum for spread of the

disease. This slows down the rate of disease spread.

4. By proper selection of crops for the mixture, soil environment can also be changed to one that is not favorable for the pathogen. Control of root rot of cotton by growing cotton with moth is an example.
5. The soil-borne pathogens are not uniformly distributed in the field soil. Generally they are randomly present as dormant structures. Activation of these dormant strutures is often dependent on contact with the host roots, the chances of which are highly reduced in mixed crop due to spacing between plants.

2. Decoy and Trap Crops

Decoy crops are non-host crops sown with the purpose of making soil-borne pathogens waste their inoculum potential. This is achieved by stimulating and activating dormant propagules of the pathogens in the absence of the host. A list of pathogens that can be decoyed in this way is given in Table 2.

Trap crops are host crops of the pathogen, sown to attract nematodes, but destined to be harvested or destroyed before the nematodes complete their life cycle. In Germany, sowing of somewhat resistant potatoes and harvesting the crop early before the potato cyst nematode matures, is recommended. Similarly, sowing of crucifers and plowing before the beet cyst nematode can develop fully is also recommended.

3. Crop Nutrition

The macronutrients, nitrogen (N), phosphorus (P), and potassium (K), are needed in large amounts by crop plants and are frequently limiting to plant growth. The micronutrients are needed in smaller quantities. Maintenance of optimum plant health with sufficient, but not excessive, levels of fertility can be beneficial in crop resistance to stress from pathogen attack.

(a) Nitrogen

Since the effect of nitrogen on plants is most pronounced, it has been studied in great detail. Fertilization with nitrogen, especially at enhanced dosages, causes new succulent vegetative growth of the plant and delays maturity. Those pathogens which attack such plant organs are, therefore, favored by high nitrogen. It is known that excess of nitrogen predisposes the host to rusts and powdery mildew of wheat, blast of rice, etc. When the nitrogen is deficient, the plant is weak, its development is incomplete and its maturity is hastened. Pathogens favored by slow growth of the host are thus favored by low nitrogen.

Studies on the effects of nitrate (NO_3-N) and ammonium (NH_4-N) forms of

Table-2 : Decoy Crops for the Reduction of Pathogen Population

Crop	Pathogen	Decoy Crop
	Fungi	
Brassicae	*Plasmodiophora brassicae*	Rye grass, *Papaver rhoeas, Reseda odorata*
Olive	*Verticillium albo-atrum*	*Tagetes minuta*
	Nematodes	
Egg plant	*Meloidogyne incognita*	*Tagetes patula*
	M. javanica	*Sesamum orientale*
Oats	*Heterodera avenae*	Maize
Soybean	*Rotylenchus* spp., *Pratylenchus* spp.	*T. minuta, Crotolaria spectabilis*
Tomato	*M. incognita, Pratylenchus alleni*	*T. patula*, Casterbean, groundnut, chrysanthemum

nitrogen have revealed that pathogen-crop interactions depend frequently on the form rather than the amount of N available. These interactions are, however, complex in nature. The effects of nitrogen as such and forms of N on the incidence and severity of some infectious diseases are summarized in Table-3 & 4.

i. Mechanisms

Suppression and stimulation of diseases by specific forms of nitrogen has been attributed to preference of N forms by plants, altered host resistance, modification of plant constituents and exudates, shift in soil and rhizosphere pH, effect on pathogens and modified microbial equilibrium.

In most aerated soils NO_3-N predominates and plants adapted to such soils grow well with NO_3-N as the sole source of nitrogen. Crops that utilize NH_4-N grow well following fumigation while those requiring NO_3-N may be adversely affected. Evidence that nitrogen affected resistance of wheat to pathogen causing take-all was demonstrated. Increased root growth permitting wheat to escape take-all has also been reported. NH_4-N reduce bacterial canker of *Prunus* by hastening periderm formation. NH_4-N has been found to increase permeability and exudation of broad bean leaf surface resulting in higher levels of sugars and amino acids. *Botrytis cinerea* infected such leaves more frequently than leaves supplied with NO_3-N. NH_4-N forms of nitrogen generally lower the pH while nitrates raise it. Nitrate nitrogen suppresses damping-off of sugarbeet (*Pythium ultimum*) but NH_4-N fails to do so. This is due to change in rhizosphere pH toward alkalinity by NO_3-N and beets being alkali-tolerant plants develop more vigor under these conditions, thereby developing resistance quickly through tissue maturity.

Table-3 : Diseases Influenced by Nitrogen Without Reference to Forms of Nitrogen

Increased		Decreased	
Disease	Pathogen	Disease	Pathogen
	Bacterial Disease		
Angular leafspot and fire blight of tobacco	*Psudomonas tabaci*	Bacterial wilt of cucumber	*Erwinia tracheiphila*
Citrus canker	*Xanthomonas campestris* pv. *citri*	Bacterial spot of peach	*Xanthomonas campestris* pv. *pruni*
Fire blight of apple	*Erwinia amylovora*	Bacterial blight of tomato and tobacco	*R. solanacearum*
Wilt of maize	*E. stewartii*	Leaf spot of peach	*Ps. pruni*
Wilt of tobacco	*R. solanacearum*	Wild fire of tobacco	*Ps. tabaci*
	Fungal Disease		
Bean root rot	*Rhizoctonia solani*	Bunt of wheat	*Tilletia* spp.
Blast of rice	*M. grisea*	Club root of cabbage	*Plasmodiophora brassicae*
Browning root rot of wheat	*Pythium* spp.	Root rot of cotton	*Phymatotrichum amnivorum*
Downy mildew of cabbage	*Peronospora parasitica*	Root rot of pea	*Aphanomyces euteiches*
Dutch elm disease	*Ophiostoma ulmi*	Root rot of wheat	*Rhizoctonia solani*
Flag smut of wheat	*Urocystis tritici*	Stem canker of soybean	*R. solani*
Graymold of grape	*Botrytis cinerea*	Sclerotium rot of sugarbeet	*Sclerotium rolfsii*
Powdery mildew of cereals	*Erysiphe graminis*		
Rusts of several crops	*Puccinia* spp.	Take-all of wheat	*Gaeumannomyces graminis*
Smut of maize	*Ustilago zeae*	Wilt of cotton	*Verticillium albo-atrum*
Wilts of tomato, melon, cabbage	*Fusarium oxysporum*		
	Viral Diseases		
Bean mosaic	*Common bean mosaic virus*	Bean mosaic	*Tobacco mosaic virus*
Lettuce mosaic	*Lettuce mosaic virus*	Ring spot of celery	*Celery ring spot virus*
Tobacco mosaic	*Tobacco mosaic virus*		
Wheat mosaic	*Barley stripe mosaic virus*		

Table-4 : Effect of Inorganic Forms of Nitrogen on Plant Diseases

Disease	Pathogen	NO_3-N	NH_4-N
	Bacterial Diseases		
Angular leaf spot of cotton	*Xanthomonas campestris* pv. *malvacearum*	D	
Canker of peach	*X. campestris* pv. *pruni*	D	
Canker of tomato	*Corynebacterium michiganense*	I	
Crown gall	*Agrobacterium tumefaciens*	I	
Ring spot of potato	*C. sepidonicum*	I	
Southern wilt of tobacco/ tomato	*R. solanacearum*	I	D
Stewert's wilt of maize	*Erwina stewartii*	I	
	Fungal Diseases		
Blast of rice	*M. grisea*	D	I
Black scurf of potato	*Rhizoctonia solani*	D	I
Brown spot of rice	*Drechslera oryzae*	I	D
Charcol rot of various crops	*Macrophomina phaseolina*	D	I
Eye spot of wheat	*Pseudocercosporella herpotrichioides*	D	I
Northern leaf blight of maize	*Drechslera turcica*	D*	I
Root rot of pea	*Aphanomyces euteiches*	D	I
Root rot of pea	*Pythium* spp.	I	D
Root rot of tobacco	*Thielaviopsis basicola*	I	D
Root rot of cotton	*Phymatotrichum omnivorum*	I	D
Root rot of bean	*Fusarium solani* f.sp. *phaseoli*	D	I
Root rot of maize	*A. euteiches*	D	I
Root rot of maize	*Pythium* sp.	I	D
Scab of potato	*Streptomyces scabies*	I	D
Stem rust of wheat	*Puccinia graminis*	I	D
Stripe rust of wheat	*P. striiformis*	I	D
Stalk rot of maize	*Fusarium* sp.	D	I
Stalk rot of maize	*Diplodia zeae*	I	D
Take-all of wheat	*Gaeumannomyces graminis*	I	D
Wilt of bean	*F. oxysporum f.sp. phaseoli*	D	I
Wilt of potato/tomato	*Verticillium albo-atrum*	I	D

Disease	Pathogen	NO_3-N	NH_4-N
Nematode Diseases			
Root knot of lima bean	*Meloidogyne incognita*	I	D
Soybean cyst nematode	*Heterodera glycinea*	I	D
Tobacco cyst nematode	*H. tabacum*		D
Viral Disease			
Cadang-Cadang of coconut	Viroid		D
Potato virus	Potato Virus X		D
Tobacco mosaic	Tobacco mosaic virus		I

Note: D= decrease, I= increase

Some nitrogen compounds are directly toxic to pathogens. Ammonia is known to be toxic to *S. rolfsii, P. omnivorum, P. brassicae, G. graminis,* and some plant parasitic nematodes. Calcium cyanamide is toxic to *S. rolfsi, Pythium, F. oxysporum,* and many nematodes are suppressed by urea due to direct toxicity. A form of nitrogen may affect virulence of a pathogen without affecting growth. Inhibition of pectolytic and cellulolytic enzymes by a specific form of N has been implicated to reduce *Rhizopus* fruit rots and susceptibility of cotton to *R. solani and V. albo-atrum.* The thallus of *F. solani f. sp. phaseoli* has been found larger and more extensive with NH_4-N than with NO_3-N. Reduction of saprophytic activity of *R. solani* with NH_4-N reduces survival and disease inducing potential.

Application of nutrients may modify microbial equilibrium in soil and thus, results in suppression of disease. NH_4-N and NO_3-N have differential effect on Fusarium root rot of bean, Aphanomyces root rot of pea, take-all of cereals and Rhizoctonia root rot of cotton. Disease severity was similar with either forms of N in sterile soil which suggests that disease control results from microbial interactions. Suppression of pea root rot in natural soil by NO_3-N was correlated with enhanced activity of bacteria and actinomycetes antagonistic to *A. euteiches.* A similar situation of antagonists (*Streptomyces*) under NO_3-N has been postulated in the control of *Poria* and *Armillaria* root rot of pine.

(b) Phosphatic Fertilizers

Results of investigations reveal that effects of phosphates on plant diseases appear to be connected with one of the two situations: (1) either the crop grows on soil markedly deficient in P, and correction of the imbalance contributes to its health or (2) maturity of the crop is somewhat advanced by liberal supply of P, and this helps it escape from biotrophs (such as downy mildews) preferring younger tissues. Effects of P on some important diseases are presented in Table-5.

Table-5 : Diseases Influenced by Phosphorus

Disease-Pathogen	Effect
Bacterial Diseases	
Bacterial wilt of cotton – *Pseudomonas cryophylli*	Increase
Black fire of tobacco- *Ps. angulata*	Decrease
Blight of lima bean - *Ps. syringae*	Decrease
Fire blight of apple - *Erwinia amylovora*	Increase
Wilt of maize - *E. stewartii*	Increase
Nematode Disease	
Pea nematode - *Pratylenchus penetrans*	Increase
Root knot of pea - *Meloidogyne incognita*	Increase
Virus Diseases	
Beans mosaic - tobacco mosaic virus	Decrease
Spinach virus - cucumber virus-I	Increase
Tomato/ tobacco mosaic – tobacco mosaic virus	Increase
Fungal Diseases	
Bunt of wheat - *Tilletia spp.*	Decrease/Increase
Club root of cabbage – *Plasmodiophora brassicae*	Increase
Damping off of pea – *Rhizoctonia solani*	Decrease
Downy mildew of cabbage - *Peronospora parasitica*	Decrease
Downy mildew of grape – *Plasmopara viticola*	Decrease
Downy mildew of lettuce- *Bremia lactucae*	Increase
Flag smut of wheat - *Urocystis tritici*	Decrease
Late blight of potato-*Phytophthora infestans*	Decrease/ Increase
Powdery mildew of cereals –*Erysiphe graminis*	Decrease/ Increase
Root rot of tobacco-*Thielaviopsis basicola*	Decrease
Root rot of cotton-*Phymatotrichum omnivorum*	Increase
Root rot of soybean-*R. solani*	Decrease
Take-all of wheat-*Gaeumannomyces graminis*	Decrease
Wilt of cotton-*F. oxysporum f.sp. vasinfectum*	Decrease/ Increase
Wilt of tomato-*F. oxysporum f.sp. lycopersici*	Decrease/ Increase

(c) Potassium

Usually high potassium (K) has been reported to reduce incidence of diseases. The mechanism by which K affects disease differ greatly, and includes both direct and indirect effects. Direct effects include reduction or stimulation of pathogen penetration, multiplication, survival, aggressiveness, and rate of establishment in the host. Indirect effects include promotion of wound healing, increase in resistance to frost injury and delay in maturity in some crops. Diseases supressed by potassium are listed in Table-6.

(d) Calcium

The principal role of calcium in host-pathogen relationship is the formation of calcium pectate in the cell walls thus making them resistant to degradation by facultative parasites: *R. solani, S. rolfsii, Botrytis cinerea* and *Erwinia amylovora*. The vascular wilts caused by *F. oxysporum* are also reduced by calcium especially in conjugation with NO_3-N. However, calcium favors black shank of potato caused by *Phytophthora nicotianae* var. *nicotianae*. It also makes potato tubers prone to attack of common scab (*Streptomyces scabies*).

4. Adjustment of Sowing Dates

The choice of sowing dates in relation to crop diseases has one principal aim, viz., to reduce to a minimum the period over which infective agent (propagules, vectors) meets susceptible host tissue. This is also the aim of operations to influence the time of flowering and fruiting, especially by pruning and by breaking of dormancy. Since most crops grow under ranges of temperature and humidity wider than those favouring pathogen development, sowing or planting at seasons which give the crop an advantage is obviously good strategy.

Bunt (*Tilletia caries*) of winter wheat can be controlled by either planting before mid September or after mid October. Cercosporella foot rot can be an important disease of winter wheat planted in the state of Washington before September 15, but almost no disease is present if planting is delayed by one month.

5. Spacing

In crowded stand of crops the proper aeration is checked, humidity and temperature of the atmosphere as well as of the soil are likely to favour the survival and spread of pathogens. The plants remain tender and weak. The soil-borne fungi can easily reach and invade the roots of healthy plants. Damping off, late blight of potato, downy mildews, etc. are some of the diseases which spread fast in close space planting. To avoid these conditions favorble for the pathogen

Table-6 : Plant Diseases Influenced by Potassium Nutrients

Disease-Pathogen	Effect
Bacterial Diseases	
Angular leaf spot of tobacco-*Pseudomonas tabaci*	D
Angular leaf spot of cucumber-*Ps. lachrymans*	D
Angular leaf spot of cotton-*Xanthomonas campestris* pv. *malvacearum*	D
Bacterial blight of lima-bean - *Ps. syringae*	D
Bacterial blight of rice-*X. campestris* pv. *oryzae*	D
Fire blight of pear-*Erwinia amylovora*	D/I
Soft rot of cabbage-*E. carotovora*	D
Wilt of tobacco-R. *solanacearum*	D
Fungal Diseases	
Ascochytosis of chickpea-*Ascochyta rabiei*	D
Blast of rice-*M. grisea*	D
Brown spot of rice-*Helminthosporium oryzae*	D
Canker of potato-*Rhizoctonia solani*	D
Daming off of beet -*Pythium ultimum*	D
Downy mildew of lettuce- *Bremia lactucae*	D
Downy mildew of grape -*Plasmopara viticola*	D
Downy mildew of cauliflower – *Peronospora parasitica*	D
Early blight of tomato- *Alternaria solani*	D
Gray mold of grape-*Botrytis cinerea*	D
Late blight of potato – *Phytophthora infestans*	D
Northern blight of corn – *Drechslera turcica*	D
Powdery mildew of cereal-*Erysiphe graminis*	D
Root rot of pea-*Aphanomyces euteiches*	D
Root rot of pine apple-*Phytophthora cinnamomi*	D
Root rot of cotton- *Phymatotrichum omnivorum*	D
Root rot of jute-*R. solani*	D
Rust of cereals-*Puccinia* spp.	D
Sheath blight of rice-*Corticium sasaki*	D

Disease-Pathogen	Effect
Stalk rot of maize- *Fusarium moniliforme* *Diplodia zeae* *Gibberela zeae*	D
Wilt of cotton-*F. oxysporum* f.sp. *vasinfectum*	D
Wilt of melon-*F. oxysporum* f.sp. *melonis*	D
Wilt of tomato-*F. oxysporum* f.sp. *lycopersici*	D
Nematode Diseases	
Pea nematode-*Pratylenchus penetrans*	I
Root knot of cucumber-*Meloidogyne incognita*	I
Root knot of lima bean-*M. incognita*	D
Sugarbeet cyst nematode-*Heterodera schachtii*	D
White tip of rice-*Aphelenchoides besseyi*	D
Viral Diseases	
Bean mosaic-Tobacco mosaic virus	D
Blotchy ripening of tomato-Tobacco mosaic virus	D
Potato leaf roll-Potato leaf roll virus	I
Spinach virus-Cucumber virus	I
Tobacco mosaic of tomato-Tobacco mosaic virus	I
Tobacco mosaic-Tobacco mosaic virus	D

Note: D= Decrease, I= Increase

proper spacing between plants must be maintained.

At 15 cm between rice plants, there is a higher percentage of seed-borne *Alternaria spp., Curvularia lunata, Drechslera oryzae* than at wider distances. Soybean seeds from narrow row (25 cm) compared to wider rows (76 cm) spacing give higher recovery of total fungi and bacteria which adversely affected the quality of seeds.

6. Water Management

Soil moisture is related to many diseases. As examples, wet soil favours club root (*Plasmodiophora brassicae)* of crucifers, silver scurf (*Helminthosporium solani*) of potatoes, and *Cercosporella* on wheat, while dry soil increases severity of white mold (*Wetzelinia sclerotiorum)* of onion, common scab (*Streptomyces scabies)* on potato and Fusarium diseases of cereals. Damping off diseases caused by *Pythium* spp. can be decreased by maintaining a dry soil surface. Using irrigation to grow crops in an otherwise dry season allows avoidance of diseases such as potato late blight and wheat stem rust in Mexico. The charcol rot fungus *Macrophomina phaseolina* attacks potato when there is a water stress. By irrigating the field stress is removed and the disease is suppressed.

Sprinkler irrigation has been studied in rather some detail with respect to foliar diseases. Generally, sprinkler irrigation increase diseases by increasing leaf wetness and by dispersing poropagules of the pathogen by water splashes just like rain water. At the same time, it has some advantages also such as washing off of inoculum from the leaf surface.

In addition to amount of irrigation, the timing of irrigation is also important. The principles governing the timing of irrigation, its begining and its frequency, from the aspect of disease management are (1) providing the crop with as uniform a water supply as possible to obviate stresses of water deficit or excess, (2) timing of giving water in relation to periods of heightened susceptibility of host to disease, and (3) minimizing hours of continuous leaf wetness, e.g., avoidance of sprinkling in continuation of dew periods.

C. Sanitation

Field and plant sanitation is a main part of disease control through cultural practices. This step is essential even if disease or pathogen free seed or propagating material has been used and other recommended cultural practices like crop rotation, alteration in date and method of sowing have been followed. The inoculum present on few plants in the field may nullify in soil or on the plant and in due course of time may be sufficient to nullify the effect of other cultural practices. Therefore, plants bearing such pathogens or plant debris introducing the inoculum in the soil should be removed as early as feasible.

In wilt disease of banana (*F. oxysporum* f.sp. *cubense*) it has been reported that so long as the dead roots and rhizomes of affected plants are present in soil the fungus continues its growth. When these plant remains are removed there is rapid decline in the population of the pathogen in the soil. Similarly, the wilt of cotton and root rot of bean, and Verticillium wilt of cotton are also reduced to some extent by removal of diseased plant debris.

1. Destruction of crop residues

Infected plant debris not only serves as source of perennation of pathogens, it also serves as a substrate for multiplication of inoculum. The destruction of this source of survival and multiplication of inoculum has been found to help in the control of many diseases. The fungi causing downy mildews of maize and pea, white blisters of crucifers, powdery mildews of pea and cereals are examples of pathogens which survive through their sexually produced oospores or perithecia in crop debris. In certain areas the linseed rust fungus, the rice blast and leaf spot fungi, and the fungus causing early blight of potato also perennate through dormant stages in diseased crop debris. Destruction of crop debris by burning immediately after harvest reduces the amount of inoculum surviving through debris.

Deep ploughing specially during hot summer in the tropics burries the debris to such depths where the pathogens are automatically destroyed. The pathogen free sub soil is brought on the surface by deep ploughing. The beneficial effect of the deep ploughing has been reported in diseases caused by *Sclerotinia sclerotiorum* and *Sclerotium rolfsii*.

Fallowing also helps in reduction of inoculum through sanitary effect of decomposition of crop debris in absence of a host. This has been demonstrated in Ascochyta blight of pea. Including fallow period in the rotation establishes types and number of microorganisms in the soil which are beneficial for the health of the crop. The principle of flood fallowing in which the land is left submerged in deep water for few weeks is also a method of cleaning the soil of pathogens.

2. Roguing

Regular removal of disease plants or plants parts from a population has been found effective in reducing spread of many diseases. This method is one of the effective recommendations in the control of virus diseases of crops. The rouging of diseased plants not only checks spread of the disease, it reduces the amount of survival structures also.

In the production of virus-free potato seed tubers and in the control of soybean mosaic virus, roguing is an important recommended practice. However, the procedure is effective only when the populations of the insect vectors of these pathogens are low or inactive. For the control of loose smut of wheat and production of disease free seed, roguing of diseased plants is always recommended in seed plots. In this disease also roguing is effective only if the smutted heads are removed before spores have been dispersed by wind. Red rot of sugarcane, wilt of cotton, smut of sugarcane, etc. are other diseases in which roguing of diseased plants is recommended for their control. Roguing is feasible and economical in small size fields and when the incidence of disease is not very high. In large size fields it is difficult to locate the diseased plants at the proper time unless a high percentage of plants are affected, and when very high percentage of plants are diseased, roguing becomes uneconomical.

3. Crop free period and crop free zone

The pathogen attacking crops of secondary importance and having a narrow host-range can be controlled by maintaining a crop-free period of definite duration. When the growers in an area agree not to grow the crops susceptible to the pathogen for a definite period, depending on longevity of the pathogen without its host, the pathogen is automatically starved out. Similarly, when the host crop is not grown in a zone surrounding the infested area, spread of the disease is checked.

On this basis, for control of bunchy top of banana, it has been suggested that a crop free belt around the area affected by the disease checks it spread provided disease planting material is also quarentined and insect vectors are unable to cross crop free zone. This method is effective for those diseases which are not seed-borne and either there is no insect-vector or if insect vectors spread then their flight range is limited.

4. Creating Barriers by Non-host or Dead host

Many diseases depend for their spread on nearness of healthy roots to infected roots. Presence of roots of non-host crops between such roots may create barrier against spread of the pathogen. This aspect has been mentioned under mixed cropping.

One of the control measure for bacterial wilt of banana (R. *solanacearum)* and spreading decline of citrus (*Radopholus similis)* is to destroy the healthy plants around the diseased plants. This checks the movement of pathogen from diseased to healthy plants that are left in the field. In potato fields spread of late blight and many virus diseases is checked if the plants are detopped early. The procedure prevents production of inoculum for secondary spread and dispersal of virus by insect vectors.

5. Control of Weeds and Insects

Another aspect of field sanitation is keeping the field free from weeds and insects. Weeds reduce the amount of nutrients available for the plants and by lowering their vitality increase their disease proneness. Excess of weeds in the field also helps in increased humidity and low temperature. In addition, many weeds may be collateral or alternate hosts of pathogens attacking the crops grown in the field. The root knot nematode infects a large number of solanaceous weeds. Wild cucurbits help the continuous infection chain of many viruses attacking cultivated cucurbits. Therefore, as a sanitary precaution destruction of these weeds is necessary.

CHAPTER XVII

HOST RESISTANCE AND IMMUNIZATION

I. HOST RESISTANCE

A. Introduction

The prevention of epidemics and ultimately the reduction of losses in yield has been of great concern. The diseases may be controlled by using chemicals; but the chemicals create hazards to human health and produce undesirable side effects on nontarget organisms. Under the existing circumstances, the use of resistant plants is one of the most attractive approaches for suppression of plant diseases. Their use requires no particular action by the cultivators during the crop growth, is not disruptive to the environment, is generally compatible with other management practices, and is sometimes singularly sufficient to suppress disease to tolerable levels.

B. Concept

All the individuals of a plant species are not equally susceptible to a pathogen. The extent to which a plant prevents the entry or subsequent growth of the pathogen within its tissues or the extent to which a plant is damaged by a pathogen is used to measure the resistance or susceptibility (Figure 1).

Resistance may be qualified by such words as high, intermediate, or low because there may be gradations between extreme resistance and extreme susceptibility. The magnitude of resistance can range from very small to very large. Even resistance that do not completely prevent pathogenesis can suppress disease adequately in populations of plants. If resistance has large enough effect to slow down pathogen reproduction rates to replacement levels, pathogen population will not increase. If resistance is of small magnitude disease can increase in the resistant plant population.

C. Classification of Resistance

1. Based on Effect of Genes

The terms major and minor genes are sometimes used instead of oligogenic polygenic when referring to gene for resistance. However, neither all oligogenes are necessarily major genes, nor all polygenes are minor genes.

2. Based on Growth Stage of Host Plant

Several terms such as seedling resistance, post-seedling resistance, adult plant resistance, etc. have invariably been used to describe the exact stage of the host growth when it shows resistance. For example, resistance of wheat to leaf rust (*P. recondita*) may be expressed at the "first leaf stage", is generally termed as "seedling resistance". The term "post seedling resistance" or "adult plant resistance" is used in the sense that a cultivar is susceptible at first leaf stage but at later stages of development it becomes resistant.

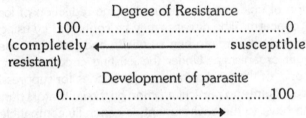

Degree of Resistance

100..0

(completely ←———————— susceptible
resistant)

Development of parasite

0..100

————————→

Figure-1 : Definition of resistance (Eenink, A.H. Proc., Symp. Induced Mutat. Against Plant Dis. Int. At. Energy. Vienna. With permission).

3. Based on Mode of Inheritance

Types of resistance have been classified into those controlled by a single gene (monogenic) and those controlled by several genes (polygenic). Monogenic resistance is often sufficiently effective to qualify as immunity. It is stable under a wide range of environmental conditions, but is usually specific for certain race virulence gene of the pathogen. In contrast, polygenic resistance, which is more sensitive to environmental fluctuations, does not result in immunity, but is more uniformly effective against variants of the pathogen. Several interrelated and complementary terms have been employed to describe the genetic concept of resistance. They are summarized in Table-1.

Table-1 : Terms Frequently Used to Express Genetic Concept of Resistance

Specific resistance	**General resistance**
Race specific	Race nonspecific
Vertical	Horizontal
Major gene	Minor gene
Monogenic	Polygenic
Multiple allele	Multiple gene
Qualitative	Quantitative
High	Low, moderate
Seedling	Adult plant
Nondurable	Durable
Hypersensitive	Nonhypersensitive
Protoplasmic	Generalized, uniform, field, partial, permanent, late rusting*, slow rusting*, tolerance

* Applicable to only rust disease

4. Cytoplasmic Resistance

There are several plant diseases in which neither vertical resistance nor horizontal resistance is controlled by genetic material contained in the cytoplasm of the cell in which genes do not normally follow the Mendelian laws of inheritance. Such resistance is sometimes referred to as cytoplasmic resistance. The two best known examples of cytoplasmic resistance occur in maize in which resistance to two leaf blights, the southern corn leaf blight caused by *Helminthosporiumn maydis* and the yellow leaf blight caused by *Phyllosticta maydis,* is conferred by characteristics present in normal cytoplasm of various types of corn but absent or suppressed in Texas male-sterile cytoplasm.

5. Resistance in Epidemiological Terms

When a host plant variety is more resistant to some races of a pathogen than to others the resistance has been termed "vertical" or "differential", whereas resistance which is evenly spread against all races is "horizontal" or uniform. Vertical and horizontal, as applied to resistance, are mathematically derived concepts, and refer respectively to the vertical and horizontal axes of a graph. This is shown in Figure 2, which indicates the foliage resistance of two potato varieties to sixteen races of *Phytophthora infestans.* Both varieties show vertical (complete) resistance to races (0), (2), (3), (4), (2,3), (2,4), (3,4), and (2,3,4), while resistance to the other eight races is horizontal and greater in Maritta than in

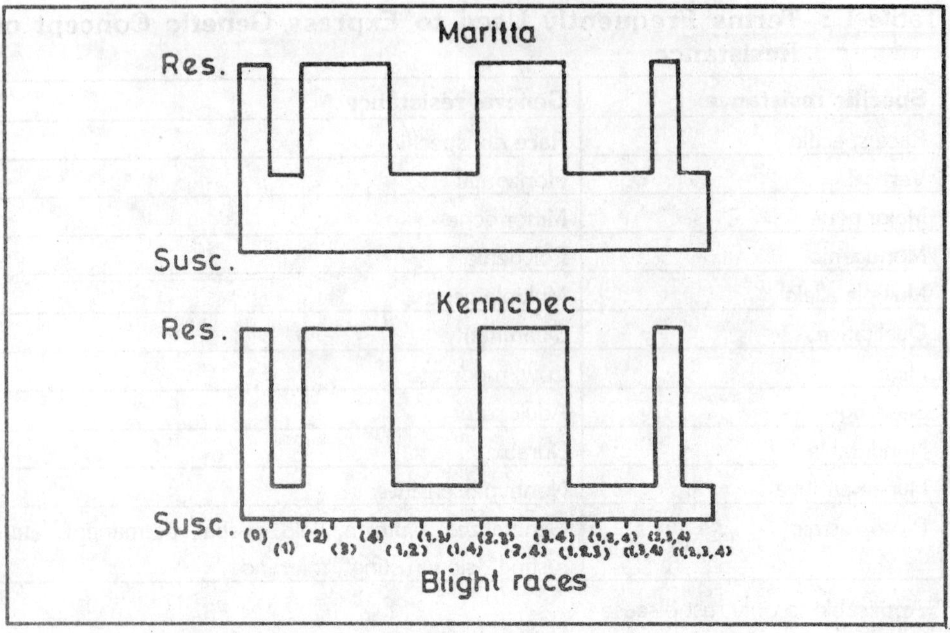
Maritta
Res.
Susc.
Kenncbec
Res.
Susc.
(0)
(1)
(2)
(4)
(1,2)
(1,4)
Blight races

Fig. 2 : Vertical and horizontal resistance in two varieties of potato

Kennebec. Vertical resiatance is probably always accompanied by some degree of horizontal resistance. It is difficult to visualize a plant completely lacking in resiatance to a pathogen. Vertical resistance operates mostly after the pathogen has penetrated the plant and that it may involve such active defense mechanisms as hypersensitivity, production of phytoalexins, etc. Although these mechanisms do not entirely prevent penetration, they do prevent the subsequent spread of the pathogen in the tissue so that the plant is, for most practical purpose remains immune. Vertical resistance is usually controlled by one or a few genes thereby the name monogenic or oligogenic.

Horizontal resistance protects against all races of a pathogen to varying extents, but the protection it affords is usually less than that given by vertical resistance against specific races. It may operate before or after infection through defense mechanisms which reduce or delay infection, colonization of the plant and/or production of spores by the pathogen. Horizontal resistance is controlled by many genes, thereby the name polygenic or multigenic resistance. Each of these genes alone may be rather ineffective against the pathogen and may play a minor role in the total horizontal resistance.

Browning *et.al.*(1977) proposed two terms "discriminatory resistance" and "dilatory resistance" and emphasized that the use of these two terms should be restricted to epidemiological concepts of resistance. A population of host is defined as having discriminatory resistance or susceptibility if it affects the epidemic by discriminating among strains, i.e. by favoring or rejecting certain components of the pathogen population. A population of host plant is defined as having dilatory resistance if it affects the epidemic by reducing the rate of development of the pathogen population. Theoretical actions of discriminatory and dilatory resistance on the components of an epidemic are given in Table-2.

Table-2 : Effect of Discriminatory and Dilatory Resistance on the Components of an Epidemic

Epidemic component	Discriminatory resistance	Dilatory resistance	Both combined
Incoming inoculum (i)	Decrease	No effect	Decrease
Rate of increase (r)	No Effect	Decrease	Decrease
Delay of onset of epidemic (t)	No Effect?	Increase	Increase
Amount of disease (x)	No Effect?	Decrease	Decrease
Outgoing inoculum (I)	No Effect?	Decrease	Decrease

D. Apparent Resistance

True resistance denotes incompatibility between the host and the parasite. It implies that so long as the incompatibility exists the host should ward off infection by the pathogen. However, under certain conditions, some very susceptible varieties or plants may remain free from infection or symptom and thus appear resistant. Apparent resistance to disease of plants known to be susceptible is generally a result of "disease escape" or "tolerance" to disease.

1. Disease Escape

The disease escaping varieties have no true genetic resistance as they are compatible with the pathogen, and can exhibit highly susceptible reaction if the pathogen can attack them at the proper time. However, due to their other genetic characters such as rapid growth, early maturity, etc. they avoid the period when the pathogen is in aggressive stage or when the environmental conditions are favorable for the pathogen. For example, varieties of pea which mature early (by January under Indian conditions) usually escape much damage from powdery mildew and rust. This disease normally becomes serious in January or later. If pods have developed before serious disease incidence the losses are considerably reduced. Varieties of raspberries susceptible to raspberry mosaic virus frequently escape the disease because the plant is a non-preferred host of the insect vector. Several leaf characters of rice cultivars like roled, narrow, dark green and slow senescence were associated with resistance to bacterial leaf blight (X. *campestris pv. oryzae*).

2. Tolerance

A plant which is attacked by a pathogen to the same degree as other plants, but suffers less damage in terms of yield or quality as a result of the attack, is said to be tolerant. Tolerance may be defined as the inherent or acquired capacity to endure disease and give satisfactory return to the growers. Browning *et.al.* (1977) in terms of epidemiological concepts of resistance said that "a population of host plants is defined as having tolerance if it is rated as susceptible visually, but is damaged less by the epidemic than another susceptible population".

Tolerance results from specific heritable characters of the host plant that allow pathogen to develop and multiply in the host while the host, either by lacking receptor sites for, or by inactivating or compensating for the irritant excretion of the pathogen, still manages to produce a good crop. Van der Plank, however, considers that tolerance is not a desirable character because it does not reduce either initial inoculum (Xo), infection rate (r) or infection time (t).

Tolerance to disease is most commonly observed in many plant virus infections in which mild strains of viruses infect plants such as potato systemically and yet cause few or no symptoms and have little effect on yield. Russell observed three different kinds of tolerance mechanisms for viral diseases.

1. The virus is able to multiply but symptoms do not appear.
2. The plants which develop disease symptoms which are as severe as those in other plants but which suffer less damage, for example sugarbeet yellows virus, such plants are referred to as disease tolerant.
3. The virus infected plants that do not show severe disease symptoms and are less damaged by infection than other plants, are true tolerant. For example, infected barley plants that are tolerant to barley yellow dwarf virus do not show pronounced yellowing or stunting of the leaves give an acceptable yield inspite of being infected.

C. Sources of Resistance

In an ongoing resistance breeding programme, the availability of genes for resistance is the first concern. The primary and secondary gene centres of cultivated plants are the best places to find genuine resistance to common diseases.

The information on the origin and evolution of cultivated plants and their wild progenitors have been collected and summarized by several workers.

The International Board of Plant Genetic Resources (IBPGR) established in 1973, has sponsored plant collections in over 20 countries or areas. All the major cereals, pulses, grasses, forage, legumes, groundnuts, vegetables, potatoes, cassava, and sweet potatoes have been included. In addition, other International Research Centres (ICRISAT, CIMMYT, ICARDA, ITTA, etc.), maintain germplasm collection.

Various sources exploited for collection of resistance genes include use of land races/ cultivated types, wild/ alien species, multiplasm, induced mutants, etc. Often genes for resistance are present in the varieties or species normally under cultivation in the area where the disease is severe, and in which the need for resistant variety is most pressing. With most diseases, a few plants remain virtually unaffected by the pathogen. These survivor plants are likely to remain disease-free because of resistance in them. Such materials are oftenly employed for breeding for disease resistance. If no resistant can be found within the local population, plants of the same species from other areas and plants of other cultivated or wild species are also checked, for resistance.

D. Failure of Resistance

There are very few varieties except some isolated cases (e.g. cabbage, yellow-wilt caused by *F. oxysporum f.sp. conglutinants*) which could maintain their complete resistance for long. In most cases even complete resistance to a disease breaks down sooner or later. There are several reasons for it. Among avoidable causes are laxity in screening, inadequate back-crossing, and such other failures on the part of the breeder. Segregation in varieties is another cause of failure. However, the most important and unavoidable cause of failure of resistance in varieties is unchecked variability of the pathogen. Mutation and hybridization cause evolution of new races/biotypes or strains of the pathogen rapidly. During screening of varieties or breeding materials, if they are not exposed to all the existing races/biotypes of the pathogen or after the release of the variety new pathotypes have appeared in the area having gene for virulence which could counteract the genes for resistance in the variety, the variety is destined to become susceptible. The new races could develop in the same area or may be introduced from some external source. Prolonged cultivation of a single genotype in the area also helps in the development of new races. Soon the inoculum build-up of these races may reach a density that could lead to an epidemic. Therefore, as far as possible, it is desirable to grow more than one genotype of the particular crop in a particular area to maintain their resistance for longer periods.

This warrants continuous breeding programme to evolve new varieties as replacement for existing ones.

G. Management of Resistant Varieties

In most cases the life of a resistant variety is limited. Complete reliance on resistant variety as the only method for reducing incidence and severity of disease is neither good for the crop variety nor for the growers. Resistance in a variety can be prolonged and efficient disease management achieved if such varieties are used as a part of an integrated control programme. In order to increase durability as well as stability of resistance and also to reduce disease damage, the use of cultural, biological and chemical control methods should be encouraged. In general, control measures are much more effective on resistant or partially resistant varieties than on susceptible ones. There must be varietal diversification in space and time using race specific or race nonspecific resistance. No single variety should be allowed to dominate cultivation. Practices like alteration of sowing dates, fertilization, irrigation, combined with fungicidal protection will prolong the life of a resistant variety by interfering with the capacity of the pathogen to develop races which overcome resistance.

II. HOST IMMUNIZATION

A. Introduction

Although plants do not possess the antigen-antibody system of man and animals, they can be systemically immunized against fungal, bacterial and viral diseases by prior inoculation/ treatment with mild strain or low doses of severe strain of the pathogens or elicitors produced by the pathogens.

B. Immunization against Microbial Diseases

Protection by immunization against several fungal and bacterial diseases has been demonstrated. Caruso and Kuc (1977) reported that in cucumbers, watermelons and muskmelons plant receiving restricted infection with *Colletotrichum lagenarium* prior to transplanting to the field had fewer and smaller lesions following challenge with high inoculum concentration of the pathogen in the field than plants that were not infected prior to transplant. The survival of protected watermelon plants was 98% as compared to 32% for controls. Tuzun *et.al.* (1984) demonstrated that imunization of tobacco plants by stem injection of sporangial suspension of *Peronospora tobacina* protected the plants against blue mold.

Above two examples and several others prove that effectiveness is not a problem for the use of immunization for the control of plant diseases in the field. So why immunization is not popular? Kuc(1987) has pointed out different advantages and disadvantages of plant immunization in relation to other plant disease control measures.

1. Advantages

1. Immunization is effective against fungal, bacterial and viral diseases.
2. It is likely that immunization is dependent upon the activation of several different mechanisms, and therefore, is probably stable. Systemic fungicides with a single metabolic site of action are generally unstable with the development of new races of pathogens.
3. Immunization is systemic and persistent. In some cases it persists for the life of annual plants.
4. Since immunization utilizes mechanisms for resistance present in plants, it may be considered natural and as safe for man and the environment as disease-resistant plants.
5. Immunization in cucurbits and tobacco is graft transmissible from root stock to scion. This could be significant since many plants are propagated by grafting.
6. Immunization in tobacco appears nonreversible. Buds from immunized plants grafted on to susceptible root stock developed into fully grown immunized tobacco plants.
7. Immunization systemically sensitizes the plant to respond, but the major expenditure of energy and the expression of resistance are generally localized and occur in the presence of pathogens, that is, when and where needed.
8. The ability to immunize susceptible plants implies that the genetic potential for resistance is present in all plants.
9. Plants can be immunized by chemicals extracted from immunized plants. This suggests the possibility of seed treatments.

2. Disadvantages/Drawbacks

1. The natural chemical signals for systemic immunization have not been characterized.
2. Immunization is not economically competitive with our present technology in modern agriculture although it appears as effective as available systemic pesticides.
3. Immunization has not received sufficient field testing to determine its stability and persistence under high natural pathogen pressure.
4. People have difficulty in accepting the reality that plants can be systemically immunized against disease.

Instead of using plant pathogens directly for immunization, possibility of using chemical elicitors, where known, produced by these pathogens to induce the host defense should be explored. At least in one case they have been found to be very effective, practical and economically sound. Metlitsky et.al. (1986) reported that a lipoglycoprotein (LGP complex) elicitor isolated from *Phytophthora infestans,* when applied as a tuber treatment before sowing at the extremely low concentration of 0.0005%, was as affective as fungicide treatment against late blight and early leaf mold, and more effective against brown patch and scab. The acquired immunity was retained not only throughout the life span of the plants but also during storage of potatoes.

C. Immunization against Viral Diseases

1. Cross Protection

In viruses, phenomenon of cross protection although first demonstrated in 1929 could be utilized for the disease control under field conditions only after Rast (1972) produced a mild mutant (MII-16) in 1972 from a common tomato strain of TMV by nitrous acid mutagenesis. Since then this mutant has been used commercially and has been applied to a high proportion of glass house grown tomato crops in the Netherlands and the United Kingdom. Successful control of TMV with an attenuated mutant (LIIA), isolated from the tomato strain of TMV in plants treated with high temperature, was also reported in Japan. Cross protection has been used on a large scale to control citrus tristeza virus, a closterovirus that is important worldwide. Naturally occurring mild strains of CTV have been demonstrated to offer protection in the field. In Brazil, the number of protected sweet orange trees exceeded 8 milions in 1980, and no breakdown in protection was observed. Recently mild strain generated from severe strain of papaya ring spot virus (PRV) by nitrous acid mutagenesis was used to protect papaya plants against infection by severe strain of PRV under field condition and a high degree of protection was obtained.

Although cross protection is a general phenomenon of plant viruses, not all plant diseases caused by viruses can be controlled by using mild strain for preimmunization. Careful attention to the selection of the best protective isolates of virus and to their introduction into the crop to be protected is essential.

For practical application of cross protection, the mild protective virus should (1) not cause severe damage to the protected plants, (2) be stable for a long period, (3) protect plants against the effects of severe strains, (4) be suitable for infecting large number of plants, (5) not affect other crops in the vicinity of the crop protected, and (6) have no synergistic reaction with other viruses.

2. Genetically Engineered Cross Protection

By using recombinant DNA techniques, plant viral genes for coat protein have been isolated, characterized, cloned and transferred to host-plant cells using Ti-plasmid as a vector. Coat protein (CP) gene got integrated with host's genome and was expressed in transgenic plants. For some viruses it has been found that transgenic plants expressing a nuclear integrated coat protein gene show a significant degree of resistance to infection by homologous virus. In several ways this phenomenon resembled cross protection: the term "genetically engineered cross-protection" was coined by the Tumer et.al. to describe this phenomenon.

Transgenic tobacco plants that accumulate the CP of the UI strain of tobacco mosaic virus (TMV) showed delay in disease development upon inoculation with the UI strain. Besides resistance to UI strains, transgenic plants also showed reduced susceptibility to more severe PV 230 strain of TMV.

As a second example of genetically engineered cross-protection it was shown that transgenic plants containing alfalfa mosaic virus (AMV) coat protein gene resist infection by AMV. Only coat protein gene is capable of providing protection to the transgenic plants. Viral genes, other than coat protein, although expressed in transgenic plants, failed to provide protection against the viral infection. Recent studies have shown that coat protein gene has multifunctional role during infection that induce encapsidation, symptom expression, and differential elicitation of resistance gene. During last 10 years coat protein genes from a number of viruses have been characterized, isolated and cloned. In next few years it is expected that a number of transgenic plants containing viral coat protein gene would be generated.

CHAPTER XVIII

CHEMICAL CONTROL

I. INTRODUCTION AND CONCEPT

The word fungicide has originated from two latin words : viz. *fungus* and *caedo* (to kill). So literally, a fungicide would be any agency (physical or chemical) which kills a fungus. However, the word is restricted to chemicals. Hence, the word fungicide should mean a chemical capable of killing fungi. However, there are a number of compounds which do not kill the fungus. They simply inhibit growth or spore germination temporarily. If the fungus is freed from such substances, it would revive. Such a chemical is called a "FUNGISTAT" and the phenomenon as "FUNGISTASIS". Some other chemicals, like certain phenanthrene derivatives and Bordeaux mixture may inhibit spore production without affecting vegetative growth. These are called "ANTISPORULANTS". There are other groups of chemicals which exhibit very poor or no antifungal activity *in-vitro* condition but provide protection to the plants either by inhibiting the penetration of host surface by the fungi or by inducing the host defense system. The former type of chemicals are termed as "ANTIPENETRANTS"and later as "ANTIPATHIC AGENTS". Even though fungistats, antisporulants, antipenetrants, and antipathic agents do not "kill" fungi, they are included under the broad term fungicide because by common usage, the word fungicide has been defined as a chemical substance which has ability to prevent damage caused by fungi to plants and their products.

Fungicide which is effective only if applied prior to fungal infection is called"PROTECTANTS". On the other hand, fungicide which is capable of eradicating a fungus after it has caused infection, and thereby "curing" the plant, is called "THERAPEUTANT"

II. FORMULATIONS

Formulation is an art which is mainly concerned with methods of presenting the active ingredient, in the most effective form, that is, with regard to storage, application, and ultimate bioligical activity. Actually the amount of fungicide that is biologically active/effective at the plant surface is so minute that for economic use the chemical must be diluted, either with a solid dust,powder, etc.) or liquid (concentrates, emulsions, suspension, etc.) before application. Water provides a cheap and effective dilution medium and, with few exceptions is used as the carrier for agricultural sprays. As majority of the fungicides are of very low water solubility they must be formulated to make them compatible with water, hence surface active agents are required to prepare water-dispersible powders, stock emulsions, or emulsifiable concentrates. In addition, surface active agents may be needed to improve the suspending, spreading, and wetting properties of the sprays. Other supplements that may be incorporated with sprays include stickers to improve the weather resistance of the deposit, and materials to improve storage stability, deposition on plant surface and penetration of plant surfaces and/or fungal cells. The type of formulation chosen for a fungicide is determined by a variety of factors of which cost and biological efficiency are the most important. Different formulations of the pesticides (including fungicides) are listed in Table 1.

III. APPLICATION OF FUNGICIDES FOR DISEASE MANAGEMENT

A. Seed Treatment

Seeds, tubers, bulbs and other propagating materials are often treated with fungicides for eradication of the pathogen propagules present on and/or in them as well as for preventing their rot in the soil after planting. The chemicals can be applied on the seed as dusts, or as thick water suspension(slurry) mixed with the seed. The seeds can also be soaked in water or solvent solution of the chemical and then allowed to dry. On the basis of their mode of action the seed treatment chemicals can be of three types:

a) Those which kill the pathogen present on the seed and do not remain active for long after the seed has been planted (seed disinfestants).

b) Those wihich act as eradicants and destroy the pathogen established in seed tissues. A number of systemic chemicals are being used for this purpose (seed disinfectants), and

c) When seeds are treated with protective fungicides, the pathogens present on the seeds are destroyed and the compound remains on seed surface for sufficiently long times (seed protectants). In treating seeds and other propagative organs, it should be ensured that enough chemical has sticked to the seeds. It should also be ensured that their viability is not lowered or destroyed.

Table 1 : Different formulations of the pesticides

Formulations	Abbreviations
Wettable powder	WP (formerly w.p)
Dustable powder	DP (formerly d.p)
Water soluble powder	SP (formerly s.p)
Powder for seed treatment	DS
Solution for seed treatment	LS
Flowable concentrate for seed treatment	FS
Water soluble powder for seed treatment	SS
Water Dispersible powder for slurry treatment (of seed)	WS
Emulsifiable concentrate	EC (formerly e.c)
Electrochargeable liquid	ED
Emulsions, water in oil	EO
Emulsions, oil in water	EW
Oil miscible liquid	OL
Hot fogging concentrate	HN
Ultra low volume liquid	UL
Aerosol generator	AE
Gas generating product	GS
Granules	GR
Water soluble granules	SG
Water Dispersible granules	WG
Suspension concentrate (flowable)	SC
Soluble concentrate	SL

B. Soil Treatment

The purpose of soil treatment with chemicals is to reduce or eradicate the inoculum density of the pathogens present in soil as *soil invaders* or as *soil inhabitants*. The chemical soil treatment can be done in any of the following ways—

1. Soil Drenching :

Fungicidal suspension or solution is applied to soil surface in quantities enough to wet 10-15 cm depth of the soil before or after planting.

2. Broadcasting :

Non-volatile fungicides (dusts, powders,granules, etc.) are mixed with soil or fertilizers and spread on the soil surface. Light ploughing or harrowing is then done to mix the chemical in sufficient depth.

3. Furrow Application :

In this method fungicides are applied either as dusts or with water to the furrow at the planting time. This method is possible in crops planted in furrows/rows such as potato and sugarcane. This method is good for the control of diseases that occur at the base of the plants. This method requires much less quantity of the chemicals per hectare than the broadcast method.

4. Fumigation:

This method is usually used to control plant parasitic nematodes. The chemicals used are generally volatile and on coming in contact with soil moisture release gases which diffuse in the soil and kill the nematodes. Special equipments (guninjectors) are required for application since diffusion is restricted to certain distance. The chemicals are injected into soil at regular distance (12-15 inch) all over the field. The depth of application is kept at 6-9 inches.

C. SPRAYING

This is the most commonly adopted method. Spraying of fungicides is done on leaves, stems, flowers, and fruits. Wettable powders are most commonly used for preparing sprays. The most common diluent or carrier is water. The dispersion of the spray is usually achieved by its passage under pressure through nozzles of sprayers. Spraying is of two types, namely, high volume and low volume. When sprays involve large quantities of liquid per unit area, they are termed high volume. Six hundred litres and above per hectare would be considered to be of the high volume category. With low volume sprays, it is usually possible to cover one hectare with about 100 litres or less.

D. DUSTING

Dusts are applied to leaves, fruits, and stems of plants as an alternative to spraying. Dry powders are used for covering host surface. Dusting is practicable only in the calmest weather and high effectivity is obtained if the dust is applied when the plant is wet with dew or rain.

E. PASTES OR PAINTS

In orchards, the wounds created during pruning and training of trees often serve as entry points of the pathogens. To protect these wounds fungicidal pastes or paints are used as a protective layer. These pastes are prepared with fungicides in a suitable carrier such as raw linseed oil, lanolin, glycerine, etc. The residual effect of the pastes lasts long enough to permit natural healing of the cut surfaces.

IV. SYSTEMICITY

A fungicide when applied on the plant surface either may remain on the surface (non-systemic) or absorbed by the plant. The latter may remain there in treated plant parts (locosystemic); may move in the direction of evapotranspiration stream (apoplastic) or may move with photosynthates to "Sink" (Symplastic) or in both directions (ambimobile).

The symplast is the living part of the plant which is enclosed by membranes, i.e. protoplasts, and plasmodesmata, including the phloem sieve cells. Long-distance transport in phloem is symplastic. The apoplast is the nonliving part of the plant, i.e., cell walls and cuticle, including xylem vessels and tracheids. Long distance movement in the transpiration stream is apoplastic.

A. Apoplastic Translocation

Apoplastic translocation, within the plant, is usually directed from roots to transpiring areas, especially leaves(Fig. 1). Fungicides are absorbed by the roots,

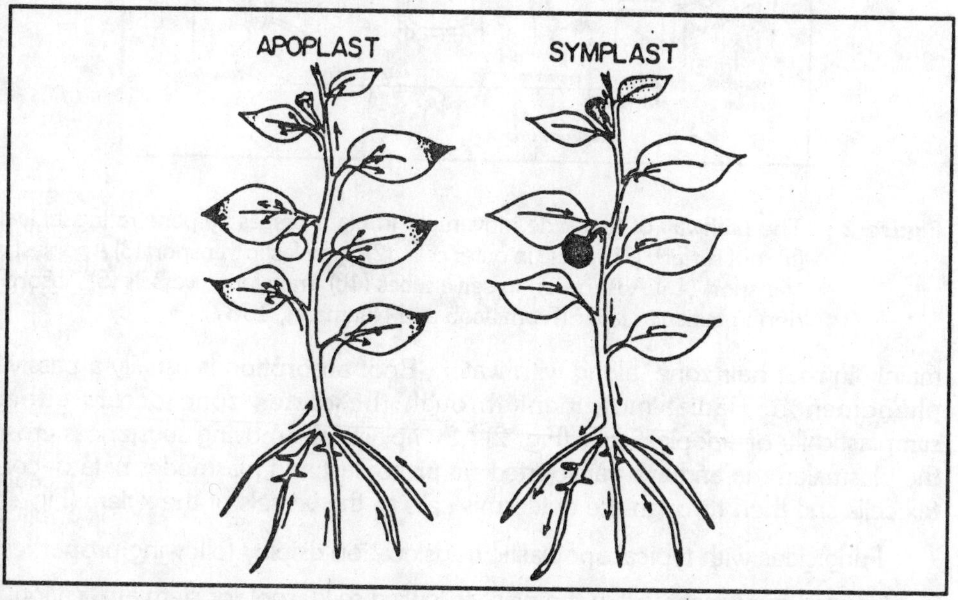
APOPLAST
SYMPLAST

Figure-1: Translocation patterns in the apoplast and symplast. Arrows indicate direction of transport, black indicates accumulation of systemic chemicals. Stippling indicates areas of lesser accumulation (from Edgington and Peterson, 1977)

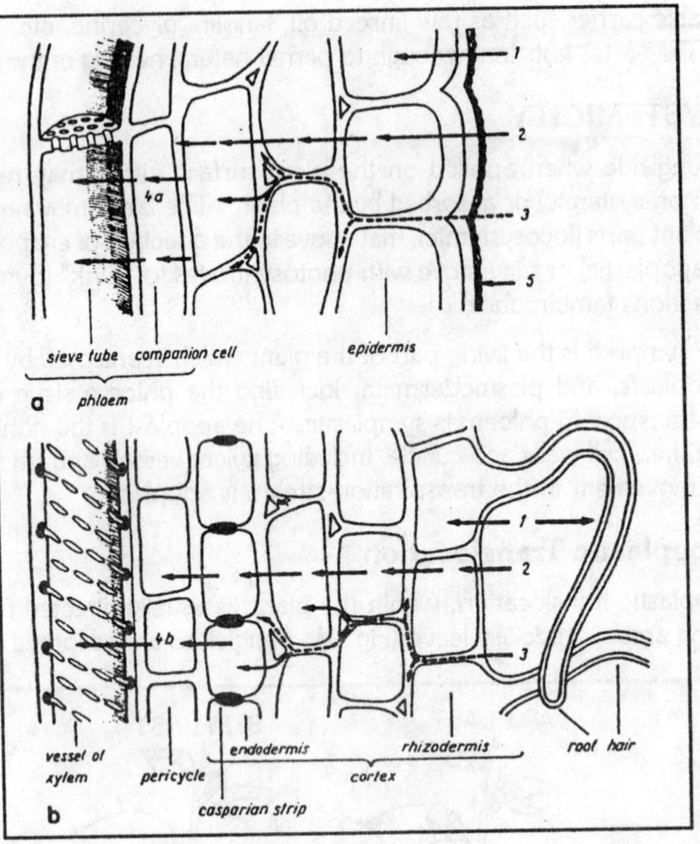
1
2
4a
3
5
sieve tube
companion cell
epidermis
a
phloem
1
2
4b
3
vessel of
xylem
pericycle
endodermis
rhizodermis
root hair
cortex
casparian strip
b

Figure-2 : The pathway of fungicide movement in plant tissues by penetrating (a) leaf (b) root surface (1) uptake in outer cells (2) Symplastic transport (3) Apoplastic transport (4a) Absorption in seive tubes (4b) Transfer to vessels (5) Absorption in cuticular layer (From Jcob and Naumann, 1987).

mainly in root hair zone, along with water. Root absorption is usually a passive phenomenon. Radial movement through the cortex zone occurs either symplastically or apoplastically (Fig. 2). Symplastically moving substances cross the plasmalemma and are transported via protoplast and plasmodesmata of cortex cells and then through the epidermis cells to the vessels of the xylem (Fig 2).

Fungicides with typical apoplastic translocation display following properties.

1. Upward movement within the plant following seed, root, or stem application.
2. Movement into various plant organs is dependent on their transpiration rate.
3. Accumulation at tips and margins of leaves. In monocotyledons, where ve-

nation is convergent palmate parallel type, metalaxyl is drawn much strongly through veins resulting in its accumulation at tips and margins.

4. In dicotyledons, because of reticulate venation, driving forces toward the periphery get drastically reduced. Thus, allowing metalaxyl to enter cytoplasm. In such cases, the fungicide is either uniformly distributed in entire lamina or accumulation at margins is quite delayed. However, majority of the fungicides with typical apoplastic translocation are accumulated at tips and margins of leaves irrespective of plant species involved.

B. SYMPLASTIC TRANSLOCATION

Symplastic movement takes place within the living parts of the cell. The movement is in the sieve tubes of the phloem tissues. This is an active movement and requires the expenditure of metabolic energy both for uptake inside the cells as well as for movement in the sieve tubes. Fully expanded photosynthesizing leaves serve as a source and roots, flowers, young growing leaves, and fruits serve as a sink. A fungicide with symplastic movement, after entry into the phloem follows the same source to sink pathway as followed by the phloem assimilates.

V. CLASSIFICATION OF FUNGICIDES

Ever since the introduction of systemic fungicides, systemicity has been the most popular criterion for the broad classification of the fungicides. They were classified as 1. NON-SYSTEMIC, and 2. SYSTEMIC. However, in this book selectivity has been preferred over systemicity and based on that fungicides are broadly classified as 1. Non-selective, and 2. Selective

A. Non selective fungicides

1. Sulfur Fungicides

Several inorganic and organic sulfur fungicides (Table 2) are in use for control of plant diseases. All sulfur fungicides are nonsystemic, and except lime sulfur, are nonphytotoxic (excluding "sulfur shy" varieties) and compatible with most of the pesticides. As far as their mode of action is concerned, elemental sulfur interferes with energy production by intercepting electron on the substrate side of cytochrome C in the mitochondrial electron transport system. The dialkyldithiocarbamates (Figure 3) are known to inhibit a multitude of enzymes; therefore, fungitoxicity probably involves concurrent inhibition of enzymes at several sites. The pyruvate dehydrogenase reaction is particularly highly sensitive to dialkyldithiocarbamates.

Table 2 : Sulfur Fungicides

Common name	Chemical name	Formulations	Uses
Sulphur	Sulphur	Elosol, Kumulus S, Fortho, Kolthior, Sodil B, Solfa, Ultranix, Imber sulfospor, Sandotox	Acaricide, fungicide, applied as dust or spray against powery mildews
Lime sulfur	Calcium polysulfide	Prepared by dissolving S (15 lbs) in suspension of calcium hydroxide (50 lbs rock lime/50 gallons of water), diluted to 10 ml/litre	Powdery mildews, scale insects, mites
Thiram TMTD	Tetramethyl thiuram disulfide	Arasan, Tersan, Pomarsol, Fernasan, etc.	wide spectrum, seed treatment, soil treatment, foliar sprays, bird, rodent repellent
Ziram	Zinc dimethyl dithio carbamate	Corozate, Cuman, Hexazir, Karbam, Milbam, Zerlate, Ziram, Zirberk, etc.	Commonly used as spray, broad spectrum
Ferbam	ferric dimethyl dithiocarbamate	Coromet, Ferbam, Ferberk, Fermate, Fermocide, Ferrodow, Hexaferb, etc	--do--
Zineb	Zinc ethylene bisdithio carbamate	Dithane Z -78, Hexathane, Lanocol, Parzate C, etc.	used as spray, brod spectrum
Maneb/ Mancozeb	Manganous ethylene bisdithiocarbamate	Dithane M-22, Manzate, MEB, MnEBD, etc.	--do--
Nabam	Sodium ethylenebis dithiocarbamate	Chembam,Dithane A-40, Dithane D-14, Parzate liquid, etc.	--do--
Vapam SMDC	Sodium methyldithiocarbamate	Vapam, VPM Chemvape 4-S, etc.	Broad spectrum, soil fungicide

2. Copper Fungicides

Ever since the discovery of Bordeaux mixture in 1885, copper fungicides predominated the field of fungicidal plant disease control for more than 50 years until synthetic organic fungicides invaded the market. Even today some of the copper compounds are used widely in many countries. Copper compounds currently in use are shown in Figure 4 and are described in Table 3. Copper is a multisite biochemical inhibitor (probably interact with-SH groups of enzymes) with little biological specificity.

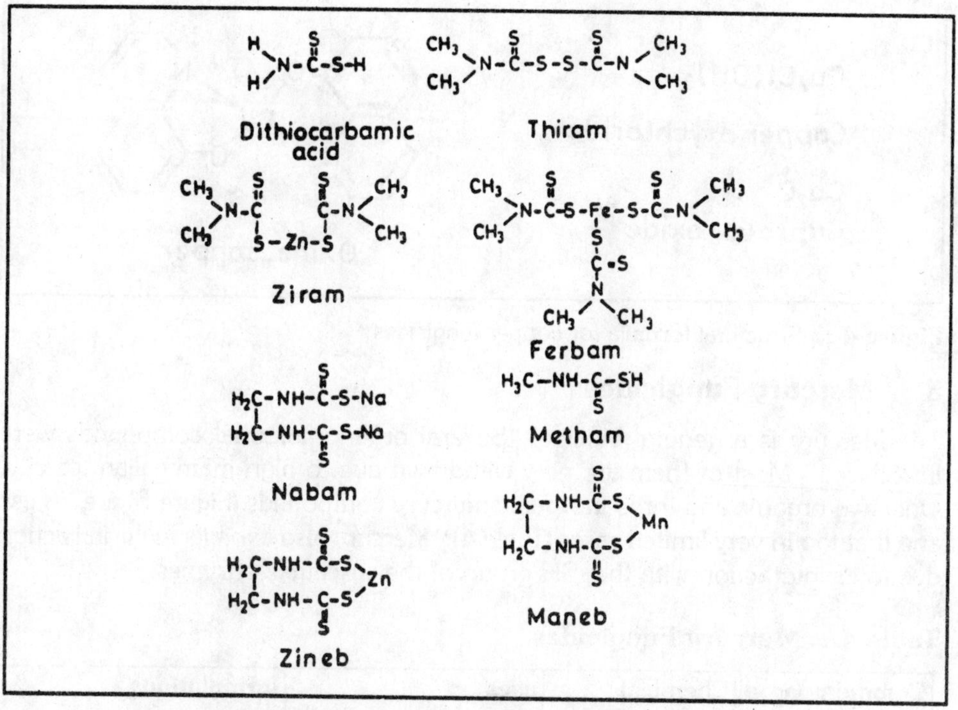
Dithiocarbamic acid
Thiram
Ziram
Ferbam
Metham
Nabam
Maneb
Zineb

Figure-3 : Structural formula for dithiocarbamate fungicides

Table 3 : Copper Fungicides

Common Name	Chemical Name	Formulations	uses
Bordeaux Mixture	Mixture with or without stabilizing agents, of calcium hydroxide and copper (II) sulfate	WP, SC, can be prepared as a tankmix using $CuSO_4$ $5H_2O$ (1.0 kg) with Ca $(OH)_2$ (1.25 kg) in 100 l water) for high volume sprays or (4kg + 2kg+ 100 l) for low volume sprays.	Protective spray fungicide, especially against Oomycetous fungi
Copper oxychloride	Dicopper chloride trihydroxide	Blimix 4%, Blitox-50, Cupramar, MicopD-06, Fytolan, Blue copper '50' etc.	Protective spray against Peronosporales fungi
Cuprous oxide	Copper (I) oxide	Fungimar, Perenox, Copper sadoz, etc.	Mainly for seed treatment, foliar application against blights, downy mildews, rusts
Oxine copper	Cupric 8 - quinolinoxide	Quinolate 400	Seed treatment in cereals

$Cu_2Cl(OH)_3$
Copper oxychloride

Cu_2O
Cuprous oxide

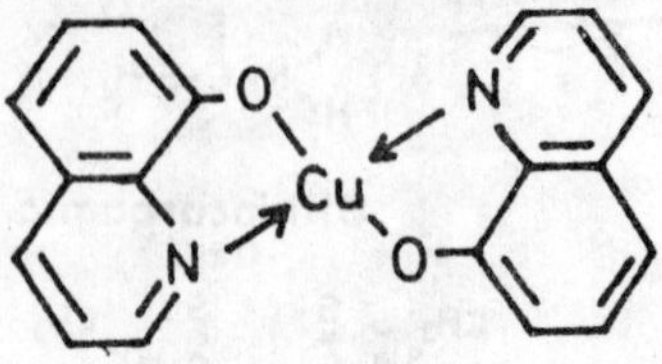

Oxine-copper

Figure-4 : Structural formula for copper fungicides

3. Mercury Fungicides

Mercury is a general biocide. Several organomercurial compounds were introduced. Most of them are now withdrawn due to high mammalian toxicity. Only two organic and three inorganic mercury compounds (Figure 5) are in use and that too in very limited cases (Table 4). Mercury also exhibits multisital action due to its interaction with the -SH group of the suscetible enzymes.

Table 4 : Mercury Fungicides

Common name	Chemical name	uses	formulations
Mercurous chloride	Mercury (I) chloride	Preplant soil application	'Cyclosan', M-C Turf fungicide DP, (40g/kg)
Mercuric chloride	Mercury (II) chloride	Turf fungicide	Mixtures: Merfusan, Mersil
Mercuric oxide	Mercuric oxide	As paint Bark, pruning cuts	Santar paint (30g/kg)
2-methoxy ethyl mercury silicate	-	Seed treatment	'Soprasan' DS
PMA	Phenyl mercury acetate	Eradicant seed treatment	Agrosan D, Ceresol, DS, WS

4. Phthalimide and Quinone Fungicides

One of the quinone derivatives, chloranil was introduced as a fungicide in 1940. Now it has been replaced by dichlone, another quinone derivative introduced in 1943. Properties of dichlone are listed in Table 5. Quinone fungicides are multisital in their mode of action. The two mechanisms which are thought to be most likely for dichlone and chloranil are 1. binding of the quinone nucleus to -SH and $-NH_2$ groups in fungal cell, and 2. interference with electron transport

$HgCl_2$

Mercuric chloride

HgO

Mercuric oxide

Hg_2Cl_2

Mercurous chloride

$CH_3OCH_2CH_2Hg$-silicate

2-methoxyethyl mercury silicate

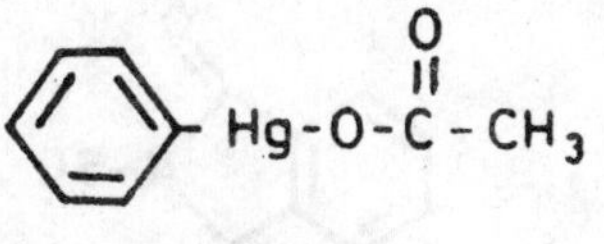

PMA

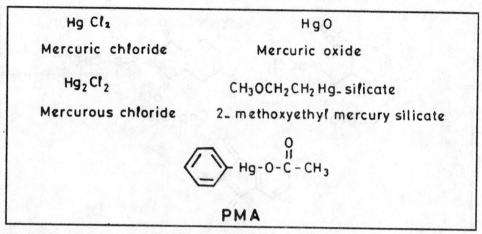

Figure-5 : Structural formula of mercury fungicides

system (Nene and Thapliyal, 1996). Among the phthalimide compounds, captan was introduced first and is still being used widely. Basically these fungicides (Figure 6, Table 5) are protective in nature but limited systemicity has been reported for captan and captafol. Phthalimides are highly reactive against thiol (-SH) groups of proteins (enzymes) and low molecular weight metabolites (cystein, glutathione, etc).

Table 5 : Phthalimide and quinone fungicides

Common name	**Chemical name**	**Formulations**	**uses**
Captan (SR 406)	N(trichloromethylthio) cyclohex-4-ene-1,2-dicarboximide	Orthocide 50 WP, Phytocape WP, etc.	As seed dresser, root dip, soil treatment, foliar sprays
Captafol (ortho 5865)	N(1,1,2,2,-tetrachloroethylthio) cyclohex-4-ene-1,2-dicarboximide	Captasor, Kenofol, Orthodifolatan, etc.	Widely used to control foliage and fruit diseases
Folpet	N-(trichloromethylthio) phthalimide	Acryptan, Phallan WP, etc.	Foliar sprays against several fungal diseases
Dichlone	2,3-dichloro-1,4-naph thoquinone	Phygon WP	used as seed dresser and foliar spreays

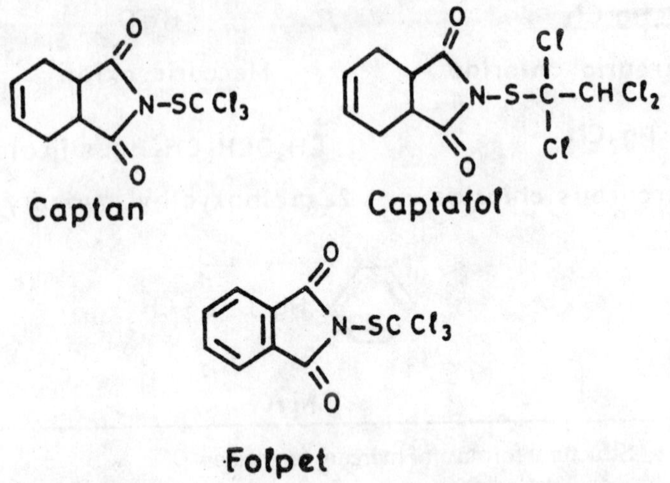
N-SC Cl3
Captan
N-S-C-CH Cl2
Cl
Cl
Captafol
N-SC Cl3
Folpet

Figure-6 : Chemical structures of phthalimide fungicides

B. Selective fungicides

1. Benzimidazoles and Related fungicides :

Benzimidazoles and thiophanates (Fig. 7) represent a group of highly effective broad spectrum systemic fungicides. Most Ascomycetes, some of the Basidiomycetes and Deuteromycetes, and none of the Phycomycetes are sensitive to these fungicides (Table 6). Their mild cytokinin like effects on some plants tend to retain chlorophyll and in some cases increase yield and delay maturity. Benzimidazoles bind with the β-tubulin subunit of microtubules of sensititve fungi and thereby inhibit formation of spindle and subsequently chromosomal separation during nuclear division.

Table 6 : Benzimidazoles and Related fungicides

Common name	Chemical name	Formulations	uses
Benomyl	Methyl 1-(butylcarbamoyl) benzimidazole-2-ylcarbamate	Benlate WP, Tersan, etc.	Seed treatment, soil treatment, sprays
Carbendazim	Methyl-2-benzimidazole carbamate (MBC)	Bavistin WP Agrozim WP Carbate, Derosol etc.	--do--
Fuberidazol	2-(2-furyl) benzimidazole	Veronit	seed treatment

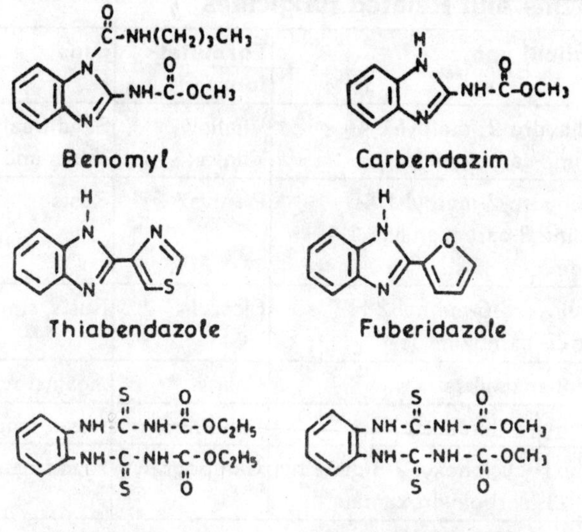
Benomyl
Carbendazim
Thiabendazole
Fuberidazole
Thiophanate
Thiophanate_methyl

Figure-7: Chemical structures of banzimidazoles and related fungicide

2. Carboxins and Related compounds

Among systemic fungicides carboxin and oxycarboxin (Fig. 8) were first to be discovered and introduced for plant disease control. These fungicides are readily absorbed by the seeds, roots, and leaves and translocated apoplastically. They are very effective mainly against Basidiomycetes, smuts, bunts, and rusts of cereals, and soil fungus *R. solani* (Table 7). Both have low animal toxicity and are quickly degraded. Their mode of action is very specific. They interact with succinate ubiquinone reductase complex resulting in inhibition of oxidation of succinate via electron transport chain.

3. Sterol Biosynthesis Inhibiting Fungicides

This group includes numerous fungicides from chemically heterogenous classes. However, all these compounds inhibit ergosterol biosynthesis in fungi and therefore, are classified as "ergosterol biosynthesis inhibitors" (EBIs) or sterol biosynthesis inhibitors(SBIs). They are effective against a wide range of fungi except Oomycetes which do not need sterol for growth. Several fungicides belonging to morpholines, piperidine, piperzine, pyridine and azole are well known sterol biosynthesis inhibitors.

Table 7 : Carboxins and Related fungicides

Common name	Chemical name	Formulations	uses
Carboxin	5,6-dihydro-2,-methyl-1,4-oxathine-carboxanilide	Vitaflow, Vitavax	Seed treatment against smuts and bunts
Oxycarboxin	5,6-dihydro-2-methyl-1,4-oxathine 3-carboxanilide 4,4 dioxide	Plantvax	Rusts
Pyracarbolid	3,4- dihydro-6-methyl-2 H-pyran-5- carboxinilide	Sicarol	Rusts, smuts, *R. solani*
Benodanil	2-iodobenzanilide	Calirus	Against rusts
Fenfuran	2-methyl-3- furanilide	Pano-ram	Rusts, smuts
Furmecyclox	methyl N-cyclohexyl-2.5-dimethyl furan -3- carbohydroxamate	Campogran	seed treatment

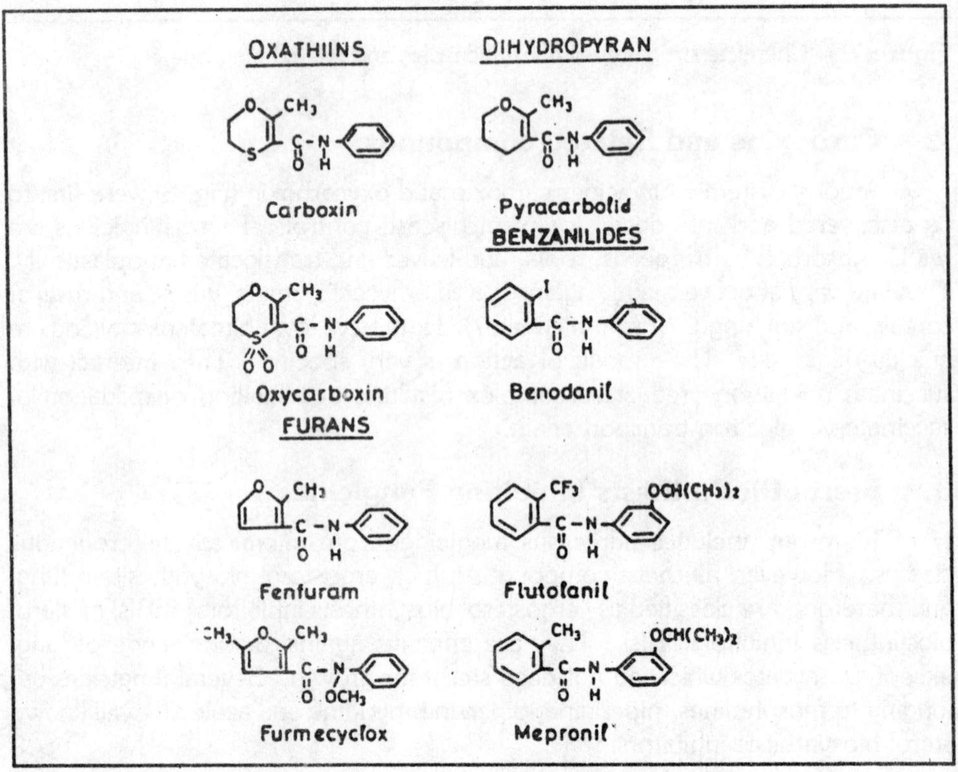
OXATHIINS
DIHYDROPYRAN
Carboxin
Pyracarbolid
BENZANILIDES
Oxycarboxin
Benodanil
FURANS
Fenfuram
Flutolanil
Furmecyclox
Mepronil

Figure-8 : Chemical structures of carboxins and related fungicides

4. Anti-oomycetes fungicides

The past decade has witnessed the introduction of five classes of fungicides controlling diseases caused by Oomycetes : 1. the carbamates (e.g. prothiocarb, propamocarb), 2. the isoxazoles (e.g. hymexazol), 3. the cyanoacetamide oximes (e.g. cymoxanil), 4. ethyl phosphonates (e.g. fosetyl-AL), and 5. phenylamides. The last class, phenylamides, covers three groups of compounds: 1 acylalanines (e.g. furalaxyl, metalaxyl, benalaxyl), 2. acylamino butyrolactones (e.g. ofurace), and 3. acylamino-oxazo lidinones (e.g. cyprofuram, oxadixyl). (Table 8). All these fungicides show comperatively high water solubility. Main practical uses and mode of application are listed in Table 8.

Table 8 : Antioomycetes fungicides.

Common Name	Pathogens on roots/stems	Foliar pathogen	Additional activity against nonoomycetes Pathogens
Hemaxazole	*Aphanomyces, Pythium*	--	*Corticium sasaki* *Fusarium* spp.
Propamocarb/ Prothiocarb	*Aphanomyces, Pythium, Phytophthora*	*Bremia, Peronospora, Pseudoperonospora*	--
Fosetyl	*Phytophthora, Pythium*	*Bremia, Plasmopara, Peronospora, Pseudoperonospora*	*Phomopsis* sp. *Guinardia* sp.
Phenylamides	*Peronosclerospora Phytophthora, Pythium, Sclerospora, Sclerophthora*	*Albugo, Bremia, Peronospora, Phytophthora, Plasmopara, Sclerospora*	--

5. Antipenetrants

There are several compounds which show no or poor fungitoxicity under *in-vitro* conditions but prevent infection by interfering with the penetration process Table 9).

Table 9 : Antipenetrants

Common name	Main uses	Formulations
Validamycin	Narrow range of fungi particularly *R. solani* (Rice, Potato and Vegetables)	Validacin, Valimon SL (30 g/l); DP (3g/kg)
Phthalide	*Magnaporthe grisea* Rice (mainly used as spray)	Rabcide, WP (300-500g/kg); DP (25g/kg), SC(200g/l)
Tricyclazole	*Magnaporthe grisea* -Rice (applied as flat drench, root soak or foliar spray)	Bean,Bim, Blascide Wp (200 or 750 g/kg); DP (10 g/kg); GR (30 g/kg)
Pyroquilon	--do--	caratop-2, caratop 5 GR (20 or 50g/kg); fungorene 50 wp (500g/kg)

6. Host Defense Inducing Compounds

Various chemicals which have been found to trigger host resistance mechanism and by doing so provide protection against diseases on prior or sometimes simultaneous application with the pathogen. They include both chemical constituents of plants or microbes (biomolecules or metabolites) and synthetic compounds which are foreign to the living system (xenobiotics). Informations pertaining to such compounds are summarized in Table 10.

Table 10 : Host Defense Inducing compounds

Chemical	Effective Against	Treated Plants' Response following infection
Probenazole or oryzemate (3-allyloxy-1,2-benzisothia-zole-1,1-dioxide)	*M. grisea* – Rice *X. campestris* pv. *oryzae*- Rice	• Augmentation of peroxidase, PAL, and catechalo – methyl transferase • Induced lignification • Accumulation of antimicrobial compounds
DDCC or WL 28325 (2,2-dichloro-3,3-dimethyl cyclopropane carboxylic acid	*M. grisea*– Rice	• Hypersensitive reaction • Accumulation of phytoalexins • Interference with host parasite recognition

Chemical	Effective Against	Treated Plants' Response following infection
PTU (phenylthiourea)	*Cladosporium cucumerinum*-cucumber	• Enhanced lignification arround penetration site • Increased PO activity
Fosetyl – AL	Peronosporales	• Induced phenolic accumulation
Metalaxyl	Peronosporales	• Hypersensitivity • Papillae formation • Accumulation of phytoalexins

7. Miscellaneous Selective Fungicides

There are several miscellaneous fungicides (Table 11) belonging to diverse chemical groups which show at least some degree of selectivity in their antifungal spectrum. This selectivity is particularly for the compounds like anilazine, fenitropan, chinomethionate, dichlofluanid, and chlorothalonil, which interact with NH_2 and /or —SH groups and hence, their mode of action is multisital.

Table 11 : Miscellaneous selective Fungicides

Common name	Chemical name	Formulations	uses
Anilazine	4,6 dichlore-N-(2chlorophenyl)-1,3,5-triazin-2-amine	Dyrene,Direz, Kemate WP (500g/kg)	*Botrytis, Septoria, Colletotrichum, Helminthosporium, Fusarium, Alternaria*
Fenitropan	(IRS,2RS)-2-nitro-1-phenyl-trimenthylenedi (acetate)	Volparox	As seed treatment for cereals, maize rice, etc
Dinocap	Isomeric reaction mixture mainly 2-(1-methylheptyl) 4,6 dinitrophenyl crotonate	Karathane WP, Sialite WP	Powdery mildews on various fruits and ornamentals
Binapacryl	2-sec-butyl-4,6 dinitrophenyl 3-methylcrotonate	Acricid, Endosan, Morocide	Powdery mildew of various fruits and crops acaricide
Chinomethionate	6-methyl-1,3-dithiolo(4,5, 6) quinoxalin-2-one	Morestan WP, Morestan2, DP	--do--
chlorothalonile	Tetrachloroisoph thalonitrile	Bravo W-75, Bravo 500	Broad range of fungal pathogens

REFERENCES

Adams, M.J. 1991. Transmission of plant viruses by fungi. *Ann. Appl. Biol.* 118: 479.

Admas, P. 1990. The potential of mycoparasites for biological control of plant diseases. *Annu. Rev. Phytopathol.* 28: 59.

Agnihotri, V.P. 1983. *Diseases of Sugarcane,* Oxford & IBH Publicing Co., New Delhi. 363.

Agrawal, V.K. and Sinclair, J.B. 1987. *Principles of Seed Pathology*, Vols., I. and II, CRC Press, Boca Raton, FL, 176.

Agrios, G.N. 1988, *Plant Pathology,* Academic Press, New York. 803.

Ainsworth, G.C. 1966. A general purpose classification of fungi. *Bibliography systematic mycology*, 1-4 common wealth mycol. Inst. Kew, Surrey.

Ainsworth, G.C. 1971. *Dictionary of Fungi* 6th Ed. C.M.I., Kew.

Ainsworth, G.C. 1973. Introductions and keys to higher texa. in: *The fungi: An advanced Treatise* Vol. IV Ainsworth G.C. Sparrow, F.K. and Sussman, A.S., Eds. Academic Press, New York, 1973.

Ainsworth, G.C. 1983. Ainsworth's & Bisby's *Dictionary of Fungi* 7 ed. Common Wealth Mycological Institute, Kew, Surrey.

Ainsworth, G.C.; Sparrow, F.K. and Sussman, A.S. 1973. *The fungi: An Advance Treatise*. Academic Press, New York.

Alcom, J.L. 1988. The taxonomy of "*Helminthospoprium*" species. *Annu Rev. Phytopathol.* 26: 37.

Alexopoulos, C.J. 1962. *Introductory Mycology.* II ed. John Wiley & Son, Inc, New York.

Alexopoulos, C.J. 1979. *Introductory Mycology.* III ed. John Wiley & Son, Inc, New York, London.

Alexopoulos, C.J. and Mims, C.W. 1983. III ed. Wiley Eastern Limited, India.

Alexopoulos, C.J.; Mims, C.W. and Blackwell, M. 1996. *Introductory Mycology.* 4th ed. Wiley, New York.

Altman, J and Campbell, C.L. 1977. Effect of herbicides on plant diseases. *Annu. Rev. Phytopathol.* 15: 361.

Andrews, J.H. 1992. Biological control in the phyllosphere. *Annu Rev. Phytopathol.* 30: 603.

Anonymous. 1994. *"Common Names of Plant Disease"* APS Press, st. Paul, Minnesota.

Apple, J.L. 1977. The theory of disease management, pp 79-102. *In*: J.G. Horsfall and E.B. Cowling (eds.), *Plant Disease-an advanced treatise* Vol. I, Academic Press, New York.

Asada, Y., Bushnell, W.R. Ouehi, S., and Vance, C.P., eds. 1982. *"Plant Infection: The physiological and biochemical basis"* springer verlag, Berlin and New York.

Bailey, J.A. and Deverall, B.J., eds. 1983. *"The Dynamics of Host Defense"* Academic Press, New York.

Bailey, J.A. 1986. *Biology and Molecular Biology of Plant Pathogen interactions,* Springer-Verlag, Berlin.

Bailey, J.A. and Mansfield. J.W., eds. 1982. *"Phytoalexins"* Wiley, New York.

Baker, K.F. and Cook, R.J. 1974. *Biological control of Plant Pathogens.* W.H. Freeman, San Fransisco, California 433p.

Baker, K.F. and Smith, S.H. 1966. Dynamics of seed transmission of plant pathogens. *Annu. Rev. Phytopathol.* 4: 311.

Baker, K.F. and Snyder, W.C. (ed.) 1965. *Ecology of Soil borne plant pathogens.* University of California Press, Berkeley, 571.

Baker, R. 1968. Mechanism of biological control of soil borne pathogens. *Annu. Rev. Phytopathol.* 6: 263.

Baldwin, J.G. 1992. Evolution of cyst and non cyst-forming Heteroderinae. *Annu. Rev. Phytopathol.* 30: 271.

Ball, A.K. 1988. Pathogenesis and host parasite specificity in *Rbizobium species In : Experimental and Conceptual Plant Pathology* Vol. II Singh R.S.; Singh, U.S.; Hess, W.M. and Weber, D.J., Eds. Garden and Breach, New York and IBH Publishing, New Delhi, 247.

Barnett, H.L. and Binder, F.L. 1973. The fungal host parasite relationship. *Annu. Rev. Phytopathol.* 11: 273.

Barnett, H.L. and Hunter, B.B. 1972. *"Illustrated Genera of Imperfect Fungi"* Burgess, Minneapolis, Minnesota.

Barr, D.J.S. 1992. Evaluation and Kingdoms of organisms from the perspective of a mycologist. *Mycologia.* 84: 1.

Barras, F., Van Gijseman, F. and Chatterjee, A.K. 1994. Extracellular enzymes and pathogenesis of sof-rot Erwinias. *Annu. Rev. Phytopathol.* 32: 201.

Bawden, F.C. 1964. *Plant Viruses and Virus diseases,* Ronald Press New York. 361.

Beachy, R.N., Loesch-Fries, S., and Tumer, N.E. 1990. Coat protein mediated resistance against plant viruses. *Annu. Rev. Phytopathol.* 28: 451.

Berger, R.D. 1977. Application of epidemiological principles to achieve plant disease control. *Annu. Rev. Phytopathol.* 15: 165.

Berkeley, F.A. 1968. *Outline classification of organisms*. Hoppkins press, Providence Massachusetts. 205.

Bird, P.M. 1988. The role of lignification in plant disease, *In : Experimental and conceptual plant pathology* Vol. III. Singh, R.S.; Singh, U.S.; Hess, W.M. and Weber, D.J. Eds., Gardon and Breach. New York and IBH Publishing, New Delhi. 523.

Bol, J.F., Linthorst, H.J.M. and Cornelissen, B.J.C., 1990. Plant Pathogenesis-related proteins induced by virus infection. *Annu. Rev. Phytopathol.* 28: 113.

Bolkan, H.A. and Reinert, W.R. 1994. Developing and implementing IPM strategies to assist farmers: An Industry approach. *Plant Dis.* 78: 545.

Booth, C. 1971. "*The Genus Fusarium*" Common Wealth Mycol. Inst., Kew, England.

Bos L. 1983. *Introduction to plant virology*, Pudoc, Wageningen. 160.

Bourke, P.M.A. 1970. Use of weather information in the prediction of plant disease epiphytotics, *Annu. Rev. Phytopathol.* 8: 345.

Bove, J.M. 1984. Wall-less prokaryotes of plants. *Annu. Rev. Phytopathol.* 22: 361.

Bowles, D.J. 1990. Defense related proteins in higher plants. *Annu. Rev. Biochem*. 59: 873.

Bradbury, J.F. 1983. The new bacterial nomenclature. What to do? *Phytopathology* 73: 1349.

Bradbury, J.F. 1986. "*Guide to Plant Pathogenic Bacteria*" CAB Int. Mycol. Inst., Kew, Surrey, England.

Brasier, C.M. 1992. Evolutionary Biology of *Phytophthora*. *Annu. Rev. Phytopathol.* 30: 153.

Brown, D.J.F., Roberston, W.M., and Trudgill, D.L. 1995. Transmission of viruses by plant nematodes. *Annu. Rev. Phytopathol.* 33: 323.

Brown, M.E. 1974. Seed and root bacterization. *Annu. Rev. Phytopathol.* 12: 181.

Brown, R.H. and Kerry, B.R., eds. 1987. "*Principles and Practices of Nematode control in crops*" Academic Press, Sydney.

Browning, J.A. Simous M.D. and Torres, E. 1977. Managing host genes: epidemiological and genetic concepts, in: *Plant Diseases An Advanced Treatise*, Vol. 1, Horsfall, J.G. and Cowling E.B. eds., Academic Press, New York. 191.

Bruchl, G.W. ed. 1975. *"Biology and control of soil borne plant pathogens"* APS Press,St. Paul, Minnesota.

Bruchl, G.W. 1991. Plant Pathology, a changing profession in a changing world. *Annu. Rev. Phytopathol.* 29: 1.

Buchanan, R.E. and Gibbons, N.E. eds. 1974. *Bergey's Manual of Determinative bacteriology, 8th Ed.* William and Wilkins, Baltimore. Md 1268.

Buczacki, S.T., ed. 1983. *Zoosporic Plant Pathogens: A Modern Perspective.* Academic Press, London.

Buddenhagen, I.W. 1965. The relation of plant pathogenic bacteria to the soil *In : Ecology of Soil Borne Plant Pathogens- Prelude to Biological Control.* Baker, K.F. and Snyder W.C. Eds. Univ. of California Press, Los Angeles. 571.

Burkholder, W.H. 1948. Bacteria as plant pathogens. *Annu. Rev. Microbiol.* 2: 389.

Burnett, J.H. 1968. *Fundamentals of Mycology,* Edward Arnold, London.

Callow, J.A. 1983. *Biochemical Plant Pathology,* Wliley, New York.

Callow, J.A., Estrada-Garcia, M.T. and Green, J.R. 1987. Recognition of non-self: the causation and avoidance of disease, *Ann. Bot.* 60. Suppl. 4, 3.

Campbell, R. 1985. *Plant Microbiology.* Edward Arnold, London, 191p.

Cambell, C.L. and Madden, L.V. 1990. *Introduction to Plant Disease Epidemiology.* Wiley, New York.

Carlile, M.J., and Watkinson, S.C. 1994. *"The Fungi".* Academic Press, San Diego.

Carlson, G.A. and Main, C.E. 1976. Economics of disease loss management. *Annu. Rev. Phytopathol.* 14: 381.

Chakravarty, A.K. and Shaw,M. 1977. A possible molecular basis for obligate host-pathogen interactions. *Biol. Rev.* 52: 147.

Charles, T.C., Jin, S. and Nester, E.W. 1992. Two component sensory transduction systems in phytobacteria. *Annu. Rev. phytopathol.* 30:463

Chattopadhyay, S.B. 1985. *Principles and Proceducers of Plant Protection,* Oxford & IBH Publishing Co. New Delhi. 547.

Chaube, H.S. 1987. *Ascochyta Blight of chickpea. Technical report,* Expt. St., G.B. Pant University of Agric. Technol. Pantnagar, India. 110.

Chaube H.S. 1989. Pathogen suppressive Soils In: *Perspectives of Phytopathology* : Agnihotri, V.P.; Chaube, H.S.; Singh, N.; Singh, U.S. and Dwivedi, T.S., eds. Today and Tomarrow's Printers, New Delhi.

Chaube, H.S. and Singh, R.A. 1984. Survival of plant pathogenic bacteria, *In: Progress in Microbial Ecology.* Mukharjee, K.G., Agnihotri, V.P. and Singh, R.P. eds. Print House (India). Lucknow. 653.

Chaube, H.S. and U.S. Singh. 1991. *Plant Disease Management: Principles and Practices,* CRC Press, Boca Raton, Ann Arbor, Boston, USA. 319p.

Chaube, H.S. Singh, U.S. and Razdan, V.K. 1987. Studies on uptake, translocation and distribution of I&C metalaxyl in pigeonpea, *Indian Phytopathol.* 40: 507.

Chaube, H.S.; Singh, U.S.; Mukhopadhyay, A.N. and Kumar, J. 1992. *Plant Diseases of International Importance.* Vol. II: Diseases of Vegetables and oilseed crops, Prentice-Hall, Englewood Cliffs, New Jersey, USA, 376p.

Chen, M. and Alexandre, M. 1973. Survival of soil bacteria during prolonged dessication. *Soil Biol. Biochem.* 5: 213-221.

Chet, I. 1987. *Trichoderma*-application, mode of action and potential as a biocontrol agent of soil borne plant pathogenic fungi *In: Innovative approaches to plant disease control,* Chet. I. Ed. John Wiley and Sons. New York. 372.

Chet, I. Ed. 1987. *Innovative Approaches to plant disease control,* John. Wiley and Sons, New York. 372.

Coakley, S.M. 1988. Variation in climate and prediction of disease in plants. *Annu. Rev. Phytopathol.* 26: 163.

Cole, R.M. 1979. Mycoplasma and spiroplasma viruses: Ultra-structure, pp. 385-410. In: M.F. Barile and S. Razinleds *The Mycoplasmas*, Vol I, *Cell Biology.* Academic Press, New York.

Cole, R.M., Tully, J.G., Popkin, T.J. and Bove, J.M. 1973. Morphology, ultrastructure and bacteriophage infection of the helical mycoplasma like organism (*Spiroplasma citri*) cultured from stubborn disease of citrus. *J. Bacteriol.* 115: 367.

Colhoun, J. 1973. Effect of environmental factors on plant disease. *Annu. Rev. Phytopathol.* 11: 343.

Collmer, A. and Keen, N.T. 1986. The role of pectic enzymes in plant pathogenesis. *Annu. Rev. Phytopathol.* 24: 383.

Cook, R.J. 1977. Management of the associated microbiota, in: *Plant Disease – An Advanced Treatise* Vol. 1. Horsfall, J.G. and Cowling, E.B. Eds. Academic Press, New York. 465.

Cook, R.J. 1981. Biological control of plant pathogens, in: *Biological control in crop production,* Papavizas G.C., Ed. Allanhold, Osmun. 461.

Cook, R.J. 1993. Making greater use of introduced microorganisms for biological control of plant pathogens. *Annu. Rev. Phytopathol.* 31: 53.

Cook, R. and Baker, K.F. 1983. *The Nature and Practices of Biological Control of Plant Pathogens,* Am. Phytopathol. Soc. St.Paul Minnesota. 539.

Commonwealth Mycological Institute 1970. *Description of Plant Viruses,* CMI, Kew Surrey, England.

Cowling, E.B. and Horsfall, J.G. 1978. Prologue: How disease develops in populations, *In: Plant Disease: An Advanced Treatise*, Vol. II. Horsfall, J.G. and Cowling E.B. Eds. Academic Press, New York. 436.

Carbett, M.K. and Sisler, H.D. 1964. *Plant Virology*, University of Florida Press, Gainesville. 527.

Crosse, J.E. 1968. Plant Pathogenic bacteria in soil, pp. 522-572. In: T.R.C. Gray and D. Parkinson (eds.) *Ecology of soil bacteria.* Liverpool, University Press, Liverpool, 522.

Cruickshank, I.A.M. 1963. . Phytoalexins. *Annu. Rev. Phytopathol.* 1: 351-374.

Daly, J.M. 1984. The role of recognition in plant disease. Annu. Rev. Phytopathol. 22: 273.

Daly, J.M. and Deverall, B.J., eds. 1983. *Toxins in Plant Pathogenesis* Academic Press, New York.

Daly, J.M. and Vritani, I. 1979. *Recognition and specificity in plant parasite interactions.* University of Tokyo Press, Tokyo.

Daniels, M.J. 1983. Mechanism of spiroplasma pathogenicity. *Annu. Rev. Phytopathol.* 21: 29.

Daniels, M.J. 1984. Molecular biology of bacterial plant pathogens. *Microbiological sciences* 1: 33.

Daniels. M.J.; Dow, J.M. and Osbourn, A.E. 1988. Molecular genetics of pathogenicity in phytopathogenic bacteria. *Annu. Rev. Phytopathol.* 26: 285.

Dasgupta, S.N. 1958. History of plant pathology and mycology in India, Burma and Ceylone. Pub. *Indian Bot. Soc.*, 118p.

Dasgupta, M.K. 1977. Concept of disease in plant pathology and its application elsewhere *Phytopathol.* Z. 88: 136.

Davis, M.J.; Lawson, R.H.; Gillaspie, A.G. Jr. and Harris R.W. 1983. Properties and relationship of two xylem limited bacteria and a mycoplasma like organism infecting bermudagrass. *Phytopathology.* 1973: 341.

Davis, M.J.; Gillaspie, A.G. Jr.; Harris; R.W. and Lawson, R.H. 1980. Ratoon stunting disease of sugarcane. Isolation of the caused bacterium. *Science* 240: 1365.

Davis, M.J.; Gillaspie, A.G. Jr.; Vidaver. A.K. and Harris, R.W. 1984. Clavibacter: a new genus containing some phytopathogenic coryneform bacteria, including *Clavibacter xyli sub. sp. xyli* sp. Nov., Sub sp. Nov. and *Clavibacter xyli sub. sp. cynodontes* sub. sp. Nov., pathogens that cause ratoon stunting disease of sugarcane and bermudagrass stunting disease, *Int. J. Syst. Bacteriol.* 34: 107.

Davis, M.J.; Purcell, A.H. and Thomson, S.V. 1978. Pierces disease of grapevine. Isolation of the bacterium. *Science,* 199: 75.

Davis, R.E. and whitcomb, R.F. 1971. Mycoplasmas, Rickettsiae, and Chlamydiae: possible relation to yellows diseases and other disorders of plants and insects. *Annu. Rev. Phytopathol.* 9: 119.

Davis, M.J.; Whitcomb, R.F. and Gillaspie, A.G. Jr. 1981. Fastidious bacteria of plant vascular tissue and invertibrates (including so called rickettsia like bacteria) in the prokaryotes: *A handbook on habitats, isolation and identification of bacteria.* Vol. 2. Starr, M.P.; Stolp, H.; Truper, H.G.; Balows, A. and schegel H.G. Eds., Springer Verlag, Berlin. 2172.

Davis, R.E., Worley, J.F., Whitcomb, R.F. Ishiyama, T. and steere, R.L. 1972. Helical Filaments produced by a mycoplasma like organism associated with corn stunt disease, *Science,* 176: 521.

David, P. 1965. Survival of microorganisms in soil, *In: Ecology of soil borne plant pathogens.* Baker, K.F. and snyder W.C. Eds. Univ. of California Press, Los Angeles. 571.

Day, P.R. 1966. Recent developments in the genetics of host parasite system. *Annu. Rev. Phytopathol.* 4: 245.

Day, P.R. 1974. *Genetics of host-parasite interaction* W.H. Freeman and Co., San Fransisco 238p.

Deward, M.A., *et.al.* 1993. Chemical control of plant diseases: Problems and prospects. *Annu. Rev. Phytopathol.* 31: 403.

De Wit, P.J.G.M. 1992. Molecular characterization of gene for gene systems in plant fungus interactions and the application of avirulence genes in control of plant pathogens. *Annu. Rev. Phytopathol.* 30: 391.

Diener, T.O. 1971. Potato spindle tuber "virus". IV. A replicating, low molecular weight RNA, *Virology.* 45: 411.

Diener, T.O. 1971. Potato spindle tuber: A plant virus with properties of a free nucleic acid. III. Subcellular location of PSTV-RNA and the question of whether virions exist in extracts or *in situ.* *Virology,* 43: 75.

Diener, T.O.; Owens, R.A. and Hammond, R.W. 1993. Viroids: The smallest and simplest agents of infectious disease. How do they make plants sick? *Virology*35: 186.

Dixon, R.A., Harrison, M.J. and Lamb, C.J. 1994. Early events in the activation of plant defense responses. *Annu. Rev. Phytopathol.* 32: 479.

Doi, Y.; Teranka, M.; Yora, K. and Asuyama, H. 1967. Mycoplasma or PLT group like microorganisms found in the phloem elements of plants infected with mulberry dwarf, potato witches broom, aster yellow, or paulownia witches broom. *Ann. Phytopathol. Soc. Jpn.* 33: 259.

Dropkin, V.H. 1988. *Introduction to plant nematotology.* Wiley, New York.

Dye, D.W. 1968. A taxonomic study of the genus *Erwinia.* I. the "Amylovona group" *N.Z. Jour. Sci.* 11: 590.

Dye, D.W. 1969a. A taxonomic study of the genus *Erwinia*.II. "The carotovora group". *N.Z. Jour. Sci.* 12: 81.

Dye, D.W. 1969b. A taxonomic study of the genus *Erwinia*. III. The "herbicola group". *N.Z. Jour. Sci.* 12: 223.

Dye, D.W. 1969c. A taxonomic study of the genus *Erwinia*. IV. A typical Erwinias. *N.Z. Jour. Sci.* 12: 833.

Dye, D.W. 1974. the problem of nomenclature of the plant pathogenic *Pseudomonads. Rev. Plant Pathol.* 53: 953.

Dye, D.W. 1975. Proposal for a reappraisal of the status of the names of plant pathogenic *Pseudomonas species. Int. J. Syst. Bacteriol* 25: 252.

Dye, D.W.; Bradburg, J.F.; Goto, M.; Hayward; A.C., Lalliott R.A. and Schroth, M.N. 1980. International standards for many pathovars of phytopathogenic bacteria and a list of pathovar names and pathotype strains. *Review of plant pathology* 59: 153.

Dye, D.W. and Kemp, W.J. 1977. A taxonomic study of plant pathogenic *Corynebacterium species N.Z. Jour. Agric. Res.* 20: 563.

Erwin, D.C.; Bartnicki-Garcia, S. and Tsaao, P.H., eds. 1983. "*Phytophthora*: 1st biology, taxonomy, ecology and pathology". APS Press, St. Paul., Minnesota.

Fahy, P.C. and Presloy, G.J. (eds.). 1983. *Plant bacterial diseases: A dignostic guide.* Academic Press, New York. 393p.

Fitt, B.D.L., Mc Cartney, H.A. and Walklate, P.J. 1989. The role of rain in dispersal of pathogen inoculum. *Annu. Rev. Phytopathol.* 27: 241.

Flor, H.H. 1942. Inheritance of pathogenicity in *Melampsora lini*, *Phytopathology,* 32: 653.

Flor, H.H., 1947. Inheritance of reaction to rust in flax, *J. Agric. Res.* 74: 241.

Flor, H.H. 1955. Host parasite interaction in flax rust – its genetic and other implications. *Phytopathol.* 45: 680.

Flor, H.H. 1956. Complimentary genic system in flax and flax rust. *Advan. Genet.* 8: 267.

Flor, H.H. 1971. Current status of the gene for gene concept. *Annu. Rev. Phytopathol.* 9: 275.

Frankel Conrat, H. 1968. *The molecular basis of virology,* Reinhold, New York. 656.

Fraser, R.S.S. 1985. Mechanisms of resistance to plant diseases, Nijhoff/Junk. Dardecht.

Fravel, D.R. 1988. Role of antibiosis in the biocontrol of plant diseases. *Annu. Rev. Phytopathol.* 26: 75.

Fritig, B. and Legrand, M. eds. 1993. *Mechanisms of Plant Defense Responses.* Kluwer, Hingham, Massachusetts.

Fry, W.E. 1987. *Principles of Plant Disease Management,* Academic Press, New York. 378.

Fry, W.E. and Thurston, H.D. 1980. The relationship of plant pathology to integrated pest management. *Bio-Science.* 665.

Fulton, R.W. 1986. Practices and precautions in the use of cross protection for plant viruses disease control. *Annu. Rev. Phytopathol.* 24: 67.

Gabriel, D.W. and Rolfe, B.G. 1990. Working models of specific recognition in plant microbe interactions. *Annu. Rev. Phytopathol.* 28: 365.

Gaunt, R.E. 1995. The relationship between plant disease severity and yield. *Annu. Rev. Phytopathol.* 33: 119.

Geargopoulos, S.G. 1987. The genetics of fungicides resistance, *In : Modern Selective Fungicides – Properties, Applications, Mechanisms of Action.* Lyr. H. Ed. Longman Group UK Ltd. London and VEB Gustav Fescher Verlag. Jena, Chap. 4.

Gibbs, A.H. and Harrison, B.D. 1976. *Plant Virology- The principles* Edward Arnold, London.. 292.

Goodman, R.N., Kiraly, Z. and Wood, K.R. 1986. *The Biochemistry and Physiology of Plant Disease.* Univ. of Missouri Press, Columbia.

Goto, M. 1992. *Fundamentals of Bacterial Plant Pathology.* Academic Press, San Diego.

Govindu, H.C. 1982. Green Revolution – its impact on plant diseases with special réference to cereals and millets. *Indian Phytopathol.* 35: 363.

Grogan, R.G. and Campbell, R.N. 1966. Fungi as vectors and hosts of viruses. *Annu. Rev. Phytopathol.* 4: 29.

Grover, R.K., 1975. Plan Pathology researches in India – an introspection and prospects. *Pesticides Annual.* 1975.

Hahlbrock, K. and Sebeel, D. 1987. Biochemical responses of plants to pathogens, *In*: *Innovative approaches to plant disease control.* Chet. I, Ed. John Weley & Sons, New York. 229.

Hanlin, R.T. 1990. *Illustrated Genera of Ascomycetes.* APS Press, St. Paul, Minnesota.

Hardison, J.R. 1976. Fire and Flame for Plant Disease Control. *Annu. Rev. Phytopathol.* 14: 355.

Hawksworth, F.G. and Wiends, D. 1970. Biology and Taxonomy of the dwarf mistletoes. *Annu. Rev. Phytopathol.* 8: 187.

Hendrix, F.F. Jr. and Campbell, W.A. 1973. *Pythiums* as plant pathogens. *Annu. Rev. Phytopathol.* 11, 77.

Henis, Y. and Katan, J. 1975. Effect of inorganic amendments and soil reaction on soil borne plant disease. *In : Biology and Control of soil borne plant diseases,* Bruebl. G.W. Ed. Am. Phytopathol. Soc. St. Paul. Minnesota. 100.

Hepting, G.H. 1974. Death of the American Chestnut. Forest History. 18: 60.

Heesett, W.B. Raski, D.J. and Gobeen, A.C. 1958. Nematode Vectors of soil borne fanleaf virus of grapevines, *Phytopathology*. 48: 586.

Hickman, C.J. 1958. *Phytophthora*, Plant Destroyer. *Trans. Brit. Mycol. Soc.* 41: 1.

Hickman, C.J. and HO, H.H. 1966. Behaviour of zoospores in plant pathogenic phycomycetes. *Annu. Rev. Phytopathol.* 4: 195.

Hildebrand, D.C., Schrorth, M.N. and Huisman, O.C. 1982. The DNA homology matrix and non-random variation concepts as basis for the taxonomic treatment of plant pathogenic and other bacteria. *Annu. Rev. Phytopathol.* 20: 235.

Hopkins, D.L. 1977. Diseases caused by leafhopper borne rickettsia like bacteria. *Annu. Rev. Phytopathol.* 15: 277.

Hopkins, D.L. 1983. Gram negative, Xylem limited bacteria in plant disease. *Phytopathology.* 73: 347.

Hornby, D. 1983. Suppressive soils. *Annu. Rev. Phytopathol.* 21: 65.

Hornby, E. *ed.* 1990. Biological control of plant pathogens. CAB Int., Oxon, U.K.

Horsfall, J.G. and cowling E.B. 1977. The sociology of plant pathology, *In: Plant Diseases – An Advanced Treatise* Vol.1. Horsfall J.G. and Cowling, E.B. Eds. Academic Press, New York. 465.

Horsfall, J.G. and Cowling, E.B. 1977. *Plant Disease: An Advanced Treatise.* Vol. 1. Academic Press, New York, 405.

Horsfall, J.G. and Cowling, E.B., eds. 1978. *Plant Disease.* Vol. 2. Academic Press, New York.

Horsfall, J.G. and Cowling, E.B. 1978. *Plant Disease.* Vol. 3. Academic Press, New York.

Horsfall, J.G. and Cowling, E.B., eds. 1980. *Plant Disease.*' Volume 5: How Plants Defend Themselves. Academic Press, New York.

Horsfall, J.G. and Dimond, A.E. 1959. Prologue: the diseased plant, In: *Plant Pathology- An Advanced Treatise.* Vol. 1. Horsfall J.G. and Dimond A.E. Academic Press Eds. New York.

Huber, D.M. 1981. The use of fertilizers and organic amendments in the control of plant disease. *In : Handbook of Pest Management in Agriculture* Vol. 1. Pimental, D. Ed. CRC Press. Boea Raton, FL. 315.

Huber, D.M. and Watson, R.D. 1974. Nitrogen form and plant disease. *Annu. Rev. Phytopathol.* 12: 139.

Huber, L. and Gillespie, T.J. 1992. Modeling leaf wetness in relation to plant disease epidemiology. *Annu. Rev. Phytopathol.* 30: 553.

REFERENCES

James, W.C. 1974. Assessment of plant diseases and losses. *Annu. Rev. Phytopathol*. 12: 27.

Joshi, L.M., Gera, S.D. and Saari, E.E. 1973. Extensive cultivation of kalyansona and disease development. *Indian Phytopathol.* 26: 370.

Joshi, L.M.; Srivastava, K.D.; Singh, D.V. and Ramanujam, K. 1980. Wheat rust epidemics in India Since 1970. Cereals Rusts Bull. 8: 17.

Kahn, R.P. 1991. Exclusion as a plant disease control strategy. *Annu. Rev. Phytopathol*. 29: 219.

Katan, J. 1981. Solar heating (solarization) of soil for control of soil borne pests. *Annu. Rev. Phytopathol.* 19: 211.

Katan, J. 1987. Soil Solarization, *In : Innovative Approaches to Plant Disease Control*, Chet. 1. Ed. John. Wiley and Sons, New York 372.

Keen, N.T. 1981. Specific recognition in gene for gene host parasite systems. *Advances in Plant Pathol.* Vol. I.

Keese, P. and Symons R.H. 1987. The structure of viroids and virusoids, in: *Viroids and viroid like pathogens.* Semaneik J.S. Ed. CRC Press, Boca Raton, FL. 177.

Kelman, A. and Cook, R.J. 1977. Plant Pathology in peoples republic of China. *Annu. Rev. Phytopathol.* 15: 409.

Kendrick, B. ed. 1971. "*Taxonomy of fungi imperfecti*". Toronto Univ. Press, Toronto.

Kerr, A. 1972. Biological control of crown gall: Seed inoculation. *J. Appl. Bacteriol*. 35: 493.

Kerr, A. 1974. Soil microbiological studies on *Agrobacterium radiobacter* and biological control of crown gall. *Soil Sci.* 118: 168.

Kerr, A. 1980. Biological control of crown gall through production of agrosin 84. *Plant Dis*. 64: 25.

Kessman, H. *et.al.* 1994. Induction of systemic acquired disease resistance in plants by chemicals. *Annu. Rev. Phytopathol.* 32: 439.

Klinkowski, M. 1970. Catastrophic plant diseases. *Annu. Rev. Phytopathol.* 8: 37.

Kommedabl, T. and Windels, C.E. 1979. Fungi: Pathogen as host dominance in disease *In : Ecology of Root Pathogens.* Kreepa S.V. and Dommirgues Y.R. Eds. Elsevier, Amsterdam. 291.

Kosuge, T. and Nester, E.W., eds. 1989. *Plant microbe Interactions: Molecular and Genetic Perspectives.* Vols. 1-3. McGraw Hill New York.

Kranz, J. 1978. Comparative anatomy of epidemics in plant disease: an advanced Treatise. Vol. II. Horsfall J.G. and Cowling, E.B. Eds. Academic Press, New York, 436.

Kranz, J. and Han, B. 1980. Systems analysis in epidemiology. *Annu. Rev. Phytopathol.* 18: 67.

Krause, R.A. and Massie, L.B. 1975. Predictive system: modern approaches to disease control. *Annu. Rev. Phytopathol.* 13: 31.

Krause, R.A., Massie, L.B. and Hyre, R.A. 1975. Blitecast: A computerized forecast of potato late blight. *Phytopathology* 65: 95.

Krieg, N.R. and Holt, J.G. (eds) 1984. *Bergey's manual of systematic bacteriology,* Vol. I. Baltimore: Williams and Walkins. 964p.

Kuc, J. 1966. Resistance of plants to infectious agents. *Annu. Rev. Microbiol.* 20: 337.

Kuc, J. 1972. Phytoalexins. *Annu. Rev. Phytopathol.* 10:207

Kuc, J. 1987. Plant Immunization and its practicability for disease control. *In* : *Innovative Approaches to plant disease control.* Chet. I Ed. John. Wiley & Sons. New York. 225.

Kuc, J. 1995. Phytoalexins, Stress Metabolism, and disease resistance in plants. *Annu. Rev. Phytopathol.* 33: 275.

Kuijt, J. 1977. Haustoria of phanerogamic parasites. *Annu. Rev. Phytopathol.* 15: 91.

Kumar, J. Chaube, H.S.; Singh, U.S. and Mukhopadhyay, A.N. 1992. *Plant Diseases of International Importance,* Vol. III. Diseases of fruit crops, Prentice Hall, Englewood Cliffs, New Jersey, USA, 456p.

Linderman, R.G. 1970. Plant residue decomposition products and their effect on host roots and fungi pathogenic to roots. *Phytopathology.* 60: 19.

Lockwood, J.L. 1977. Fungistasis in soil, *Biol. Rev.* 52: 1.

Lomonossoff, G.P. 1995. Pathogen-derived resistance to plant viruses. *Annu. Rev. Phytopathol.* 33: 323.

Lyr.H. 1977. Mechanisms of action of fungicides. *In* : *Plant Disease.* Vol. 1. Horsfall J.G. and Cowling, E.B. Eds. Academic Press. New York. 239.

Lyr. H. 1987. Aromatic hydrocarbon Fungicides. In: *Modern Selective Fungicides - Poperties, Applications, Mechanisms of Action.* Lyr. H., Ed. Longman Group UK Ltd. London and VEB Gustav Eischer Verlag Jena. Chap. 5.

Lyr. H. 1987. Selectivity in modern systemic fungicides, *In*: *Modern Selective Fungicides - Properties, Application, Mechanisms of Action,* Lyr. H., Ed. Longman Group U.K. Ltd. London VEB Gustav Eischer Verlag. Jena Chap. 2.

Macdonald, D. 1979. Some interactions of plant parasitic nematodes and higher plants in *Ecology of Root Pathogens*, Kreepa, S.V. and Dommergues, Y.R., Eds., Elsevier 1979. 281.

Madden, L.V. and Hughes, G. 1995. Plant Disease incidence : Distribution, heterogeneity and temporal analysis. *Annu. Rev. Phytopathol.* 33: 529.

REFERENCES

Maramorosch, K. 1966. Transmission of plant viruses by vectors, pp. 243-250. In: S.P. Raychaudhari (ed.) *Plant disease problems. Indian Phytopath.* Soc., New Delhi.

Maramorosch, K. 1969. *Viruses, Vectors and Vegetation.* Interscience Press, New York.

Maramorosch, K. and Raychaudhari, S.P. 1981. Mycoplasma diseases of trees and shrubs. Academic Press, New York.

Mankau, R. 1980. Biological control of nematode pests by natural enemies. *Annu. Rev. Phytopathol.* 18: 415.

Mathys, G. 1975. Thoughts on quarantine problems. *EPPO Bull.* 5(2). 55.

Mathys, G. and Bakor, E.A. 1980, An appraisal of the effectiveness of quarantines. *Annu. Rev. Phytopathol.* 18: 85.

Mathews, R.E.F. 1970. Plant Virology, Academic Press, New York, 778.

Mathews, R.E.F. 1991. Plant Virology. 3rd Ed. Academic Press, San Diego.

McGee, D.C. 1955. Epidemiological approach to disease management through seed technology. *Annu. Rev. Phytopathol.* 33: 445.

Mc Intyre, J.L. and Sands, D.C. 1977. How disease is diagnosed, *In: Plant Disease: An advanced treatise,* Vol. 1, Horsfall, J.G. and Cowling, E.B. Eds. Academic Press, New York. 35.

Mc New, G.L. 1950. Outline of a new approach in teaching plant pathology. *Plant Dis. Rep.* 34: 106.

Mc New, G.L. 1960. The nature, origin and evolution of parasitism in *plant pathology - An advbanced treatise* Vol. II. Horsfall, J.G. and Dimond A.E. Eds. Academic Press, New York. 16.

Mc New, G.L. 1972. Concept of pest management, pp. 119-125. *In* : Pest control strategies for the future. National Academy of Science, Washington.

Mc Gregor, R.C. 1978. People placed pathogens; the emigrant pests, *In: Plant Diseases: an advanced treatise,* Vol. II. Horsfall J.G. and cowling E.B. Eds. Academic Press, New York: 383.

Mehrotra, R.S. 1980. *Plant Pathology,* Tata McGraw-Hill New Delhi. 771.

Mehta, P.R. 1963. Plant Pathology in India- past, present and prospects. *Indian Phytopath.* 16: 1-7.

Mclouk, H.A. and Shokes, F.M. eds. 1995. *Peanut Health Management.* A P S Press, St.Paul, Minnesota.

Menzies, J.D. 1963. Survival of microbial plant pathogens in soil. *Bot. Rev.* 29: 79.

Meredith, D.S. 1973. Epidemiological consideration of plant diseases in the Tropical environment, *Phytopathology* 63: 1446.

Miller, P.R. 1967. Plant disease epidemics- their analysis and forecasting, in paper : presented at the F.A.O. Symp. Crop Losses. F.A.O. Rome. 330.

Miller, P.R. and O'Brien, M.J. 1957. Prediction of plant diseases epidemics. *Ann. Rev. Microbiol.* 11: 77.

Mount, M.S. and Lacey, G.H. eds. 1982. *Phytopathogenic Prokaryotes.* Vols 1 and 2. Academic Press, New York.

Mukhopadhyay, A.N. 1987. Biological control of soil-borne plant pathogens by *Trichoderma spp. Indian J. Mycol. Plant Pathol.* 17: 1.

Mukhopadhyay, A.N. and Pavgi, M.S. 1973. Cytology of chlamydospore development in *Protomyces macrosporus* unger, Cytologia 38: 467.

Mukhopadhyay, A.N. and Singh U.S. 1985. Recent thoughts on plant disease control 1. Crop protection through host immunization by the chemical. *Pesticides.* 14: 14.

Mukhopadhyay, A.N. and Singh, U.S. 1985. Recent thoughts in plant disease control. II. Fungal resistance to fungicide. *Pesticides.* 14, 23.

Mukhopadhyay, A.N.; Kumar, J. Chaube, H.S. and Singh, U.S. *Plant Diseases of International Importance,* Vol IV. Diseases of Sugar, Forest, and Plantation crops, Prentice Hall, Engkewood Cliffs, New Jersey, USA, 376p.

Musselman, L.J. 1980. The biology of striga, orobancha and other root parasitic weeds. *Annu. Rev. Phytopathol.* 18: 463.

National Academy of sciences, 1968. Plant disease development and control, National academy sciences, Washington, D.C.

Neergard, P.A. 1980. A review on quarantine for seed. *In: Golden Jubilee Commemoration Volume* National Academy of Science, New Delhi.

Nelson, P.E. and Dicky, R.S. 1970. Histopathology of plants infected with vascular bacterial pathogens. *Ann. Rev. Phytopathol.* 8: 259.

Nelson, P.E., Toussoun, T.A. and Cook, R.J., eds. 1981. "*Fusarium: Diseases, Biology, and Taxonomy*". Pensylvania State Press, University Park.

Nelson, R.R. 1978. Genetics of horizontal resistance in plant diseases. *Annu. Rev. Phytopathol.* 16: 359.

Nene, Y.L. and Singh, S.D. 1976. Downy Mildew and ergot of pearl millet. *PANS* 22: 366.

Nene, Y.L. and Thapliyal, P.N. 1979. *Fungicides in Plant Disease Control,* 2nd ed. Oxford and IBH Publishing Co. New Delhi, 507.

New, P.B. and Kerr. A. 1972. Biological control of crown gall: Field measurement and glasshouse experiments. *J. Appl. Bacteriol.* 35: 279.

Nicholson, R.L., and Hammerschmidt, R. 1992. Phenolic compounds and their role in disease resistance. *Annu. Rev. Phytopathol.* 30: 369.

Nienhaus, F. and Sikora, R.A. 1979. Mycoplasmas, Spiroplasmas and Rickettsia like organisms as plant pathogens. *Annu. Rev. Phytopathol.* 17: 37.

REFERENCES

Noble, M. and Richardson, M.J. 1968. Annotated list of seed borne diseases. CMI *Phytopathol.* Papers 8: 1-191.

Okabe, N. and Goto, M. 1963. Bacteriophages of plant pathogens. *Annu. Rev. Phytopathol.* 1: 397.

Ordish, G. and Dufour, D. 1969. Economic bases for protection against plant diseases. *Annu. Rev. Phytopathol.* 7: 31.

Owens, R.G. 1963. Chemistry and physiology of fungicidal action. *Annu. Rev. Phytopathol.* 1: 77.

Padmanabhan, S.Y. 1973. The great Bengal Famine. *Annu. Rev. Phytopathol.* 11: 11.

Palleroni, N.J. and Doudoroff, M. 1972. Some properties and taxonomic subdivision of the genus *Pseudomonas. Ann. Rev. Phytopathol.* 10: 73.

Palti, J. 1981. *Cultural practices and infectious crop diseases.* Springer-Verlag, Berlin. 343p.

Palu Kaitis, P. and Zaittin, M. 1984. A model to explain the "Cross protection phenomenon shown by plant viruses and viroids in plant microbe interactions molecular and genetic perspectives. Kosuge, T.; Nester, E.W., Eds. Macmillan, New York, 420.

Panopoulos, N.J. and Peet, R.C. 1985. The molecular genetics of plant pathogenic bacteria and their plasmids. *Annu. Rev. Phytopathol.* 23: 381.

Papavizas, G.C. 1981. *Biological control in crop production.* Allanhold, Osmun. 461.

Papavizas, G.C. 1985. *Trichoderma* and *Gliocladium*: biology, ecology and potential for biocontrol, *Annu. Rev. Phytopathol.* 23: 23.

Papavizas, G.C. and Lumsden, R.D. 1980. Biological control of soil-borne fungal propagules. *Annu. Rev. Phytopathol.* 18: 389.

Park, D. 1960. Antagonism- the background to soil fungi, pp 148-159. In: D. Parkinson and J.S. Waid (eds.), *Ecology of soil fungi.* University Press, Liverpool.

Park, D. 1963. The ecology of soil borne fungal disease. *Annu. Rev. Phytopathol.* 1: 241.

Parlevliet, J.E. and Zadoks, J.C. 1977. Integrated concept of disease resistance - new view including horizontal and vertical resistance in plants, *Euphytica,* 26: 5.

Patil, S.S. 1974. Toxins produced by phytopathogenic bacteria. *Annu. Rev. Phytopathol.* 12: 259.

Patil, S.S. *et.al.* 1991. "*Molecular strategies of pathogens and hosts*" Springer-Verlag, New York.

Paxton, J.D. 1988. Phytoalexins in plant parasite interaction, in: *Experimental and Conceptual Plant Pathology* Vol. III Singh, R.S.; Singh, U.S.; Hess,

W.M. and Weber, D.J.; Eds. Gardon and Breach, New York and IBH Publishing New Delhi. 537.

Person, C. and Sidhu, G. 1971. Genetics of host parasite relationship in proceedings of the symposium in mutation breeding for disease resistance, Int. Atomic Energy, Vienna. 31.

Pezian, M.A. Koltunow, A.M. and Krake, L.R. 1988. Isolation of three viroids and circular RNA from grape vines, *J. Gen. Virol.* 69: 413.

Pontecorvo, G. 1956. The parasexual cycle in fungi. *Annu. Rev. Microbiol.* 10: 343.

Parter, I.J. and Merriman, P.R. 1983. Effect of solarization of soil on nematode and fungal pathogens at two sites in victoria, *Soil Biol. Biochem.* 15: 39.

Powell, N.T. 1971. Interactions between nematodes and fungi in disease complexes. *Annu. Rev. Phytopathol.* 9: 253.

Pringle, R.B. and Scheffer, R.P. 1964. Host specific plant toxins. *Annu. Rev. Phytopathol.* 2: 133.

Prusiner, S.B. 1982. Novel proteinaceous infectious particles cause scrapie. *Science* 216: 136.

Prusiner, S.B. 1984. Prions: Novel infectious pathogens. *Advances virus Res.* 29:1.

Prusky, D. 1988. Hypersensitivity: an overview, in: *Experimental and conceptual plant pathology*. Vol. III. Singh, R.S.; Singh, U.S.; Hess, W.M. and Weber, D.J. Eds. Gordon and Breach. New York and IBH Publishing, New Delhi. 485.

Pullman, G.S.; De Vay, J.E. and Carber, R.H. 1981. Soil solarization and thermal death: A logarithmic relationship between time and temperature for four soil borne pathogens. *Phytopathology.* 71: 959.

Pullman, G.S.; De Vay, J.E.; Garber, R.H. and Weinhold, A.R. 1979. Control of soil borne pathogens by plastic tarping of soil, *In: Soil borne pathogens.* Schippors B. and Gams, W., Eds. Academic Press, New York, 686.

Purcell, A.H. 1982. Insect vector relationships with prokaryotic plant pathogens. *Ann. Rev. Phytopathol.* 20: 397.

Raychaudhari, S.P. 1967. Development of mycological and plant pathological researches, education, and extension work in India. *Rev. Appl. Mycol.* 46: 577.

Raychaudhari, S.P. 1972. History of plant pathology in India. *Annu. Rev. Phytopathol.* 10: 21.

Raychaudhari, S.P. and Verma, J.P. 1977. Therapy by heat, radiation and meristem culture, in: *Plant diseases: an advanced treatise* Vol. 1. Horsfall J.G. and Cowling, E.B. Eds. Academic Press, New York. 177.

REFERENCES

Reddy, D.B. 1977. The international plant protection convention for southeast asia and pacific region. *Plant Prot. Bull.* 25: 157.

Reddy, P.P. 1983. *Plant Nematology*, Agricole publishing, New Delhi. 287.

Riesner, D. 1991. Viroids: From thermodynamics to cellular structure and function. *Mol. Plant Microbe Interact.* 4; 122.

Robinson, R.A. 1969. Disease resistance terminology. *Rev. Appl. Mycol.* 48: 593.

Robinson, R.A. 1976. *Plant Pathosystems*. Springer-Verlag, Berlin. 184 pp.

Rodriguez-Kabana, R. and Curl, E.A. 1980. Nontarget effects of pesticides on soil borne pathogens and disease. *Annu. Rev. Phytopathol.* 18: 311.

Rotem, J. 1994. "*The Genus Alternaria*" APS Press, St. Paul, Minnesota.

Rotem, J. and Palti, J. 1969. Irrigation and Plant Diseases. *Annu. Rev. Phytopathol.* 7: 267.

Rudolph, K. 1976. Non-specific toxins, in: *Physiological plant pathology.* Hutefuss, R. and Williams, P.H. Eds. Springer-Verlag, Berlin, 270.

Ryals, J. Uknes, S. and Ward, E. 1994. Systemic acquired resistance. *Plant Physiol.* 104: 1109.

Samules, G.J. and Seifert, K.A. 1995. The impact of molecular characters on systematics of filamentous ascomycetes. *Annu. Rev. Phytopathol.* 33: 37.

Sasser, J.N. and Jenkins, W.R. (ed.) 1960. *Nematology*. University of North Carolina. Press. Chapel Hill.

Sayre, R.M. and Walter, D.E. 1991. Factors affecting the efficacy of natural enemies of nematodes. *Annu. Rev. Phytopathol.* 29: 149.

Schaad, N.W. 1979. Serological identification of plant pathogenic bacteria. *Annu. Rev. Phytopathol.* 17: 123.

Schaad, N.W. (ed) 1980. *Laboratory guide for identification of plant pathogenic bacteria. Amer. phytopath. Soc., St. Paul, Minn.*

Schafer, J.F. 1971. Tolerance to plant disease. *Annu. Rev. Phytopathol.* 9: 235.

Schafer, W. 1994. Molecular mechanisms of fungal pathogenicity to plants. *Annu. Rev. Phytopathol.* 32: 461.

Scheffer, R.P. 1976. Host specific toxins in relation to pathogenesis and disease development. *In : Physiological plant pathology*. Heitefuss, R. and Williams, P.H., Eds. Springer Verlag. Berlin. 247.

Scheinpflug, H. and Kuck, K.H. 1987. Sterol biosynthesis inhebiting piperazine, pyridine, pyrimidine and azole fungicides, *In: Modern selective fungicides, Properties, Application, Mechanisms of action*. Lyr. H.Ed. Longman Group U.K. Ltd. London, and VEB Gustax Fisher Verlag, Jena, Chap. 13.

Schippers, B.; Bakker, A.W. and Bakker, P.A.H. 1987. Interactions of deleterious and beneficial rhizosphere microorganisms and the effect of cropping practices. *Annu. Rev. Phytopathol.* 25: 339.

Schneider, R.W. Ed. 1982. Suppressive soils and plant diseases. Am. Phytopathol. Soc. St. Paul, MN.

Schuster, M.L. and Coyne, D.C. 1974. Survival mechanisms of phytopathogenic bacteria. *Ann. Rev. Phytopathol.* 12: 199.

Scott, K.J. and Chakravorty, A.K. eds. 1982. "*The Rust Fungi*" Academic Press, London.

Seem, R.C. 1984. Disease incidence and severity relationships. *Annu. Rev. Phytopathol.* 22: 133.

Shaw, C.G., 1950. The genera of the Peronosporaceae. *Phytopathology.* 40: 25.

Shaw, C.G. 1970. Morphology and physiology of downy mildews. Significance in toxonomy and pathology. *Indian Phytopath.* 23: 364.

Shaw, C.G. 1978. *Peronospora species* and other downy mildews of the Graminae. *Mycologia* 70: 594.

Shephard, M.C. 1987. Screening for fungicides. *Annu. Rev. Phytopathol.* 25: 189.

Shoemaker, R.A. 1981. Changes in taxonomy and nomenclature of important genera of plant pathogens. *Annu. Rev. Phytopathol.* 19: 297.

Sigee, D.C. 1992. "*Bacterial Plant Pathology:* Cell and molecular aspects" Cambridge Univ. Press, New York.

Sikora, R.A. 1992. Management of the antagonistic potential in agricultural ecosystems for the biological control of plant parasitic nematodes. *Annu. Rev. Phytopathol.* 30: 245.

Singh, D.P. 1986. *Breeding for resistance to diseases and insect pests.* Narosa Publishing House, New Delhi and Springers Verlag, Berlin.

Singh, N. and Singh, R.S. 1984. Significance of organic amendment of soil in biological control of soil borne plant pathogens *In* : Progress in *Microbial Ecology.* Mukharjee, K.G.; Agnihotri, V.P. and Singh, R.P. Eds. Print House (India), Lucknow. 533.

Singh, R.P. 1988. Pathogenesis and host parasite relationship in viroids, *In*: *Experimental and conceptual plant pathology* Vol. 2. Singh, R.S.; Singh U.S.; Hess, W.M. and Weber, D.J., Eds. Gordon and Breach Science Publishers, New York. 599.

Singh, R.P. and Clark, M.C. 1971. Infectious low molecular weight ribonucleic acid, Biochem, Biophys, Res. Common 44: 1077.

Singh, R.S. 1984. *Introduction to principles of plant pathology.* Oxford & IBH Publishing Co. New Delhi, Bombay, Calcutta. 534.

Singh, R.S. and Chaube, H.S. 1968. The occurrence of *Sclerospora sacchari* and its oospores on maize in India. *Labdev J. Sci. Technol.* B6: 197.

Singh, R.S. and Singh, U.S. 1988. Pathogenesis and host-parasite specificity in plants. *In : Experimental and conceptual plant pathology,* Vol. III, Singh R.S.; Singh, U.S.; Hess, W.M. and Weber, D.J. Eds. Gordon and Breach, New York and IBH Publishing, New Delhi. 139.

Singh, R.S.; Singh, U.S.; Hess, W.M. and Weber, D.J. 1988. *Experimental and conceptual plant pathology.* Vol. III, Gordon and Breach, New York and IBH Publishing, New Delhi.

Singh, R.S. and Sitaramaiah, K. 1971. Control of plant parasitic nematodes with organic amendments of soil, PANS 16: 287.

Singh, R.S. and Sitaramaiah, K. 1973. Control of plant parasitic nematodes with organic amendments of soil, *G.B. Pant Univ. Expt. St. Res. Bull.* 6: 289.

Singh, U.S. 1987. Uptake, translocation, distribution and persistance of 14C-Metalaxyl in maize (*Zea-mays*) *Z. pflanzenkr. Pflanzenschutz.* 94: 478.

Singh, U.S. 1989. Studies on the systemicity of [14c] metalaxyl in cowpea (Vigna unguiculata [L] walp), *Pestic Sci.* 25: 145.

Singh, U.S. and Singh, R.S. 1988. Philosophy of defense in plants, *In : Experimental and Conceptual plant pathology.* Vol. III. Singh, R.S.; Singh, U.S.; Hess, W.M. and Weber, D.J., Eds. Gordon and Breach, New York and IBH Publishing, New Delhi, 459.

Singh, U.S.; Singh, R.P. and Kohmoto, K. eds. 1995. "*Pathogenesis and Host specificity in plant diseases*: Histochemical, biochemical, genetic and molecular bases". Pergamon/Elsevier, Tarrytown, New York.

Singh, U.S. and Tripathi, R.K. 1982. Physio-chemical and biological prosperities of metalaxyl II. Absorption by excised maize roots, *Indian J. Mycol. Plant Pathol.* 12: 295.

Singh, U.S. and Tripathi, R.K. 1982. Physico—chemical and biological properties of metalaxyl, I. Octanol number, absorption spectrum and effect of different physico-chemical factors on stability of metalaxyl. *Indian J. Mycol. Plant Pathol.* 12: 287.

Singh, U.S.; Mukhopadhyay, A.N.; Kumar, J. and Chaube, H.S. 1992. *Plant Diseases of International Importance*, Vol. I: Diseases of cereals and pulses, Prentice Hall, Englewood Cliffs, New Jersey, USA.

Sisler, H.D. and Ragsdale, N.N. 1987. Disease control by non-fungitoxic compounds *In : Modern selective fungicides-properties, application, mechanisms of action*. Lyr. H.Ed. Longman Group U.K. Ltd. London and VEB Gastav Eischu Verlag, Jena Chap. 23.

Slykhuis, J.T. 1976. Virus and virus like diseases of cereal crops. *Annu. Rev. Phytopathol.* 14: 189.

Smulex, R.W. 1975. Forms of nitrogen and the pH in the root zone and their importance in root infection *In : Biology and control of soil borne pathogens*, Breach, G.W. Ed. Am. Phytopathol. Soc. St. Paul, Minnesota. 52.

Smith, K.M. 1972. *A text book of plant virus diseases*, 3rd eds. Longmans Group, New York. 652.

Sparrow, F.K. 1973. Mastigomycotina (Zoosporic Fungi) pp 61-73, In. G.C. Ainsworth; F.K. sparrow and A.S. Sussman (Eds.) *The Fungi*, Vol. IV B. Academic Press, New York.

Sparrow, F.K. 1976. The present status of classification of biflagellate fungi. pp. 213-22. In., E.B. Gareth Jones (ed.) *Recent Advances in Aquatic Mycology*, John Wiley, New York.

Spencer, D.M., ed. 1978. *The Powdery Mildew"* Academic Press, New York.

Spencer, D.M. ed. 1981. *The Downy Mildew"* Academic Press, New York.

Stakeman, E.C. and Hassar, J.C. 1957. *Principles of plant pathology*. Ronald Press, New York. 581.

Staples, R.C. (ed.) 1981. *Plant Disease Control*. Wiley Interscience, New York.

Starr, M.P., ed. 1983. "Phytopathogenic bacteria". Selections from *The Prokaryotes*. Springer-Verlag, Berlin and New York.

Starr, M.P. 1959. Bacteria as plant pathogens. *Ann. Rev. Microbiol.* 13: 211.

Starr, M.P. 1984. Landmarks in the development of phytobacteriology. *Annu. Rev. Phytopathol.* 22: 169.

Starr, M.P. and Mandel, M. 1969. DNA base composition and taxonomy of phytopathogenic and other enterobacteria. *J. Gen. Microbiol.* 56: 113.

Starr, M.P., Stolp, H.; Truper, H.S., Balows, A. and Schlege. H.C. 1981. *The prokaryotes*, Vol. I and II, Springer Verlag, Berlin.

Staskawicz, B.J.; Ausabel, F.M.; Baker, B.J. *et.al.* 1995. Molecular genetics of plant disease resistance. *Science* 268: 661.

Staub. T. and Sozzi, D. 1984. Fungicide resistance : a continuing challenge *Plant Dis.* 68: 1026.

Stover, R.H. 1962. The use of organic amendments and green manures in the control of soil borne phytopathogens, *Recent Prog. Microbiol.* 8: 267.

Stover, R.H. 1986. Disease Management Strategies and the survival of the banana industry. *Annu. Rev. Phytopathol*. 24: 83.

Sussman, A.S. 1965. Darmancy of soil microorganisms in relation to survival. *In: Ecology of soil borne plant pathogens prelude to biological control* Baker, K.F. and Snyder, W.C. Eds. Univ. of California Press, Los Angles. 571.

Swenson, K.G. 1968. Role of aphids in the ecology of plant viruses. *Annu. Rev. Phytopathol.* 6: 351.

Swings, J.G. and Civerolo, E.L. 1993. *Xanthomonas*. Chapman & Hall, London.

Talbot, P.H.B. 1971. *Principles of Taxonomy*. Macmillan Press, London.

Tarr, S.A.J. 1981. *The principles of plant pathology,* Mcmillan publishers Ltd. London and Basingstoke, 632.

Taylor, A.G. and Harman, G.E. 1990. Concepts and technologies of selected seed treatments. *Annu. Rev. Phytopathol.* 28: 321.

Teakle, D.S. 1969. Fungi as vectors and hosts of viruses in *Viruses vectors and vegetation*. Maramorosch, K. Ed. Wiley, New York. 23.

Teakle, D.S. 1980. *Fungi in vectors of plant pathogens.* Harris, K.F. and Maramorosch. K. Eds. Academic Press, New York. 417.

Thirumalachar, M.J. 1969. Morphological basis for the characterization and separation of the genera *Phytophthora, Sclerophthora and Sclerospora. Indian Phytopath.* 22: 155.

Thomson, W.T. 1993. "*Agricultural Chemicals,* Book IV: Fungicides". 1993-1994 Revision. Thomnson, Fresno, California.

Thurston, H.D. 1973. Threatening plant diseases. *Annu. Rev. Phytopathol.* 11: 27.

Thurston, H.D. 1991. "*Sustainable Practices for Plant Disease Management in Traditional Farming Systems*". Westview, Boulder, Colorado.

Tinline, R.D. and Mac Neill, B.H. 1969. Parasexuality in plant pathogenic fungi. *Annu. Rev. Phytopathol.* 7: 147.

Tojamis, E.C. Papavizas, G.C. and Cook, R.J. eds. 1992. "*Biological control of plant diseases: Progress and challenges for the future*". Proc. NATO Advanced Res. Workshop, May 19-24, 1991. Cape sounion, Athens, Greece, Plenum, New York and London.

Tomiyama, K. 1963. Physiology and biochemistry of disease resistance in plants. *Annu. Rev. Phytopathol.* 1: 295.

Vance, C.P., Kirk, T.K. and Sherwood, R.T. 1980. Lignification as a mechanism of disease resistance. *Annu. Rev. Phytopathol.* 18: 259.

Van der Plank, J.E. 1963. *Plant Diseases: Epidemics and Control*. Academic Press, New York.

Van der Plank, J.E. 1968. *Disease Resistance in plants.* Academic Press, New York. 206.

Van der Plank, J.E. 1975. *Principles of Plant Infection*. Academic Press, New York.

Van der Plank, J.E. 1978. *Genetics and molecular basis of plant pathogenesis.* Springer-Verlag, Berlin, 167p.

Van Loon, L.C. 1985. Pathogenesis related proteins. *Plants Mol. Biol.* 4: 111.

Van Gundy, S.D. 1972. Non-chemical control of nematodes and root infecting fungi, pp. 317-329. In: *Pest control strategies for the future.* National Academy of sciences, Washington.

Vidaver, A.K. 1976. Prospect for control of phytopathogenic bacteria by bacteriophages and bacteriocins. *Ann. Rev. Phytopathol.* 14: 451.

Veech, J.A. and Dickson, D.W. eds. 1987. *Vistas on nematology.* Soc. of Nematologists, Hyattsville, Maryland.

Waggoner, P.E. 1960. Forecasting epidemics in *plant pathology.* Vol. 3. Horsfall J.G. and Dimond, A.E. Eds. Academic Press, New York. 291.

Waggoner, P.E. 1965. Microclimate and plant disease. *Annu. Rev. Phytopathol.* 3: 103.

Walker, J.C. 1969. *Plant Pathology.* 3rd edn. Mc Graw Hill, New York.

Wallace, H.R. 1978. Disposal in time and space. Soil pathogens, *In :* Plant Disease : *An advanced treatise.* Vol. 2. Horsfall J.G. and Cowling E.B. Eds. Academic Press, New York. 181.

Wallace, H.R. 1989. Environment and plant health-A nematological perception. *Annu. Rev. Phytopathol.* 27: 59.

Walter, J.M. 967. Hereditary resistance to disease in tomato. *Annu. Rev. Phytopathol.* 5: 131.

Waterhouse, G.M. 1964. The genus *Sclerospora, Diagnosis* and a key. *Misc. Publ. No.* 17, *CMI*, 30pp.

Waterhouse, G.M. 1967. Key to *Pythium,* Pringshein Mycol. Pap. No. 109. CMI, *15pp.*

Waterhouse, G.M. 1968. *The genus Pythium Pringsheim.* Diagnosis and figures from the original papers. Mycol. Papers No. *110. CMI, 71pp.*

Waterhouse, G.M. 1970. "The genus *Phytophthora*" 2nd Ed. Commonw. Mycol. Inst. Misc. Publ. 122.

Waterhouse, G.M. 1970. Taxonomy in *Phytophthora, Phytopathology.* 60: 1141.

Waterhouse, G.M. 1973. Peronosporales, pp 165-183. In G.C. Ainsworth and A.S. Sussman (eds.) *The Fungi*, Vol. IV. B. Academic Press New York.

Waterson, A.P. and Wilinson L. 1978. *An introduction to the History of virology.* Cambridge Univ. Press, Cambridge 237.

Webster, J. 1970. *Introduction to fungi.* Cambridge Univ. Press.

Webster, R.K. 1974. Recent advances in the genetics of plant pathogenic fungi. *Annu. Rev. Phytopathol.* 12: 331.

Weller, D.M. 1988. Biological control of soil borne pathogens in the rhizosphere. *Annu. Rev. Phytopathol.* 26: 379.

Wells, J.M., Raju, B.C.; Hung, H.Y.; Weisburg, W.G.; Mandel Co-Paul, L. and

Brenner, D.J. *Xylella fastidiosa* gen. Nov. Sp. Nov. Gram negative, Xylem Limited, Fastidious Plant bacteria related to *Xanthomonas* spp., Int. J. Syst. Bacteriol, 37, 136.

Wheeler, B.E.J. 1976. *Diseases in crops.* Edward Arnold, London.

Wheeler, H. 1975. *Plant Pathogenesis.* Springer Verlag, Berlin 104p.

Wheeler, H. and Luke, H.H. 1963. Microbial toxins in plant diseases. *Annu. Rev. Microbiol.* 17: 223.

Whitcomb, R.F. 1980. The genus *Spiroplasma. Annu. Rev. Microbiol.* 34: 677.

Whitcomb, R.F. and Tully, J.G. (eds.) 1979. *The Mycoplasmas* Vol. 3. *The mycoplasmas of plants and insects.* Academic Press, New York. 351p.

Whitney, P.J. 1976. *Microbial plant pathology.* Hutchinson, London.

Whittaker, R.H. 1969. New Concepts of kingdoms of organisms. *Science,* 163: 150.

Williams, R.J. 1984. Downy mildews of tropical cereals. *Adv. Plant Pathol.* 2: 2.

Wood, R.K.S. 1960. Pectic and cellulolytic enzymes in plant disease. *Annu. Rev. Plant Physiol.* 11: 247.

Wood R.K.S. 1967. *Physiological Plant Pathology.* Blackwell Scientific Publications, Oxford.

Yamada, T. 1993. The role of auxins in plant disease development. *Annu. Rev. Phytopathol.* 31: 253.

Yarham, D.J. 1979. The effect on soil borne diseases of changes in crop and soil management, pp. 371-383. In: B. schippers and W. Gams (eds.), *Soil borne plant pathogens.*

Young, J.M., Takikawa, Y.; Garden, L. and Stead, D.E. 1992. Changing concepts in the taxonomy of plant pathogenic bacteria. *Annu. Rev. Phytopathol.* 30: 67.

Young, R.A. 1968. *Plant Disease Development and Control.* Publ. 1596. Nat. Acad. Sci. Washington, D.C.

Zadoks, J.C. and Sehein, R.D. 1979. *Epidemiology and plant disease management.* Oxford Univ. Press, New York, 42.

Zitter, T.A. and Simons, J.N. 1980. Management of viruses by alteration of vector efficacy and by cultural practices. *Ann. Rev. Phytopathol.* 18: 289.

Index

C

INDEX

E

INDEX